Forming Technologies and Mechanical Properties of Advanced Materials

Forming Technologies and Mechanical Properties of Advanced Materials

Editors

Tomasz Trzepieciński
Valentin Ştefan Oleksik

MDPI • Basel • Beijing • Wuhan • Barcelona • Belgrade • Manchester • Tokyo • Cluj • Tianjin

Editors

Tomasz Trzepieciński	Valentin Ştefan Oleksik
Rzeszow University of Technology	Lucian Blaga University of Sibiu
Poland	Romania

Editorial Office
MDPI
St. Alban-Anlage 66
4052 Basel, Switzerland

This is a reprint of articles from the Special Issue published online in the open access journal *Materials* (ISSN 1996-1944) (available at: https://www.mdpi.com/journal/materials/special_issues/Metal_Forming_Mechanical).

For citation purposes, cite each article independently as indicated on the article page online and as indicated below:

LastName, A.A.; LastName, B.B.; LastName, C.C. Article Title. *Journal Name* **Year**, *Volume Number*, Page Range.

ISBN 978-3-0365-7282-6 (Hbk)
ISBN 978-3-0365-7283-3 (PDF)

Contents

About the Editors

Tomasz Trzepieciński

Tomasz Trzepieciński is currently an associate professor in the Department of Manufacturing Processes and Production Engineering of the Rzeszow University of Technology. After he finalized his PhD thesis, he took over the assistant professor position in the Department of Materials Forming and Processing. In 2013, Dr. Trzepieciński competed his habilitation thesis on the "Selected Aspects of Tribological Evaluation of Plastically Deformed Sheets". He has many years of research experience in metal forming processes, plasticity, the tribological behaviour of metallic materials, the numerical modeling of plastic working processes, and the application of artificial neural networks and genetic algorithms to optimize forming processes. His research areas also include advanced manufacturing technology, the processing of advanced materials, and the analysis of contact phenomena in sheet metal forming.

Valentin Ştefan Oleksik

Valentin Ştefan Oleksik is currently a full professor in the Department of Industrial Machines and Equipment of the Lucian Blaga University of Sibiu. After he finalized his PhD thesis, he became a lecturer in the same department. In 2016, Dr. Oleksik completed his habilitation thesis on "Studies and Research on the Incremental Forming Process". Dr. Oleksik has many years of research experience in metal forming processes, plasticity, and the numerical simulation of sheet metal forming processes. His research areas also include biomechanics and unconventional forming technologies.

Preface to "Forming Technologies and Mechanical Properties of Advanced Materials"

The aim of this book is to present the latest achievements in various modern forming processes and the latest research related to the computational methods for forming technologies. This publication includes research articles focusing on new developments in the formation of advanced materials including optimisation of the equal-channel angular pressing parameters, the single-point incremental forming process of polymeric and aluminium sheets, the mechanical and structural properties of titanium alloy subjected to impact-oscillatory loading, friction and lubrication in sheet metal forming, optimisation of the thermomechanical process of Ni-based superalloys, numerical and experimental analysis of titanium sheet forming, investigation of the size effects on the elongation of titanium alloy foils, optimisation of the cutting performance of cutting tools, and optimisation of the parameters of the shear spinning process. One review article summarises the development of the single-point incremental forming technique in titanium and its alloys. The authors truly believe that this book will help the research community to enhance their understanding of the present status and trends in modern material forming processes.

Tomasz Trzepieciński and Valentin Ştefan Oleksik
Editors

materials

Article

Optimizing the ECAP Parameters of Biodegradable Mg-Zn-Zr Alloy Based on Experimental, Mathematical Empirical, and Response Surface Methodology

Majed O. Alawad [1], Abdulrahman I. Alateyah [2,*], Waleed H. El-Garaihy [2,3,*], Amal BaQais [4,*], Sally Elkatatny [3], Hanan Kouta [5], Mokhtar Kamel [3] and Samar El-Sanabary [5]

1 Materials Science Research Institute, King Abdulaziz City for Science and Technology (KACST), Riyadh 12354, Saudi Arabia
2 Department of Mechanical Engineering, College of Engineering, Qassim University, Unaizah 56452, Saudi Arabia
3 Mechanical Engineering Department, Faculty of Engineering, Suez Canal University, Ismailia 41522, Egypt
4 Department of Chemistry, College of Science, Princess Nourah Bint Abdulrahman University, Riyadh 11671, Saudi Arabia
5 Department of Production Engineering and Mechanical Design, Port Said University, Port Fuad 42526, Egypt
* Correspondence: a.alateyah@qu.edu.sa (A.I.A.); w.nasr@qu.edu.sa (W.H.E.-G.); aabaqeis@pnu.edu.sa (A.B.); Tel.: +966-055-313-3322 (A.I.A.)

Citation: Alawad, M.O.; Alateyah, A.I.; El-Garaihy, W.H.; BaQais, A.; Elkatatny, S.; Kouta, H.; Kamel, M.; El-Sanabary, S. Optimizing the ECAP Parameters of Biodegradable Mg-Zn-Zr Alloy Based on Experimental, Mathematical Empirical, and Response Surface Methodology. *Materials* **2022**, *15*, 7719. https://doi.org/10.3390/ma15217719

Academic Editors: Tomasz Trzepieciński, Mingchun Zhao and Valentin Ştefan Oleksik

Received: 27 September 2022
Accepted: 31 October 2022
Published: 2 November 2022

Abstract: Experimental investigations were conducted on Mg-3Zn-0.6Zr alloy under different ECAP conditions of number of passes, die angles, and processing route types, aimed at investigating the impact of the ECAP parameters on the microstructure evolution, corrosion behavior, and mechanical properties to reach optimum performance characteristics. To that end, the response surface methodology (RSM), analysis of variance, second-order regression models, genetic algorithm (GA), and a hybrid RSM-GA were utilized in the experimental study to determine the optimum ECAP processing parameters. All of the anticipated outcomes were within a very small margin of the actual experimental findings, indicating that the regression model was adequate and could be used to predict the optimization of ECAP parameters. According to the results of the experiments, route Bc is the most efficient method for refining grains. The electrochemical impedance spectroscopy results showed that the 4-passes of route Bc via the 120°-die exhibited higher corrosion resistance. Still, the potentiodynamic polarization results showed that the 4-passes of route Bc via the 90°-die demonstrated a better corrosion rate. Furthermore, the highest Vicker's microhardness, yield strength, and tensile strength were also disclosed by four passes of route Bc, whereas the best ductility at fracture was demonstrated by two passes of route C.

Keywords: severe plastic deformation; equal channel angular pressing; biodegradable Mg-Zn-Zr alloy; ultrafine-grained structure; corrosion behavior; Response Surface Methodology; genetic algorithm; optimization

1. Introduction

Due to their exceptional properties of high specific strength and low density, which suit the needs of the transportation and lightweight structural industries, magnesium (Mg) alloys have been a source of huge attraction [1,2]. In addition, Mg alloys show a great opportunity in biomedical applications. They look promising to be adopted as bone implant material because of their remarkable advantages compared to traditional biomedical material [3–5]. Mg alloys have suitable mechanical properties, good biocompatibility, and extraordinary biodegradable nature [6–8]. In addition, Mg alloys are very close in density and elastic modulus to their natural bone counterparts [9]. Moreover, degradation in the body fluid of Mg alloys is one of the most desirable properties of a material used for medical purposes. In addition, Mg alloys' have high biodegradability in the human body, as they fade after the surgery, which leads to the absence of the need for other surgery to remove the implant [10,11]. Furthermore, Mg is non-toxic and has exceptional biocompatibility; even Mg may possess fortunate impacts on the growth and adhesion of new bone cells [12,13].

However, Mg alloys have limitations in clinical applications because of their high corrosion rate in the high chloride physiological systems and rapid degradation rate [14]. Thus, the mechanical integrity damage occurs before the complete curing of bone tissues; Moreover, throughout the corrosion process, a detrimental impact on the Mg alloy as a biodegradable implant is also caused by the development of hydrogen gas bubbles [15–17]. Consequently, many approaches from various alloying designs and surface modifications were developed to ameliorate Mg alloys' corrosion resistance and mechanical properties [15,18–23]. Additionally, other attempts were made to adjust the degradation rate by removing impurities and controlling the impurities content ratio [24,25]. Thus, it is crucial for Mg alloys adopted in biomedical applications to possess suitable mechanical properties during their service lifetime, along with being safe bio-alloying elements to guarantee non-toxicity, biocompatibility, and cytocompatibility. Therefore, improving the mechanical and corrosion characteristics of magnesium alloys is of the utmost priority. [19,26–29].

Several studies were performed to improve the mechanical and corrosion properties of Mg alloys, such as effective alloying elements. It was reported that alloying aluminum (Al) resulted in potential toxicity since Al was found to be the reason for Alzheimer's. Likewise, some rare-earth elements, such as yttrium, were also found to cause liver toxicity [6,30]. Consequently, Al-free Mg alloy is recommended for use with humans [31]. On the other hand, alloying zinc elements (Zn) showed the potential to enhance deformability and D%. In addition, Zn is a vital nutritive element for the body, and its ions can be absorbed effortlessly with no harm done to substantial organs. Likewise, zirconium (Zr) with small contents was reported to be a biocompatible alloying element [5,28,29,32]. Furthermore, Gu et al. discovered that adding Zr and Zn to Mg alloys improved strength, corrosion resistance, and cytocompatibility compared with pure Mg [28,29,32,33].

On the other hand, grain refining is an effective method for improving the mechanical characteristics of Mg alloys [19,28,29,34,35]. It was applied to improve the homogenous nanoscale distributions that claimed uniform corrosion behavior. Therefore, these properties could be attained when using severe plastic deformation (SPD) to reach ultra-fine-grained materials (UFG) [36–47]. Additionally, it was stated that the largest positive pitting potentials were associated with the coarse-grained reference alloys and that the ultra-fine-grained Mg alloys induced an enhancement in polarization resistance [48]. Consequently, the Mg-Zn-Zr alloy series (ZKxx) could be a superior alternative in medical implants; it promises a future choice for metallic biodegradable materials [19,28,29,49]. Furthermore, in the modern era of digital databases, generating functional outputs in mathematical forms has become a critical demand [44].

ECAP is the SPD technique that refines metallic materials' grains the most effective out of all the other SPD approaches and hence improving both the mechanical and electrochemical properties [34,35,50–52]. In the ECAP technique, the materials are compelled to exit a die with two parallel channels having an alike cross-sectional profile that intersects at an internal channel angle of ϕ and an angle of curvature of Ψ, as illustrated in Figure 1 [28]. The ECAP route type, in addition to the number of processing passes, has a substantial impact on the mechanical characteristics, crystallographic texture, and microstructural development of the processed billets [53]. The most common ECAP route types are A, Bc, and C [28,53]. In route A, the sample is processed through multiple passes without rotation between the subsequent passes. In contrast, in route Bc, the sample is revolved 90° in the same direction about its longitudinal axis after each pass [28,46,53]. In route C, the sample rotated 180° about the extrusion direction after each pass [53,54]. The imposed equivalent strain (ε_{eq}) can be calculated in terms of the number of ECAP passes (N), die channel angle (ϕ), and the corner curvature angle (Ψ) as shown in Equation (1) [53].

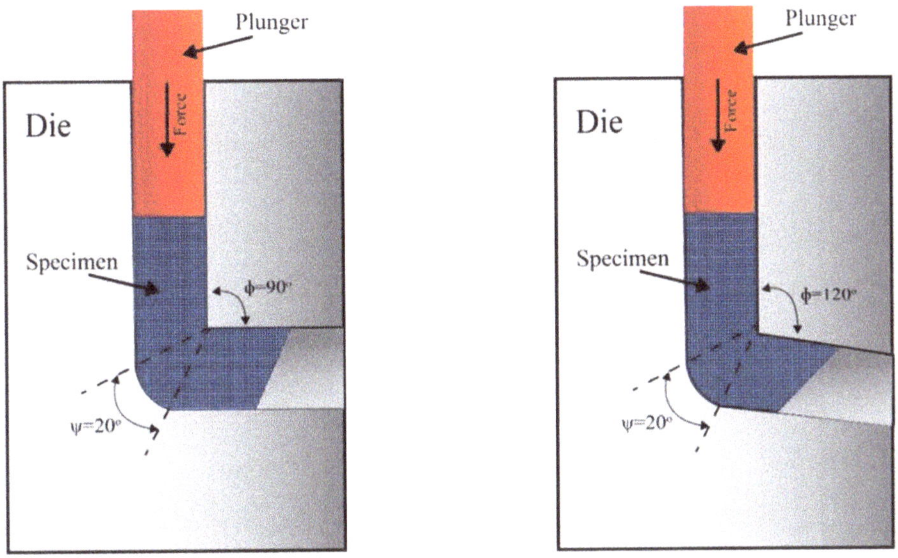

(a) (b)

Figure 1. Schematic of the ECAP dies with internal channel angle of (**a**) 90° and (**b**) 120°.

A, Bc, and C are the most prevalent ECAP route types [28,53]. Unlike route Bc, which rotates the sample by 90 degrees across its longitudinal axis after each pass, route A processes the sample in successive passes without rotating it [28,46,53]. After each pass along route C, the sample is rotated 180 degrees around the direction of extrusion [53,54]. Equation (1) illustrates how to determine the imposed equivalent strain (ε_{eq}) using the number of ECAP passes (N), die channel angle (ϕ), and corner curvature angle (Ψ) [53].

$$\varepsilon_{eq} = \frac{N}{\sqrt{3}}\left[2\cot\left(\frac{\varphi+\psi}{2}\right) + \psi\, cosec\left(\frac{\varphi+\psi}{2}\right)\right] \quad (1)$$

On the other hand, the expansion and advancement of useful mathematical insights is a fundamental necessity in the current digital database era [44]. RSM is an effective multivariate statistical method based on an empirical collection of statistical and mathematical instruments that are utilized to create, alter, and finally optimize processes. RSM works by correlating the real and modeled behavior of a response output to several effective input factors based on their own and interaction effects. RSM technique is capable of modeling and optimizing experiments. In addition, GA might be used in optimization to avoid local optimum solutions [55].

Many researchers were optimizing the ECAP conditions using RSM. Daryadel [56] validated the finite element simulation of the ECAP process of AA7075 with copper casing by examining thirty-one tests built by RSM to investigate the ECAP process parameters. The tests were focused on the highest required force and strain, where the main impacts of four chosen significant input factors (friction coefficient, casing thickness, channel, and corner angle) were studied. Consequently, an analysis of variance (ANOVA) of the process variables was conducted to analyze the obtained regression models. Based on the ANOVA analysis, it was assumed that the channel angle affected the response the most; thus, it was the most effective ECAP input parameter. Moreover, the copper casing thickness didn't show any significant effect on the resultant force response. Likewise, the strain response was affected by channel and corner angle input parameters; conversely, the friction coefficient and copper thickness showed an insignificant effect on the strain response. Finally, the performed optimization reached the optimum predicted ECAP condition aiming at maximizing the forming force and minimizing the strain. The obtained optimum values for channel angle and corner angles were 93.64° and 0°, respectively. Alateyah et al. [57] used RSM, ANOVA, GA, and RSM-GA to optimize the ECAP parameters of pure Mg, and they reported that ECAP processing using a die with $\phi = 90°$ through 4-passes of route Bc was the most significant parameters in grain reining and Vicker's microhardness values. Furthermore, ECAP processing using an ECAP die with $\phi = 120°$ through two passes of route Bc displayed the highest TS, while 4-passes of route C using the 120°-die showed the best D% at fracture. Saleh et al. [58] utilized RSM to optimize the wear resistance of AZ91during ECAP processing using a rotary ECAP die. The RSM findings revealed that the AZ91 wear resistance increased applied load, sliding time, and sliding speed. Furthermore, they reported that the AZ91 wear resistance was improved by increasing the number of processing passes.

As a result, this study's objective is to statistically analyze the ECAP performance through tests that were conducted to determine how the ECAP process parameters affected the ZK30 alloy's mechanical properties and corrosion performance. Experimental investigations were conducted on Zk30 alloy under different ECAP conditions of the ECAP die angles, a number of passes, and processing route types, aiming at reaching the optimum performance characteristics. A complete analysis of the influence of the ECAP conditions on microstructural evolution, mechanical properties, and corrosion performance was presented. The experimental investigation was designed based on RSM that was adopted to classify the optimum ECAP parameters through the analysis of the effect of different ECAP conditions on the numerical responses. Furthermore, analysis of variance (ANOVA) and second-order regression models was obtained to evaluate the optimum ECAP parameters; consequently, GA was used to optimize the ECAP conditions. At last, the optimization of the ECAP responses was enhanced by creating a hybrid RSM-GA, and the subsequent conditions were assessed via GA.

2. Methodology

2.1. The Experimental Design Matrix

The most widespread ECAP process parameters reported in previous studies were the number of passes, ECAP die angle, and the type of processing route [53]. The number of passes (one, two, and four passes), ECAP die angle (90° and 120°), and type of processing route (A, Bc, and C) were the levels of the ECAP parameters employed in this investigation, as shown in Table 1.

Table 1. ECAP parameters and corresponding levels.

ECAP Parameters	Parameters Levels		
	−1	0	1
Number of passes	1	2	4
ECAP die angle	90	120	
Processing route type	A	Bc	C

In this study, RSM was used to provide a design for the combination of the levels of ECAP parameters. Sixteen runs were performed and examined for several ECAP responses, namely, grain size, corrosion response, hardness, and tensile characteristics. Three factors were investigated with a minimum number of experiments using the RSM technique to model a second-order response surface.

2.2. Material and Experimental Procedure

The current study employed a commercial ZK30 alloy (Mg-3Zn-0.6 Zr, wt%). ZK30 billets measuring 20 mm in diameter and 60 mm in length were annealed at 430 °C for 16 h. Two cylindrical channels with an interconnection were used in the ECAP dies, which had internal angles of 90° and 120° and an external die angle of 20°. The ECAP process was applied to the as-annealed (AA) billets under various conditions at a ram speed of 10 mm/min and a temperature of 250 °C. Various routes (A, Bc, and C) were used, as were different passes (one pass (1P), two passes (2P), and four passes (4P)).

The microstructural evolution was studied using a longitudinal cross-section from the center of the ZK30 alloy. The samples were ground incrementally on a grinding wheel spinning at 150 rpm using silicon-carbide sandpaper. Then the samples were polished using diamond suspensions of particle sizes 3 μm, then 1 μm mixed with yellow DP-lubricant. All samples were to have scratch-free surfaces, as seen using a microscope. To that end, a final polishing step was conducted; a 0.05-micron colloidal silica formula was used to provide the final polish. Samples were then etched in a solution of 100 mL ethanol, 5 mL acetic acid (95%), 6 g picric acid, and 10 mL water for 50 s. Finally, to remove the top amorphous layer, the samples were flat ion milled for 30 min using a flat ion milling system. The milling parameters were a grazing angle of 5°, a specimen rotational speed of 0.425 s^{-1}, and a beam energy of 2 keV [28,29].

The microstructure evolution of the ZK30 biodegradable alloy was investigated using a SU-70 SEM equipped with an EBSD accessory which was used to characterize the microstructural and crystallographic texture evolution as well. The samples investigated by the SEM and EBSD were sectioned from the central longitudinal plane of the ECAPed billets parallel to the pressing direction. The axes of the reference system coincide with the extrusion ECAP direction (ED). The SEM operated at 15 kV and 1.5 nA. The EBSD data were collected in 100 nm increments from the top surface ED plan using HKL Flamenco Channel 5 software (Hitachi, Ltd., Tokyo, Japan) [28,29].

A three-electrode corrosion cell was used to evaluate the corrosion properties of the ECAPed ZK30-Mg alloy. 20 × 30 mm rectangular samples were cleaned with acetone, then rinsed in deionized water after being ground with various silicon-carbide papers up to 4000 grit. A platinum mesh was utilized as a counter electrode; however, the working electrode was an ECAPed ZK30 sample, and the reference electrode was a saturated calomel electrode (SCE). At room temperature, corrosion tests were conducted on ringer lactate corrosive agents. A Luggin capillary was employed to ensure measurement precision and to reduce ohmic drop. An SP-200 Potentiostat was used to record the measurements. Furthermore, a potential scan rate of 0.2 mVs^{-1} using the polarization technique was also used to confirm the steady-state situation. With an open circuit potential and a potential window of ±250 mV, linear potentiodynamic polarization was carried out. At open-circuit potential (E_{corr}), electrochemical impedance spectroscopy (EIS) was used with a sinusoidal voltage of ±10 mV and a frequency range of 10 MHz to 100 kHz.

Furthermore, Vicker's microhardness tests (Hv) were carried out using a digital microhardness tester (Qualitest Canada Ltd, Alberta, Canada) before and after the various ECAP operations, beginning at the sample's periphery and progressing into the center. A 0.5 kg applied stress was used for 15 s during the microhardness testing. The average outcomes are determined across a minimum of five equispaced indentations. Additionally, utilizing 100 kN universal testing equipment (Instron 4210, Norwood, MA, USA), the room temperature tensile characteristics of ZK30 ECAPed samples were assessed at a strain rate of 10^{-3} s^{-1}. The chosen tensile samples were cut to dimensions in accordance with the E8M/ASTM standard and taken from the middle of the ZK30 ECAPed samples. For each processing condition, three tensile samples were examined.

3. Response Surface Methodology-Based Experiments

3.1. Regression Model

RSM is a very effective tool in most engineering problems that are adopted for model formulation, analysis, design, and enhancement of an optimization process. The interaction between one or more input parameters can also be evaluated using RSM designs. RSM consists of three main steps that were applied in this study. The first step is concerned with the setup of the experimental technique for navigating the process or input factors domain. ECAP independent variables domain was defined as the number of passes, ECAP die angle, and processing route type. The second step focuses on the development of the appropriate model. The models were formed by regression modeling between the input factors and the process responses of grain size, corrosion response, and tensile characteristics. The last step is about getting a three-dimensional surface of the response to fully visualize how the response is behaving as a result of the impact of input variables, which are usually referred to as parameters. Using optimization procedures, the three-dimensional surfaces were used to identify the effect of the ECAP process factors that provide the most suitable output for the ECAP responses [59,60].

A regression model provides a function that describes the relationship between a process response (Y) and one or more independent variables, which is obtained by best fitting into first, second, or more-order polynomial equations. Polynomial models of linear, two-way interaction and full quadratic models were adopted in this study to evaluate the obtained experimental data. Equation (2) represents the second-order polynomial mathematical models for the ECAP independent variables and the output response of (Y).

$$Y = f\,(\text{No. of passes, Die angle, Processing route type}\,) + \epsilon \quad (2)$$

where Y is the output response, f represents the ECAP process independent input factors, and ε is the random error distributed about the response Y [61].

Analysis of variance (ANOVA) is adopted to investigate the ECAP process independent input factors of the number of passes, ECAP die angle, dummy variable x_1, and dummy variable x_2, and identify which ones of these parameters are significantly impacting the output responses of mechanical properties and corrosion performance [62]. The obtained experimental data were thoroughly studied and analyzed using Stat-Ease Design Expert software (version 13.0.5, Stat-Ease, Inc., Minneapolis, MN, USA). It is a very powerful and efficient computer package used widely in practice for industrial and scientific purposes aiming at designing and optimizing complex systems [63,64]. Design expert provides several types of regression transformation forms, such as linear, square root, natural logarithm, power, and many others.

The adopted ECAP process input parameters are the number of passes, die angle, and processing route type, as shown in the design of experiments which contains 16 runs (Table 2). Special consideration is placed regarding the processing route type parameter due to the variable's nature as a categorical type, which is classified into route A, route Bc, and route C.

Table 2. Design of experiment of ECAP parameters.

Run	A: No. of Passes	B: Die Angle	C: Processing Route Type
1	1	120	Bc
2	2	120	A
3	4	90	C
4	2	120	C
5	2	90	Bc
6	2	120	A
7	2	90	Bc
8	4	120	Bc
9	4	120	C
10	2	120	Bc
11	1	120	C
12	4	90	Bc
13	1	90	A
14	4	90	A
15	4	90	A
16	1	90	C

One of the adopted techniques to transform a categorical variable into a numerical one is dummy coding. It is based on a binary coding system as it provides all of the crucial information about group membership using only zeros and ones. It is one of the adopted methods for employing variables of categorical predictor nature in various types of prediction models of linear and other regressions. Dummy variables are used to denote a category variable that was transformed through dummy coding. To build dummy variables that are exhaustive and mutually exclusive and relate to a specific category variable with K classifications, a series of K-1 dummy variables is required [65]. As illustrated in the matrix in Equation (3), the dummy variables for the category variables route A, Bc, and C were coded as x_1 and x_2.

$$\begin{matrix} \\ A \\ Bc = \\ C \end{matrix} \begin{matrix} x_1 \quad x_2 \\ \begin{bmatrix} 1 & 0 \\ 0 & 1 \\ 0 & 0 \end{bmatrix} \end{matrix} \quad (3)$$

where route type A has $x_1 = 1$ and $x_2 = 0$, if route type Bc then $x_1 = 0$ and $x_2 = 1$, and if route type C then $x_1 = 0$ and $x_2 = 0$.

The finest regression models that could be statistically significant were found after numerous iterations of regression transformation forms and research into interactions between independent variables of ECAP parameters.

3.2. Genetic Algorithm

A genetic algorithm (GA) is a widely used method in various engineering and science-based applications. GA is characterized by its smart, effective, and inexpensive way of tackling real-life optimization problems. GA provides optimum algorithms along with a random number of generations in each individual. On the other hand, a common convention algorithm adopts a predetermined strategy for establishing the following generation and only generates a single point. Every generation evaluates an individual's fitness functions. The convergence of the results is ensured by GA, which adopts certain criteria to arrive at a value aiming at a global minimum for a fitness function [60,66].

4. Results and Discussion

The experimental findings [28,29] of the different studied ECAP process factors were used for the development of a qualitative and quantitative assessment strategy to examine how the ECAP parameters affected the ZK30 alloy regarding the evolution of grain size, electrochemical response, and mechanical characteristics of the ZK30 alloy. Several trials of regression transformation form and interactions of independent variables were tested thoroughly for modeling the output responses of the ECAP. The generated models in the experimental investigation were proven to show statistical significance.

4.1. Experimental Results and RSM

4.1.1. Microstructural Evolution

EBSD was adopted to assess the grain structure of the ZK30. Figure 2 shows the orientation maps of AA-ZK30 after different ECAP processes, all relative to the ED [28,29]. Based on the results of the experiments, Table 3 depicts the average grain size of ZK30 alloy billets processed by AA and ECAP. As seen in Figure 2a, the orientation map of AA-ZK30 revealed equiaxed coarse grains and certain regions with fine grains. Grain refinement can be observed after the first ECAP pass, utilizing die angles of 90° and 120°, as revealed in Figure 2b,f, respectively. However, the die angle of 120° produces a coarser grain size compared with 90° as the latter experienced a higher strain relative to the first. Furthermore, augmenting the number of passes to four using different routes of A (Figure 2c), Bc (Figure 2d), and C (Figure 2f) resulted in further refinement. However, route Bc showed finer and more homogeneous grain distribution, as shown in Figure 2d. Furthermore, by inspection of Figure 2d, it is clear that 4Bc processing using the 90°-die leads to increasing areas of ultrafine grain structure (UFG) despite the existence of minor areas of coarser grains. Accordingly, the existing coarse grains of the 4Bc condition resulted in increasing the average grain size up to 1.94 μm as displayed in Table 3. In addition, it is clear from Figure 2 and

Table 3 that in terms of grain refinement, route Bc is the most efficient. On the other hand, processing through multiple passes indicated that the die angle had an insignificant effect on the average grain size since processing through 4-Bc with the 90°-die and 120°-die reduced grain size by 92.7% and 92.8%, respectively, as relative to the AA equivalent.

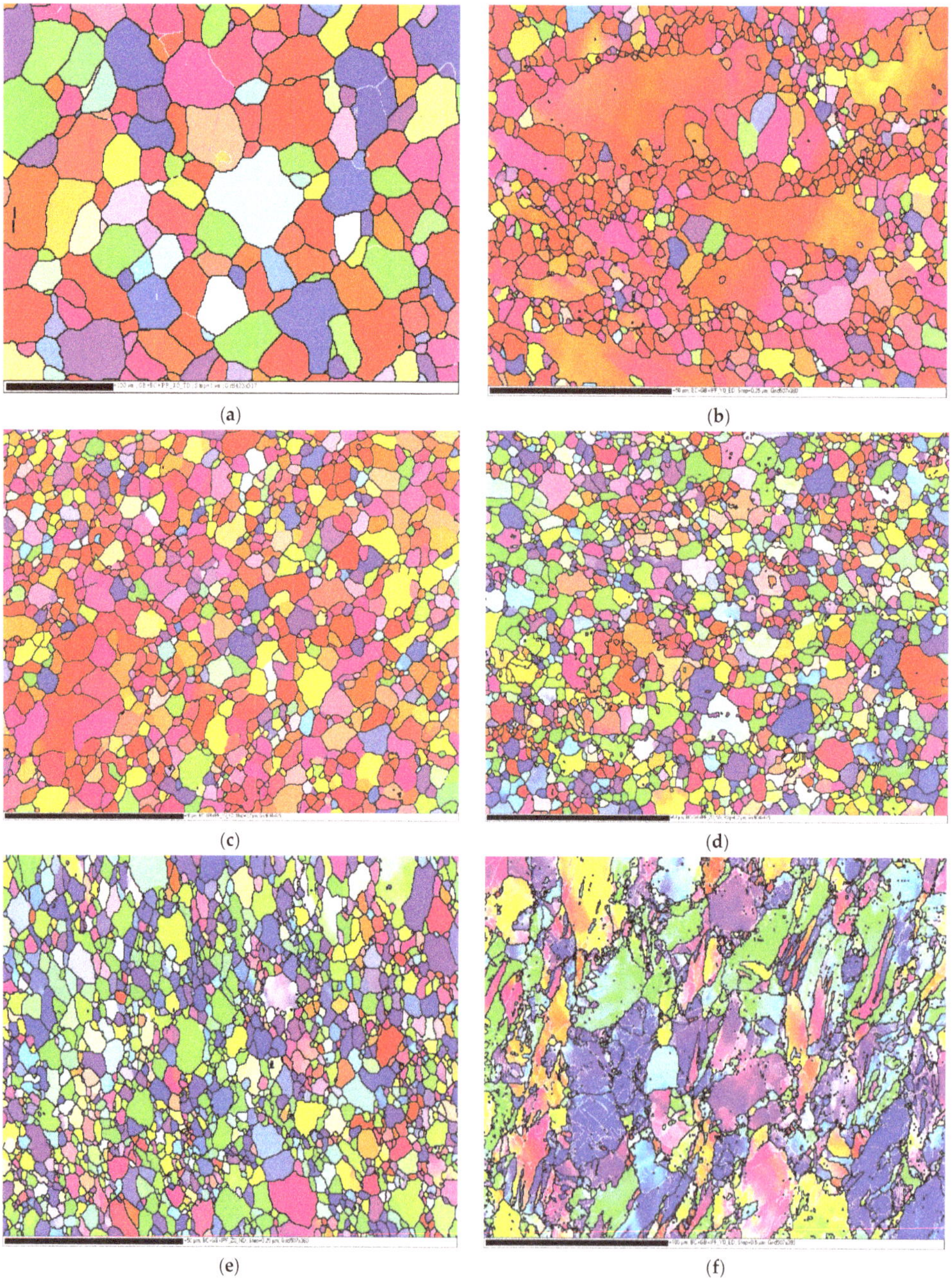

(a) (b) (c) (d) (e) (f)

Figure 2. *Cont.*

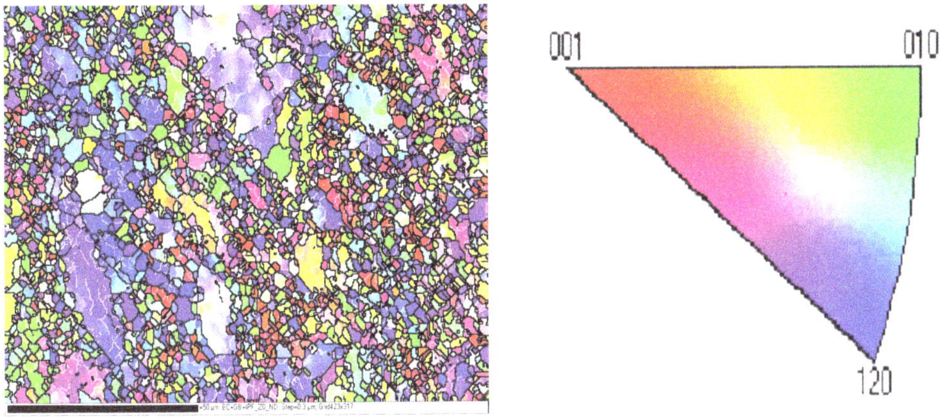

(**g**) (**h**)

Figure 2. EBSD orientation maps for the AA- ZK30 (**a**) and after the ECAP processing through 1-P (**b**), 4-A (**c**), 4-Bc (**d**), 4-C (**e**) using the 90°-die and 1-P (**f**), 4-Bc (**g**) using the 120°-die and the inverse pole figure (IPF) coloring triangle is shown in (**h**), red: [001]; blue: [120]; and green: [010].

Table 3. Grain size data of the AA and ECAPed Mg-Zn-Zr billets. All units are in μm.

	AA	90°-Die				120°-Die	
		1P	**4A**	**4Bc**	**4C**	**1P**	**4Bc**
Min	3.39	1.13	0.23	0.23	0.28	2.24	0.76
Max	76.73	38.10	14.53	11.76	12.73	35.22	17.86
Average	26.69	3.24	2.89	1.94	2.25	5.43	1.92
St. Deviation	14.74	2.42	1.92	1.54	1.60	4.22	1.09

The predicted inverse model of ZK30 grains size performed from the ECAP process is presented in Equation (4), whereas the ANOVA results are shown in Table A1 (Appendix A).

$$1/\text{Grain Size} = 0.617014 - 0.0643169 \times \text{No. of Passes} - 0.00418436 \times \text{Die Angle} + 0.0693044 \times x_1 - 0.0448229 \times x_2 + 0.00129203 \times \text{No. of Passes} \times \text{Die Angle} - 0.0415678 \times \text{No. of Passes} \times x_1 + 0.0250216 \times \text{No. of Passes} \times x_2 \quad (4)$$

Referring to Table A1, the coefficient of determination (R^2) of grain size is 0.9857, and the adjusted R^2 is 0.9732, which is close to and within 0.2 of the predicted R^2 value of 0.9155. Therefore, the obtained high values of R^2, adjusted R^2, and predicted R^2 for grain size indicate that the created model is desirable. The model terms A, B, C, AB, AC, and AD all have *p*-values that are lower than 0.05, implying that they are significant.

Similarly, the grain size model is significant with *p*-values less than 0.05, which designates that altering an input ECAP parameter would significantly affect the grain size [67], indicating that this model is satisfactory at a 95% confidence level [68]. The number of passes of the ECAP process has the greatest impact on grain size, followed by ECAP die angle, and finally, dummy variables x_1 and x_2. The adequate precision is 29.85, which is greater than four, implying that there is an adequate signal and the model could be used for navigating the design space [69].

Figure 3 is a comparison between actual experimental data and predicted values of grain size for ZK30 samples calculated by the regression model for a course of 16 iterations. It could be deduced from the figure that the bulk of the anticipated findings matches the actual experimental data extremely well with a narrow slight deviation. In addition, it indicates that the obtained regression model is adequate and could be useful for predicting the optimization of ECAP parameters for the best grain size.

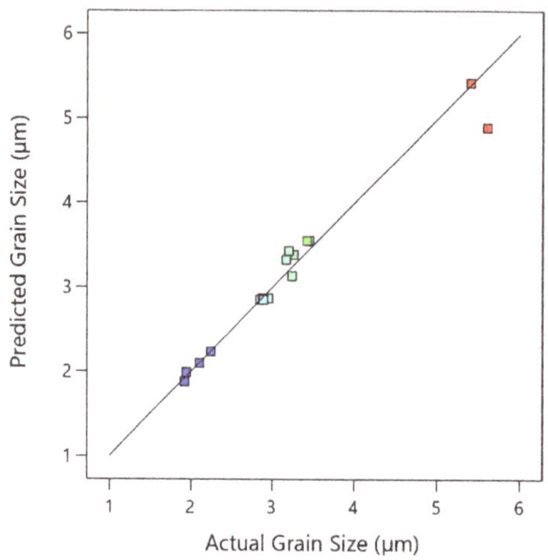

Figure 3. Predicted versus actual values of grain size, where the blue points are for minimum output value and gradually changed to red points for maximum output value.

Figure 4 illustrates three-dimensional interaction viewgraphs on the impact of the ECAP parameters on the output response of average grain size. It shows response surface plots for interaction viewgraphs between two variables, ECAP die angle and number of ECAP passes, while fixing the processing route variable. For route A, the increase of ECAP die angle resulted in increasing the grain size. Changing the number of passes has affected the grain size minimally. The minimum optimum grain size is 2.89 μm at route A, which is obtained at four passes and a 90° ECAP die angle. The die angle of the ECAP process at route Bc is proportional to grain size; moreover, the ECAP grain size shrank as the number of passes increased. In this context, the minimum optimum grain size is 1.92 μm at route Bc, which is obtained at four passes and 120° ECAP die angle. Likewise, the effect of die angle and number of passes at route C on grain size is similar to those obtained by route Bc. The minimum optimum grain size is 2.1 μm at route C, which is obtained at four passes and 120° ECAP die angle.

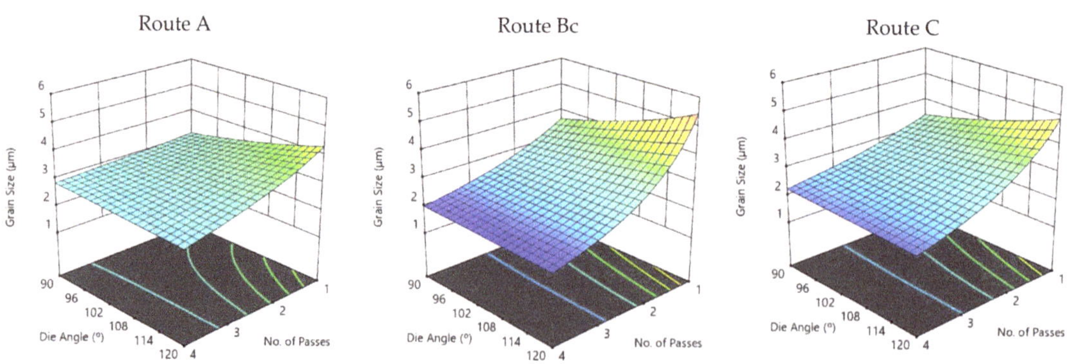

Figure 4. Three-dimensional plot of grain size with ECAP die angle and number of passes at routes A, Bc, and C.

4.1.2. Corrosion Behavior

The corrosion response of ZK30 was explained by corrosion rate and corrosion resistance. The predicted inverse model of corrosion rate is presented in Equation (5) and the linear quadratic model of corrosion resistance is presented in Equation (6).

$$1/\text{Corrosion Rate} = 16.1243 - 8.08576 \times \text{No. of Passes} - 0.0752773 \times \text{Die Angle} + 0.145218 \times x_1 - 40.4689 \times x_2 + 0.0303579 \times \text{No. of Passes} \times \text{Die Angle} + 16.7947 \times \text{No. of Passes} \times x_2 + 0.333308 \times \text{Die Angle} \times x_2 + 0.932053 \times \text{No. of Passes}^2 - 0.134907 \times \text{No. of Passes} \times \text{Die Angle} \times x_2 \quad (5)$$

$$\text{Corrosion Resistance} = 863.142 - 869.61 \times \text{No. of Passes} + 10.6176 \times \text{Die Angle} - 24.5789 \times x_1 - 2116.51 \times x_2 - 3.14036 \times \text{No. of Passes} \times \text{Die Angle} + 612.013 \times \text{No. of Passes} \times x_2 + 18.0897 \times \text{Die Angle} \times x_2 + 210.811 \times \text{No. of Passes}^2 - 5.19565 \times \text{No. of Passes} \times \text{Die Angle} \times x_2 \quad (6)$$

Table A2 (Appendix A) presents the analysis of variance (ANOVA) results for ZK30 corrosion characteristics after ECAP processing. The individual model coefficients, interaction, and quadratic terms, as well as the appropriate p-value from the ANOVA for corrosion rate and resistance, are shown in Table A2. In the case of corrosion rate, the model terms A, D, AB, AD, A^2, and ABD all have p-values that are lower than 0.05, implying that they are significant. In the case of pitting corrosion resistance, the model terms A, B, AB, and A^2 all have p-values that are lower than 0.05, implying that they are significant. Similarly, both corrosion rate and resistance models are significant with p-values less than 0.05, which designates that altering an input ECAP parameter significantly affects the corrosion rate and corrosion resistance quality criteria [67], indicating that these models are satisfactory at a 95% confidence level [68]. The number of passes of the ECAP process, factor A, has the greatest impact on corrosion rate and corrosion resistance. The adequate precision is 34.17 and 11.85 for corrosion rate and corrosion resistance, respectively, which is greater than four, implying that there is an adequate signal and the model could be used for navigating the design space [69]. The coefficient of determination (R^2) values is 0.991 and 0.9456 for corrosion rate and resistance, respectively. Additionally, the adjusted R^2 of the corrosion rate is 0.9775, which is close to and within 0.2 of the predicted R^2 value of 0.9846. In addition, the adjusted R^2 of corrosion resistance is 0.864, which is close to and within 0.2 of the predicted R^2 value of 0.8798. Therefore, the obtained high values of R^2, adjusted R^2, and predicted R^2 for corrosion rate and resistance indicate that the created model is desirable. The corrosion rate's lack of fit p-value is 0.6, which is more than 0.05, indicating an insignificant lack of fit and a good model [67].

Electrochemical experiments were carried out on the biodegradable ZK30 ECAPed at various process settings, as well as on the AA, as found in [28,29]. The measurements were carried out using ringer lactate electrolytic solution, as it mimics the human body fluids. Figure 5 illustrates the potentiodynamic polarization curves (a) and Nyquist plots (b) of ZK30 for the different ECAP process parameters.

The Tafel plot is a reliable method of corrosion resistance investigation [52]. As illustrated in Figure 5a, the 1P using the 90°-die condition showed a significant reduction in corrosion potential relative to the AA counterpart, along with a notable noble corrosion current shift toward the lower current density (I_{corr}). Additional ECAP processing passes, 4P using the 90°-die, through different routes resulted in an additional drop of corrosion current compared to the 1P, except for the 90°_4C. In addition, increasing the die angle up to 120° using route Bc resulted in significant corrosion I_{corr} reduction compared to 90°_4Bc. The I_{corr} reduction could be considered a dependable indicator for decreasing the corrosion rate. However, increasing the die angle to 120° (120°_4Bc) resulted in shifting the corrosion potential E_{corr} to more negative values.

The EIS results, Nyquist plots, were similar; however, the semicircle diameters of were dissimilar. The semicircle diameter is significantly connected to the charge resistance and, consequently, the corrosion rate. Consequently, the largest semicircle diameter represents the best corrosion resistance [70]. As shown in Figure 5b, the response of the AA- ZK30 billets was the smallest semicircle compared to the ECAPed billets. Moreover, the ECAP processing through 1P using the 90°-die showed a substantial rise in the semicircle diameter. However, further ECAP processing passes, four passes, caused an increase in the semicircle diameter compared to 1P, which might be attributable to strain buildup, which increases dislocation density [28,70,71]. On the other hand, the different routes have a substantial impact on the corrosion resistance, as shown in Figure 5b. Using route C resulted in the smallest semicircle diameter; however, route A shows an insignificant increase in the semicircle diameter compared to the 4Bc counterpart. In addition, processing through the ECAP die with 120°, 4Bc increases the semicircle diameter compared to the sample processed through 4Bc using the 90°-die which indicated higher corrosion resistance. The 120°-die's improved corrosion resistance compared to the 90°-die can be ascribed to an improvement in dislocation density during ECAP processing via the 90°-die because of the increased plastic strain, as reported earlier in the literature [34,72]. Accordingly, increasing the dislocation density resulted in decreasing the corrosion resistance.

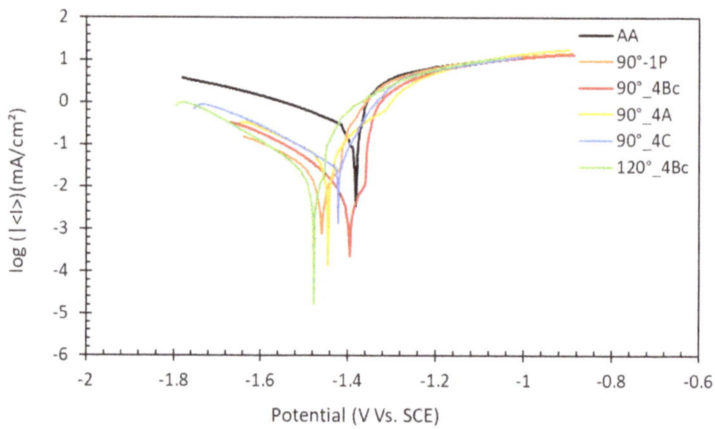

(**a**)

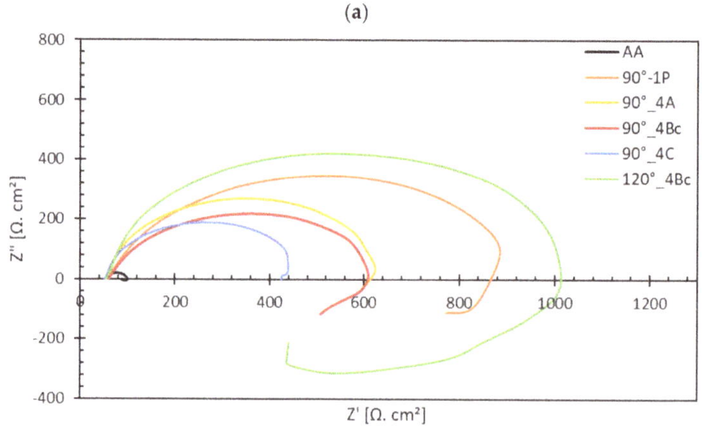

(**b**)

Figure 5. Corrosion measurements (**a**) potentiodynamic polarization curves, and (**b**) Nyquist plot of AA and ECAPed billets processed via various ECAP conditions.

Figure 6 is a comparison between actual experimental data and predicted values of corrosion rate and corrosion resistance of the ZK30 samples calculated by the regression model, for a course of 16 iterations. It could be deduced from the figure that the bulk of the anticipated findings matches the actual experimental data extremely well. In addition, it indicates that the obtained regression model is adequate and could be useful to predict the optimization of ECAP parameters for the best grain size. Figure 7 illustrates three-dimensional interaction viewgraphs on the effect of the ECAP parameters on the output response of corrosion rate and resistance. It shows response surface plots for interaction viewgraphs between two variables, ECAP die angle and number of ECAP passes, while fixing the processing route variable.

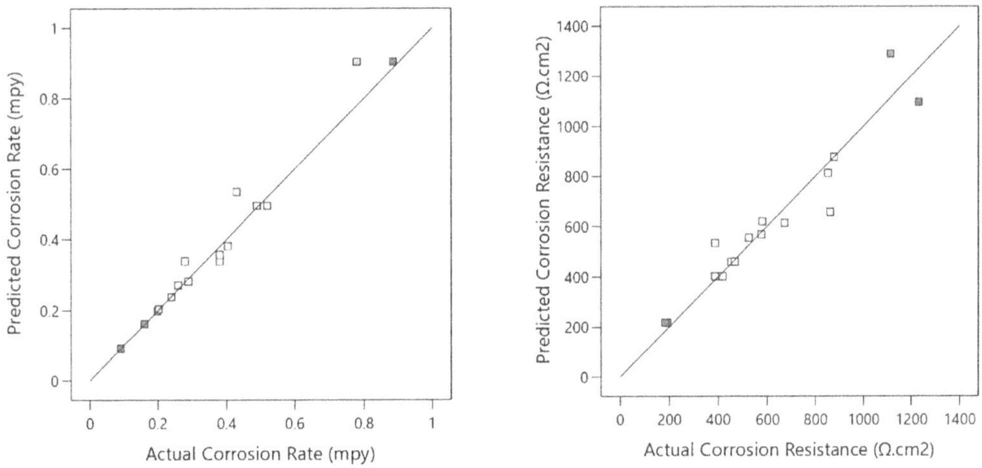

(**a**) (**b**)

Figure 6. Predicted versus actual values of the ECAP corrosion rate (**a**) and corrosion resistance (**b**), where the blue points are for minimum output value and gradually changed to red points for maximum output value.

For route A, the increase of the ECAP die angle resulted in increasing the corrosion rate. The change of the number of passes has affected the corrosion rate minimally. The minimum optimum corrosion rate is 0.198 mils per year (mpy) at route A, which is attained at one pass and a 90° ECAP die angle. The number of passes of the ECAP process at route Bc is inversely proportional to corrosion rate; moreover, the ECAP grain size decreased as the number of passes increased. Additionally, the ECAP die angle affects slightly the corrosion rate. In this context, the minimum optimum corrosion rate is 0.091 mpy at route Bc, which is obtained at four passes and 90° ECAP die angle (Figure 7a). Similarly, there is a minor effect at route C of die angle and number of passes on corrosion rate. The aforementioned results suggest that the improved corrosion rate following ECAP processing might be attributable to the obtained fine grain size (Figure 2), which is consistent with the potentiodynamic polarization findings.

Regarding corrosion resistance (Figure 7b), the corrosion resistance decreased as the ECAP number of passes at route A increased. Altering the die angle affects the corrosion resistance minimally. The maximum optimum corrosion resistance is 878 $\Omega\cdot cm^2$ at route A, which is attained at one pass and a 90° ECAP die angle. The corrosion resistance showed a decline with augmenting the number of passes nearly up to two passes, then it improved with augmenting the number of ECAP passes at route Bc. The ECAP die angle has a minor effect on corrosion resistance. In this context, the maximum optimum corrosion resistance 1232 $\Omega\cdot cm^2$ at route Bc, which is attained at one pass and 120° ECAP die angle. Similarly, the effect of die angle and number of passes at route C on corrosion resistance is similar to those obtained by route Bc. The maximum optimum corrosion resistance is 1114 $\Omega\cdot cm^2$ at route C, which is attained at one pass and 120° ECAP die angle.

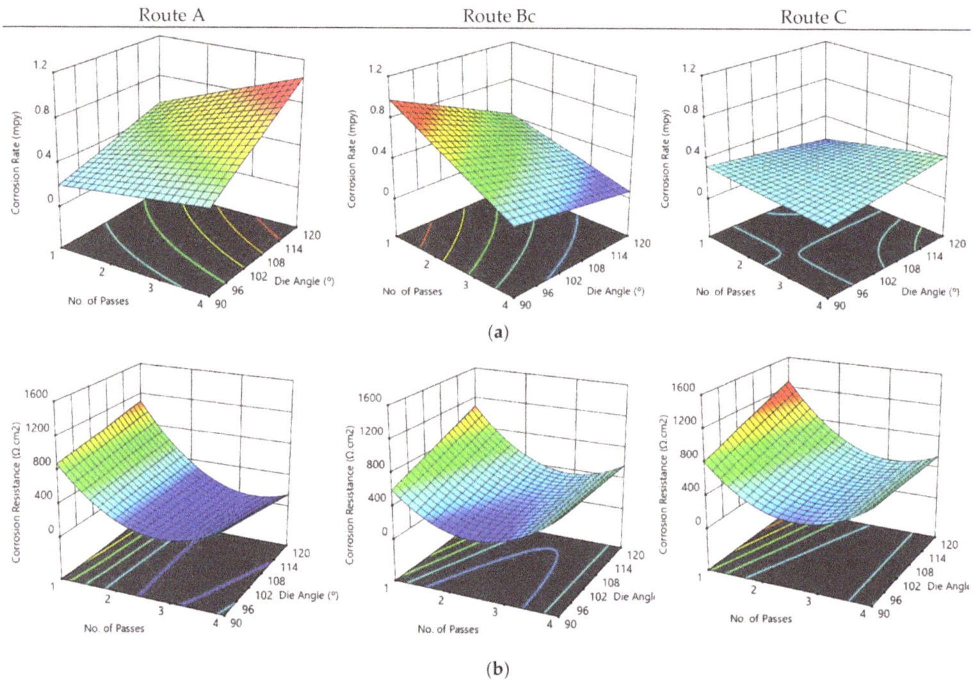

Figure 7. Three-dimensional plot of corrosion rate (**a**) and corrosion resistance (**b**) with ECAP die angle and number of passes at routes A, Bc, and C.

4.1.3. Mechanical Properties

Hardness Distribution

The inverse square root predicted models of hardness response at the center and edge of the ECAP specimen of ZK30 are presented in Equations (7) and (8).

$$1/\text{Sqrt (Hardness at center)} = +0.125954 - 0.009301 \times \text{No. of Passes} - 5.33485 \times 10^{-6} \times \text{Die Angle} - 0.021424 \times x_1 - 0.001141 \times x_2 + 0.000240 \times \text{Die Angle} \times x_1 + 0.001192 \times \text{No. of Passes}^2 \quad (7)$$

$$1/\text{Sqrt (Hardness at Edge)} = +0.101197 + 0.002123 \times \text{No. of Passes} + 0.000094 \times \text{Die Angle} - 0.016928 \times x_1 + 0.000252 \times x_2 - 0.000035 \times \text{No. of Passes} \times \text{Die Angle} - 0.000680 \times \text{No. of Passes} \times x_2 + 0.000180 \times \text{Die Angle} \times x_1 \quad (8)$$

The analysis of variance (ANOVA) results for ZK30 of the ECAP parameters on hardness response is represented in Table A3. The individual model coefficients of hardness response, interaction, and quadratic terms, as well as the appropriate p-value from the ANOVA for hardness at the center and edge, are shown in Table A3 (Appendix A).

In the instance of hardness at the edge, the p-value of every model term is lower than 0.05, indicating that every model term is significant. In contrast, the hardness at the center case has model terms for A, B, C, BC, and A^2 that are smaller than 0.05, suggesting that these model terms are significant. Similarly, both hardness at the center and edge models are significant with p-values less than 0.05, which designates that altering an ECAP parameter significantly affects both the hardness at the center and edge quality criteria [67], indicating that these models are satisfactory at a 95% confidence level [68]. The number of passes of the ECAP process has the greatest impact on both the hardness at the center and edge, followed by the ECAP die angle. The adequate precision values are 24.5 and 26.68 for the hardness at the center and edge, respectively, which is more than four, implying that there is an adequate signal and the model could be used for navigating the design space [69]. The coefficient of determination (R^2) values is 0.984 and 0.9825 for the hardness at the center and edge, respectively. Additionally, the adjusted R^2 of the hardness at the center is 0.9743, which is close to and within 0.2 of the predicted R^2 value of 0.9481. In addition, the adjusted R^2 of the hardness at the edge is 0.9671, which is close to and within 0.2 of the predicted R^2 value of 0.9245. Therefore, the obtained high values of R^2, adjusted R^2, and predicted R^2 for both the hardness at the center and edge indicate that the created model is desirable.

The relationships between the actual experimental data and the predicted response values calculated by the regression model of the hardness of the ZK30 at the center and edge are shown in Figure 8 for a course of 16 iterations. It could be deduced from the figure that the bulk of the anticipated findings matches the actual experimental data extremely well. Additionally, it indicates that the obtained regression models are adequate and could be useful to predict the optimization of ECAP parameters for the best hardness at the center and edge.

Figure 9 illustrates three-dimensional interaction viewgraphs on the effect of the ECAP parameters on the output response of hardness at the center and edge. It shows response surface plots for interaction viewgraphs between two variables, ECAP die angle and number of ECAP passes, while fixing the processing route variable. It can be seen that there is a minor effect of the ECAP die angle on hardness at the center of the specimen at routes A, Bc, and C. However, the hardness at the center showed an increase when the ECAP number of passes increased. The maximum hardness at the center at route A is attained at 87 HV with four passes and a 90° die angle. Likewise, the maximum hardness at the center at route Bc is attained at 90 HV with four passes and a 120° die angle. Regarding route C, the maximum hardness at the center is attained at 87.6 HV with four passes and 120° die angle (Figure 9a).

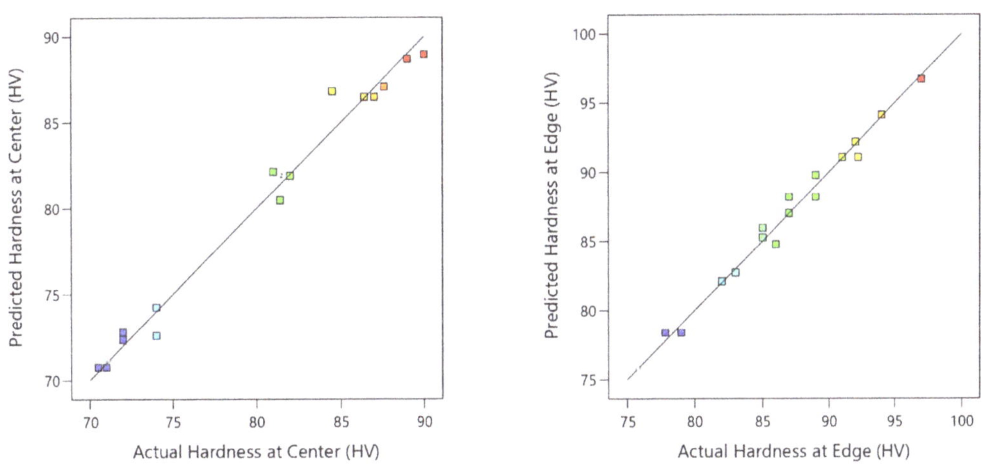

(a)

(b)

Figure 8. Predicted versus actual values of the ECAP hardness at the center (**a**), and the edge (**b**), where the blue points are for minimum output value and gradually changed to red points for maximum output value.

Regarding route A, the increase in the ECAP number of passes resulted in increasing the hardness at the edge; on the other hand, the hardness at the edge decreased with the increase in the ECAP die angle. The maximum optimum hardness at the edge at route A is attained at 92.2 HV with 4-passes, and 90° ECAP die angle. Regarding routes Bc and C, it is noticed that there is a minor effect of the ECAP die angle on hardness at the edge; in addition, a proportional effect between hardness at the edge and the number of passes is observed (Figure 9b). As a result, the maximum optimum value of hardness at the edge for routes Bc and C is 97 HV, and 92 HV, respectively, which is obtained at four passes and 120° ECAP die angle.

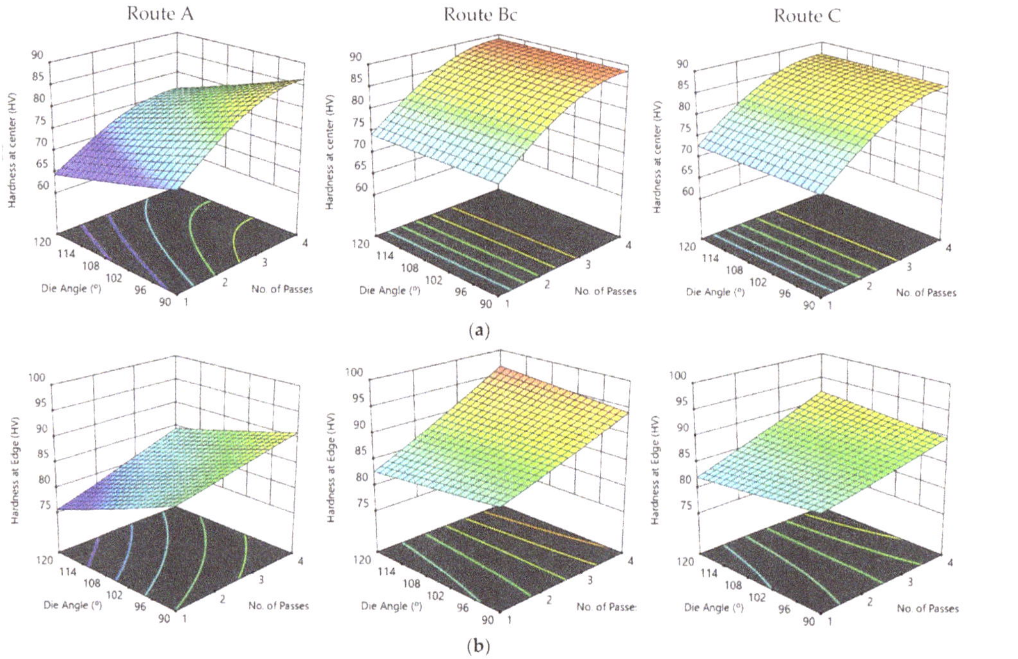

Figure 9. A three-dimensional plot of hardness at the center (**a**) and edge (**b**) with the ECAP die angle and number of passes for routes A, Bc, and C.

From the experimental and aforementioned findings, the central areas clearly exhibited lower hardness values relative to the areas at the peripheries, which could be attributed to the friction between the internal die walls and the ZK30 billets. This finding agrees with a previous study in literature [35]; in addition, increasing the number of ECAP passes resulted in increasing the hardness distribution homogeneity at both peripheral and central areas. Moreover, increasing the die angle up to 120° in both regions resulted in decreasing the hardness, which could be attributed to the decrease in the plastic strain [35]. Furthermore, increasing the number of passes resulted in hardness improvement, which could be argued to be the result of strain hardening [57,73].

Tensile Properties

The calculated tensile responses of the ZK30 specimen are yield strength (YS), tensile strength (TS), and ductility percentage (D%). Equations (9)–(11) represent the three models of tensile responses.

$$\text{YS} = 134.224 - 13.603 \times \text{No. of Passes} - 0.402982 \times \text{Die Angle} - 1.34108 \times x_1 + 3.98405 \times x_2 + 0.0973018 \times \text{No. of Passes} \times \text{Die Angle} + 0.939662 \times \text{No. of Passes}^2 \quad (9)$$

$$\text{TS} = 446.709 - 25.5164 \times \text{No. of Passes} - 1.26603 \times \text{Die Angle} - 9.29292 \times x_1 - 4.91157 \times x_2 + 0.293265 \times \text{No. of Passes} \times \text{Die Angle} + 2.6981 \times \text{No. of Passes} \times x_2 \quad (10)$$

$$\text{D\%} = 25.4808 + 2.07884 \times \text{No. of Passes} + 0.082289 \times \text{Die Angle} - 3.08458 \times x_1 + 0.894089 \times x_2 + 1.49867 \times \text{No. of Passes} \times x_1 - 1.51498 \times \text{No. of Passes} \times x_2 - 0.623325 \times \text{No. of Passes}^2 \quad (11)$$

The analysis of variance (ANOVA) results for ZK30 of the ECAP parameters on characteristic tensile responses are represented in Tables A4 and A5 (Appendix A). The individual model coefficients of characteristic tensile responses, interactions, and quadratic terms, as well as the appropriate p-value from the ANOVA for YS, TS, and D%, are shown in Tables A4 and A5. In the case of YS, the *p*-values of A, B, D, and AB are less than 0.05, indicating that these model terms are significant. In the case of TS, the *p*-values of A, B, C, and AB are less than 0.05, indicating that these model terms are significant. In the case of the D% percentage, the *p*-values of A, B, D, AC, AD, and A^2 are less than 0.05, indicating that these model terms are significant. However, the other model terms with *p*-values greater than 0.05 are insignificant. Similarly, the YS, TS, and D% percentage models are significant with *p*-values less than 0.05, which designates that changing an input ECAP parameter has a significant impact on the YS, TS, and D% quality criteria [67], indicating that these models are satisfactory at a 95% confidence level [68]. The ECAP die angle has the greatest impact on YS, TS, and D percentages followed by the number of ECAP passes.

The adequate precision values are 12.5, 19, and 29.4 for YS, TS, and D%, respectively, which is greater than four, implying that there is an adequate signal and the model could be used for navigating the design space [69]. The coefficient of determination (R^2) values is 0.9321, 0.97, and 0.9848 for the YS, TS, and D%, respectively. Additionally, the adjusted R^2 of YS is 0.886, which is close to and within 0.2 of the predicted R^2 value of 0.742. In addition, the adjusted R^2 of TS is 0.95, which is close to and within 0.2 of the predicted R^2 value of 0.906. Moreover, the adjusted R^2 of D% is 0.97, which is close to and within 0.2 of the predicted R^2 value of 0.94. Therefore, the obtained high values of R^2, adjusted R^2, and predicted R^2 for the YS, TS, and D% indicate that the created model is desirable. The YS, TS, and D% lack of fit *p*-values is greater than 0.05, indicating an insignificant lack of fit and a good model [67].

The relationships between the actual experimental data and the predicted response values calculated by the regression model of the ZK30's YS, TS, and D% are shown in Figure 10 for a course of 16 iterations. It could be deduced from the figure that the bulk of the anticipated findings matches the actual experimental data extremely well., especially in the case of YS and TS. Additionally, it indicates that the obtained regression models are adequate and could be useful for predicting the optimization of ECAP parameters for the best YS, TS, and D percentages.

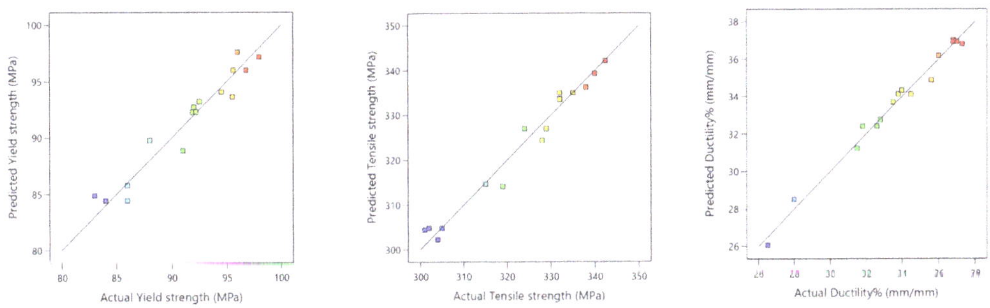

(a) (b) (c)

Figure 10. Predicted versus actual values of the ECAP YS (**a**), TS (**b**), and D% (**c**), where the blue points are for minimum output value and gradually changed to red points for maximum output value.

Figure 11 illustrates three-dimensional interaction viewgraphs on the effect of the ECAP parameters on the output response of YS, TS, and D%. It shows response surface plots for interaction viewgraphs between two variables, ECAP die angle and number of ECAP passes, while fixing the processing route variable. It is observed that the ECAP die angle and number of passes showed a similar effect on YS at different routes. It is seen that for routes A, Bc, and C, the YS increases with the decrease in the ECAP die angle and the increase in the number of passes. The maximum YS at route A is attained at 92.2 MPa with four passes and a 90° die angle. Moreover, the maximum YS at route Bc is attained at 98 MPa with four passes and a 120° die angle. Regarding route C, the maximum YS is attained at 95.5 MPa with four passes and a 90° die angle (Figure 11a). The TS of the ECAP process showed an alike trend for routes A, Bc, and C. The values of the maximum TS are 329 MPa, 342.4 MPa, and 338 MPa for routes

A, Bc, and C, respectively, which were attained with four passes and 90° ECAP die angle (Figure 11b). For routes A, Bc and C, the D% decreased with the decrease of the ECAP die angle. The maximum D% at routes Bc and C is attained at 36.19 and 36.81, respectively, with one pass and 120° die angle. Moreover, the maximum D% at route A is attained at 34.79 with two passes and a 120° die angle (Figure 11c).

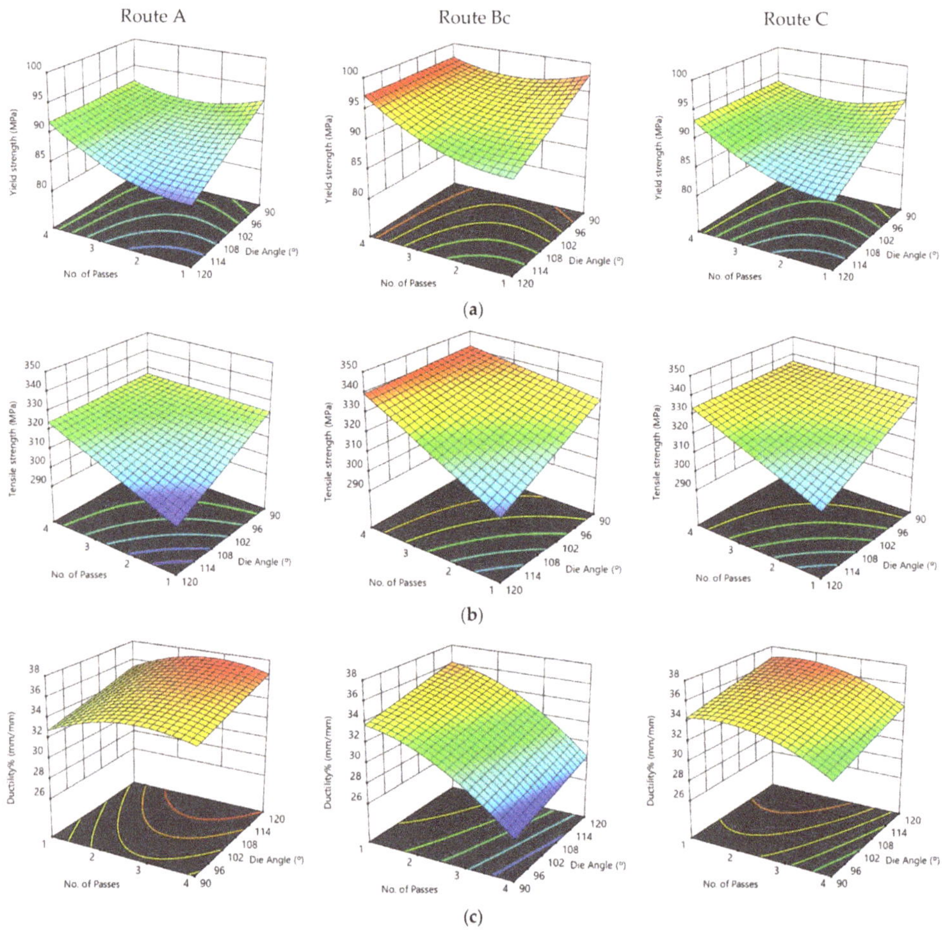

Figure 11. Three-dimensional plot of YS (**a**), TS, (**b**) and D% (**c**) with the ECAP die angle and number of passes at routes A, Bc, and C.

From the tensile results, it can be observed that ECAPed ZK30 billets displayed a momentous improvement in YS and TS without showing a substantial drop in D% in comparison with the AA counterparts. Furthermore, it was shown that the processing conditions at 90° ECAP die angle, four passes, and route Bc revealed the best YS. Consequently, route Bc showed to be the furthermost efficient route in enhancing the YS because of the substantial grain size decrease, as shown in Figure 2. The UFG obtained through ECAP processing via multiple passes could be associated to be the main reason for hindering the dislocation motion [74,75]. Therefore, the grain refining mechanism is the most efficient strengthening mechanism, which led to an enhancement in the mechanical properties. Additionally, the adoption of the 90° die angle caused a significant improvement in the YS, which could be attributed to the higher plastic strain. Furthermore, the shear strain accumulation resulting from the ECAP passes of up to four passes could be assigned to the dislocation density growth, which hinders the dislocation mobility [76]. Moreover, the D% reduction after the ECAP processing could be associated with grain refinement. Additionally, better D% was observed at route Bc with a 120° die angle and four passes, compared with route Bc with a 90° die angle and four passes, which could be assigned to imposing lower strain as reported in [29]. In the same context, route Bc exhibited the highest grain refinement; therefore, it shows lower D% compared to the remaining studied route types of A and C. Consequently, route Bc could be considered the most effective route type in this perspective.

4.2. Genetic Algorithm Results

4.2.1. Optimization of Grain Size

A minimization was considered for the grain size response presented in Equation (4), which was set to be the objective function using GA and subjected to the ECAP boundary conditions of number of passes, ECAP die angle, dummy variable x_1, and dummy variable x_2. It is presented as follows:

Minimize grain size (number of passes. ECAP die angle, x_1, x_2)

Subjected to ranges of ECAP conditions:

$1 \leq$ No. of passes ≤ 4 (pass);

$90 \leq$ die angle ≤ 120 (°);

Dummy variable $x_1 \in [0, 1]$;

Dummy variable $x_2 \in [0, 1]$.

The optimization technique of GA was done through MATLAB, where the performance of fitness value and the results of the run solver view displayed the minimum possible grain size subjected to the ECAP boundary conditions. The best value of grain size by GA is 1.8759 μm, which was attained at route Bc with four passes and 120° ECAP die angle, as shown in Figure 12a. The grain size value of RSM compared with the GA technique is 1.882 μm and 1.875 μm, respectively.

A hybrid RSM-GA was performed to enhance the obtained GA results of grain size response. The starting population of hybrid RSM-GA was based on RSM optimum ECAP conditions of four passes, 120° ECAP die angle, 0 for the dummy variable x_1, and 1 for the dummy variable x_2. The minimum optimum grain size value obtained from the hybrid RSM-GA is 1.875 μm, which was better than its counterpart response obtained by RSM at route Bc with four passes and 120° ECAP die angle, as shown in Figure 12b.

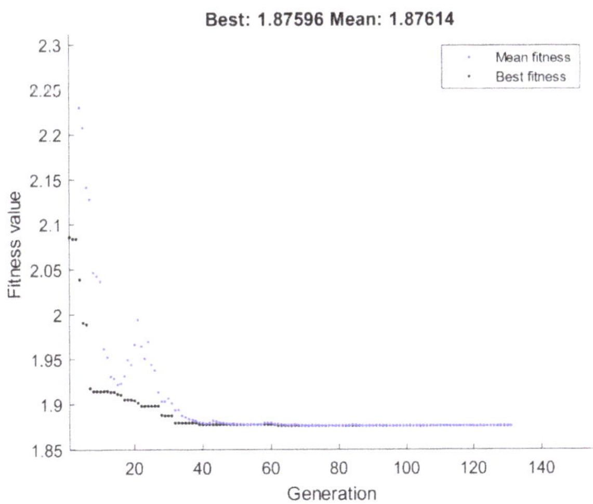

(**a**)

Figure 12. *Cont.*

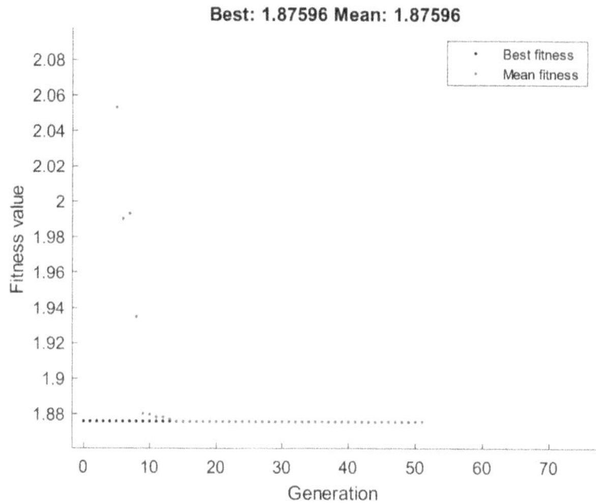

(b)

Figure 12. Optimum grain size by GA (**a**) and hybrid RSM-GA (**b**).

4.2.2. Optimization of Corrosion Response

The optimization of corrosion response by GA is shown in Figure 13. A minimization was considered for the corrosion rate response presented in Equation (5), which was set to be the fitness function and subjected to the ECAP boundary conditions of number of passes, ECAP die angle, dummy variable x_1, and dummy variable x_2. The best value of corrosion rate by GA is 0.0909 mpy, which was attained at route Bc with four passes and 90° ECAP die angle, as shown in Figure 13a. The corrosion rate values of RSM compared with the GA technique are 0.091 mpy and 0.090 mpy, respectively.

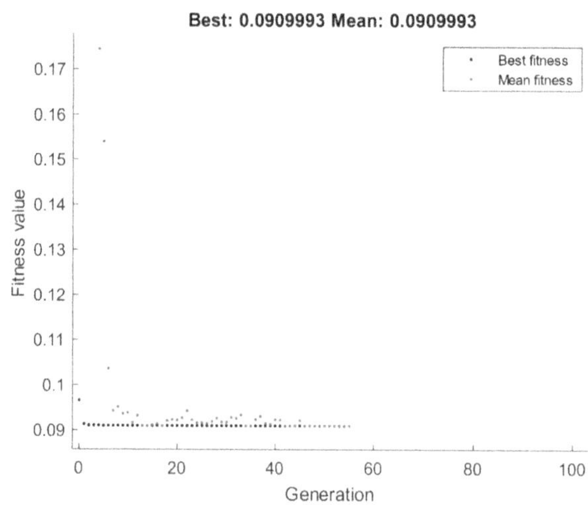

(a)

Figure 13. *Cont.*

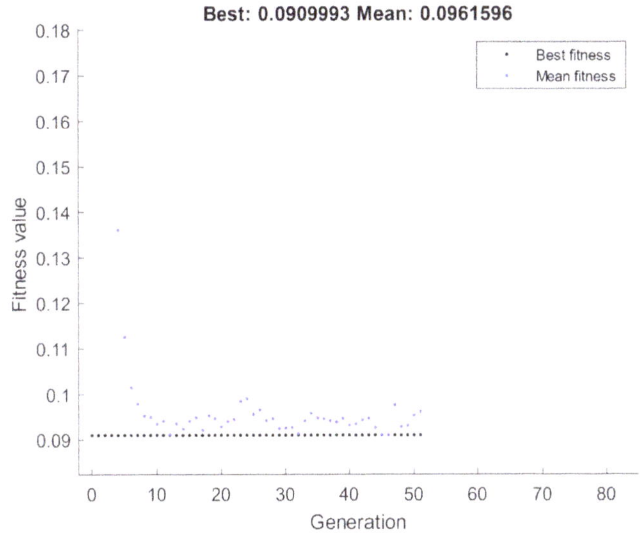

(**b**)

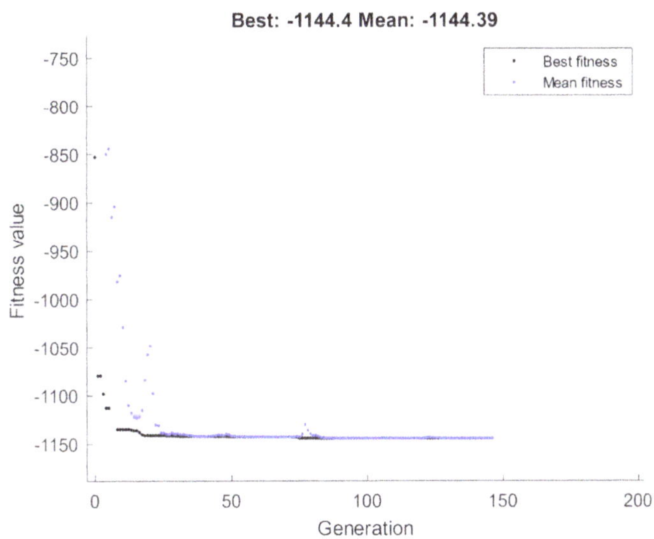

(**c**)

Figure 13. *Cont.*

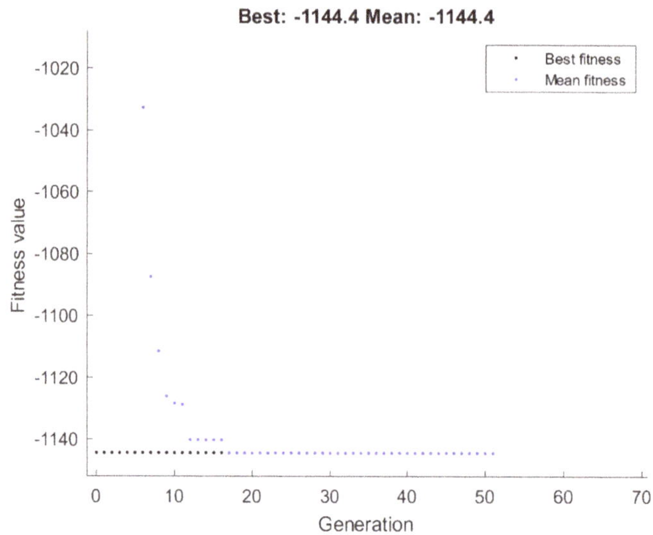

(d)

Figure 13. Optimum corrosion rate (**a**,**b**) and corrosion resistance (**c**,**d**) by GA (**a**,**c**) and hybrid RSM-GA (**b**,**d**).

Conversely, a maximization was considered for the corrosion resistance response presented in Equation (6), which was set to be the fitness function and subjected to the ECAP boundary conditions of number of passes, ECAP die angle, dummy variable x_1, and dummy variable x_2. The best value of corrosion resistance by GA is 1144 $\Omega \cdot cm^2$, which was attained at route Bc with one pass and 120° ECAP die angle, as shown in Figure 13c. The corrosion resistance value of RSM compared with the GA technique is 1149 $\Omega \cdot cm^2$ and 1144 $\Omega \cdot cm^2$, respectively.

A hybrid RSM-GA was performed to enhance the obtained GA results of corrosion response. The starting population of hybrid RSM-GA was based on RSM optimum ECAP conditions of corrosion rate and resistance. The minimum optimum corrosion rate obtained from the hybrid RSM-GA (Figure 13b) is 0.090 mpy, which is better than its counterpart response obtained by RSM at route Bc with four passes and 90° ECAP die angle. Moreover, the maximum optimum corrosion resistance obtained from the hybrid RSM-GA is 1144 $\Omega \cdot cm^2$ which is better than its counterpart response obtained by RSM at route Bc with one pass and 120° ECAP die angle, as shown in Figure 13d.

4.2.3. Optimization of Hardness Response

The optimization of hardness response by GA is shown in Figure 14. A maximization was considered for the hardness at the center and edge responses presented in Equations (7) and (8), which were set to be the fitness functions and subjected to the ECAP boundary conditions of number of passes, ECAP die angle, dummy variable x_1, and dummy variable x_2. The best values of hardness at the center and edge by GA are 88.936 HV and 96.7 HV, respectively, which were attained at route Bc with four passes and 120° ECAP die angle, as shown in Figure 14a,c.

A hybrid RSM-GA was performed to enhance the obtained GA results of hardness at the center and edge responses. The maximum optimum hardness at the center and edge values obtained from the hybrid RSM-GA are 88.936 HV and 96.7 HV, respectively, which were better than its counterpart responses obtained by RSM at route Bc with four passes and 120° ECAP die angle, as shown in Figure 14b,d.

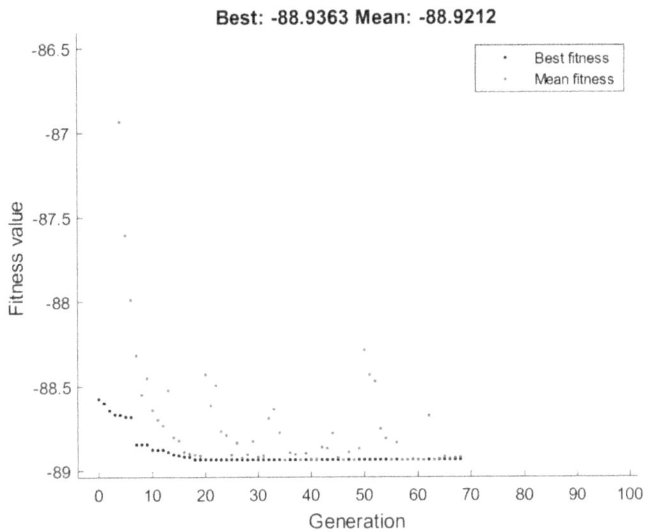

(a)

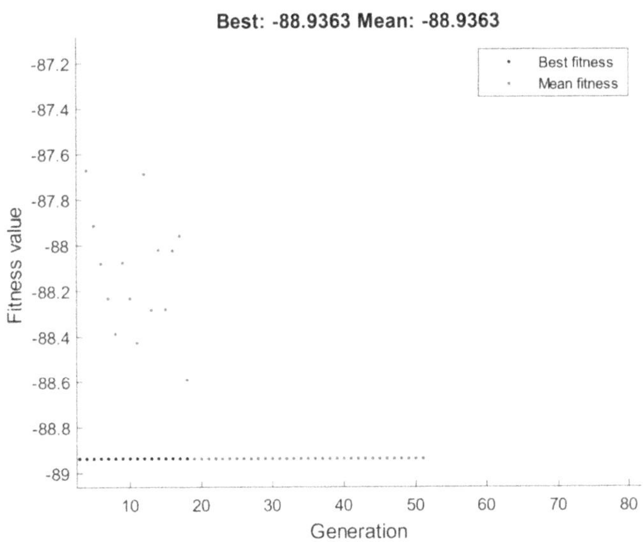

(b)

Figure 14. *Cont.*

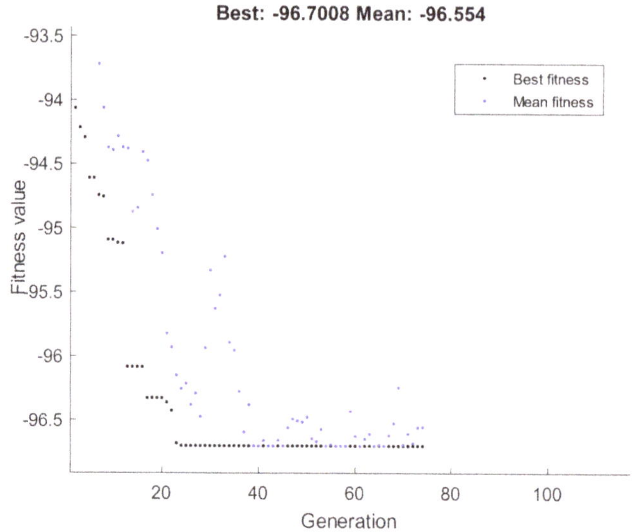

(**c**)

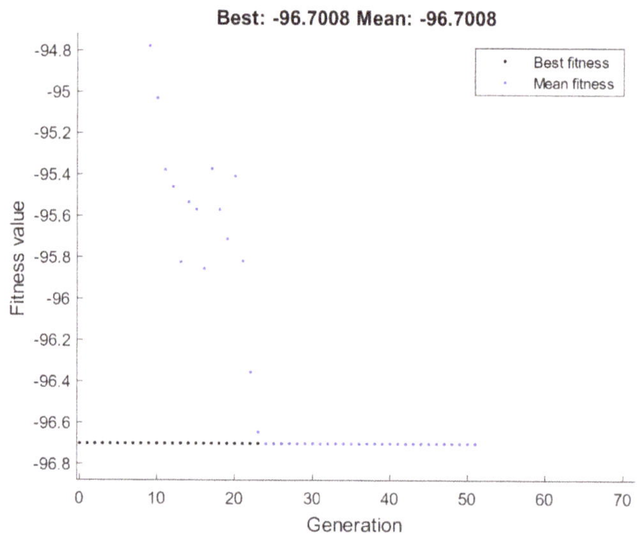

(**d**)

Figure 14. Optimum hardness at center (**a**,**b**) and edge (**c**,**d**) by GA (**a**,**c**) and hybrid RSM-GA (**b**,**d**).

4.2.4. Optimization of Tensile Response

The optimization of tensile response by GA is shown in Figure 15. A maximization was considered for the YS and TS responses presented in Equations (9) and (10), which were set to be the fitness functions and subjected to the ECAP boundary conditions of number of passes, ECAP die angle, dummy variable x_1, and dummy variable x_2. The best values of YS and TS by GA are 97.58 MPa, and 342.157 MPa, respectively, which were attained at route Bc with four passes and 90° ECAP die angle, as shown in Figure 15a,c.

In addition, a maximization was considered for the D% percentage response presented in Equation (11), which was set to be the fitness function and subjected to the ECAP boundary conditions of number of passes, ECAP die angle, dummy variable x1, and dummy variable x2. The best value of D% by GA is 36.19, which was attained at route Bc with one pass and 120° ECAP die angle, as shown in Figure 15e.

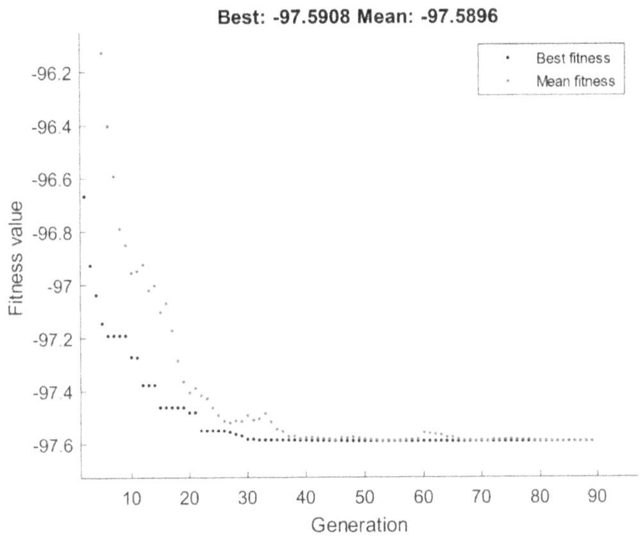

(a)

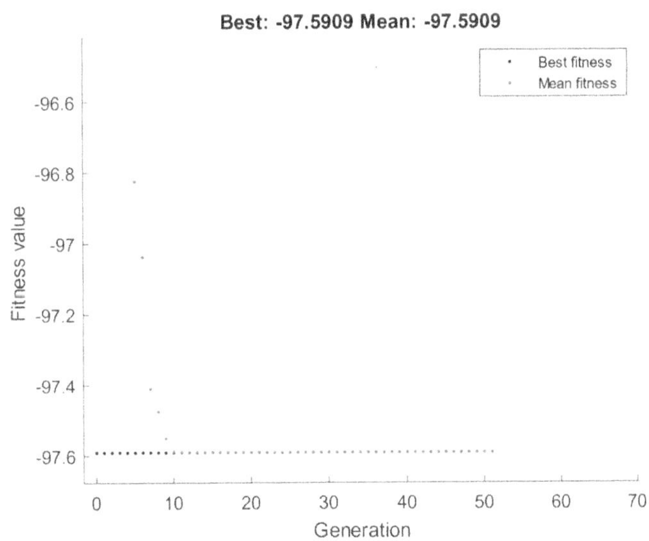

(b)

Figure 15. *Cont.*

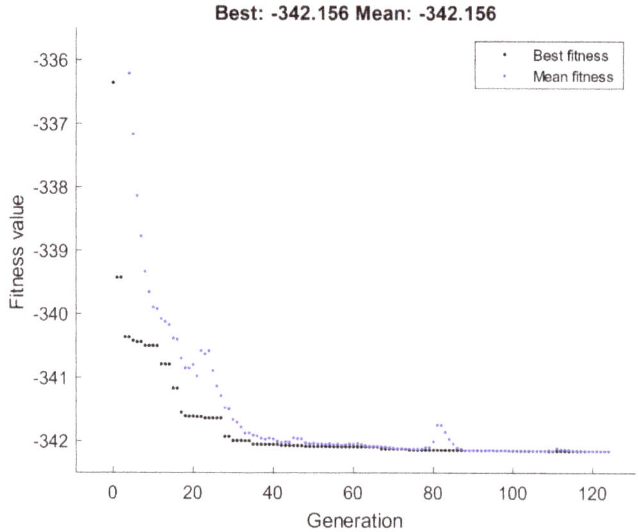

(c)

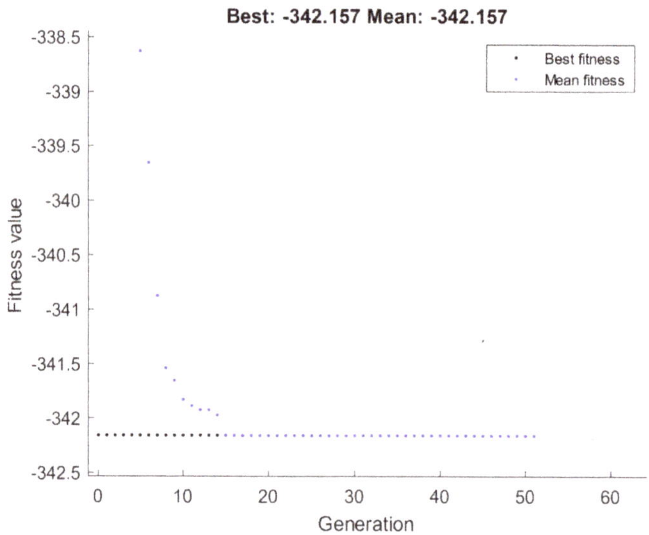

(d)

Figure 15. *Cont.*

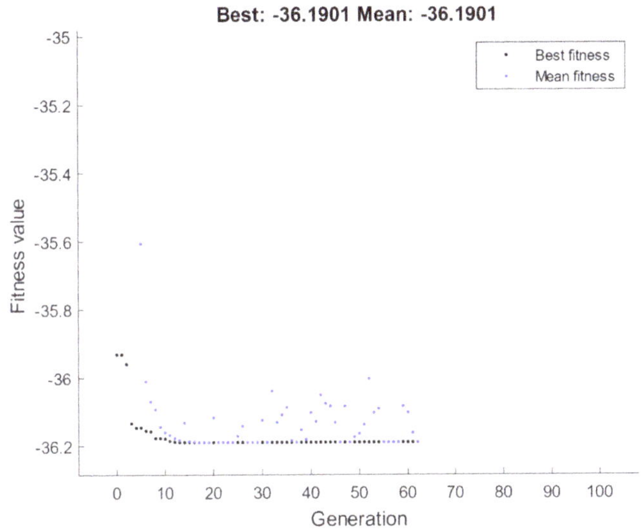

(**e**)

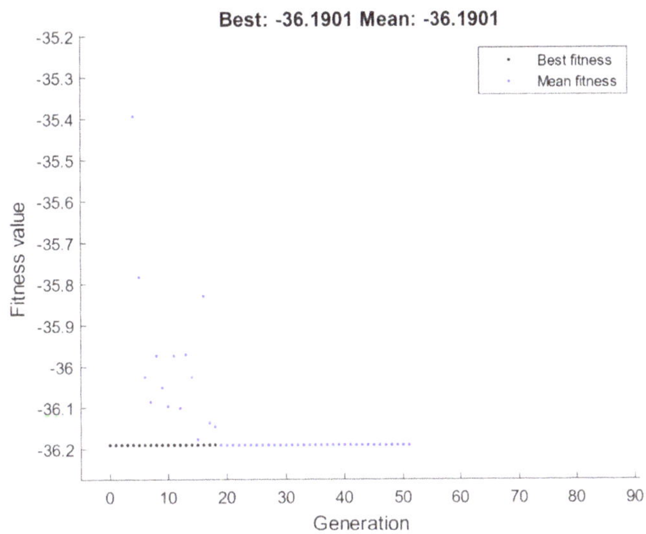

(**f**)

Figure 15. Optimum tensile response YS (**a**,**b**), TS (**c**,**d**), and D% (**e**,**f**) by GA (**a**,**c**,**e**) and hybrid RSM-GA (**b**,**d**,**f**).

A hybrid RSM-GA was performed to enhance the obtained GA results of YS and TS and D% percentage responses. The maximum optimum YS and TS values obtained from the hybrid RSM-GA are 97.59 MPa and 342.157 MPa, respectively, which were better than their counterparts obtained by RSM at route Bc with four passes and 90° ECAP die angle, as shown in Figure 15b,d. Moreover, the maximum optimum D% obtained from the hybrid RSM-GA is 36.19, which is better than its counterpart response obtained by RSM at route Bc with one pass and 120° ECAP die angle, as shown in Figure 15f. Table 4 summarizes the comparison of ECAP response values at experimental, RSM, GA, and hybrid RSM-GA cases.

Table 4. Summary of the results of the ECAP process.

Response		Experimental	RSM	GA	RSM-GA
Grain Size (μm)	Value	1.92	1.8821	1.8759	1.8759
	Cond.	4 passes, 120°, Route Bc	4 passes, 117°, Route Bc	4 passes, 120°, Route Bc	4 passes, 120°, Route Bc
Corrosion rate (mpy)	Value	0.091	0.09109	0.0909	0.0909
	Cond.	4 passes, 90°, Route Bc	4 passes, 90°, Route Bc	4 passes, 90°, Route Bc	4 passes, 90°, Route Bc
Corrosion resistance ($\Omega\cdot cm^2$)	Value	1232	1149	1144	1144
	Cond.	1 pass, 120°, Route Bc	1 pass, 120°, Route Bc	1 pass, 120°, Route Bc	1 pass, 120°, Route Bc
Hardness at center (HV)	Value	90	88.9517	88.936	88.936
	Cond.	4 passes, 120°, Route Bc	4 passes, 120°, Route Bc	4 passes, 120°, Route Bc	4 passes, 120°, Route Bc
Hardness at edge (HV)	Value	97	96.7099	96.7008	96.7008
	Cond.	4 passes, 120°, Route Bc	4 passes, 120°, Route Bc	4 passes, 120°, Route Bc	4 passes, 120°, Route Bc
YS (MPa)	Value	98	98.0049	97.5896	97.5909
	Cond.	4 passes, 120°, Route Bc	1 pass, 90°, Route Bc	4 pass, 90°, Route Bc	4 pass, 90°, Route Bc
TS (MPa)	Value	342.4	342.156	342.157	342.157
	Cond.	4 passes, 90°, Route Bc	4 passes, 90°, Route Bc	4 passes, 90°, Route Bc	4 passes, 90°, Route Bc
D% (mm/mm)	Value	37.3	36.19	36.19	36.19
	Cond.	1 pass, 120°, Route C	1 pass, 120°, Route Bc	1 pass, 120°, Route Bc	1 pass, 120°, Route Bc

4.3. *Validation of GA*

In this part, the optimal ECAP parameters of the various responses presented in this context of the grain size, corrosion response, hardness properties, and tensile response are displayed. The presented optimal ECAP parameters of the ECAP die angle, number of passes, and processing route types were chosen based on earlier studies in the literature of Mg-alloys that suggested the adoption of the ECAP die angle from 70° to 135° and a number of passes from one to twelve [75,77–79]. Table 5 shows the optimal conditions of the ECAP process for the various responses by GA and hybrid RSM-GA.

Table 5. Validated ECAP response based on previous studies.

Response		GA	RSM-GA
Grain Size (μm)	Value	0.6139	0.6139
	Cond.	12 passes, 135°, Route Bc	12 passes, 135°, Route Bc
Corrosion rate (mpy)	Value	0.0069	0.0069
	Cond.	12 passes, 70°, Route Bc	12 passes, 70°, Route Bc
Corrosion resistance ($\Omega\cdot cm^2$)	Value	21,019.5	21,019.5
	Cond.	12 pass, 70°, Route Bc	12 pass, 70°, Route Bc
Hardness at center (HV)	Value	89.0707	89.0707
	Cond.	4 passes, 135°, Route Bc	4 passes, 135°, Route Bc
Hardness at edge (HV)	Value	178.73	178.73
	Cond.	12 passes, 135°, Route Bc	12 passes, 135°, Route Bc
YS (MPa)	Value	213.51	213.509
	Cond.	12 passes, 135°, Route Bc	12 passes, 135°, Route Bc
TS (MPa)	Value	472.153	472.153
	Cond.	12 passes, 135°, Route Bc	12 passes, 135°, Route Bc
D% (mm/mm)	Value	37.42	37.42
	Cond.	1 pass, 135°, Route Bc	1 pass, 135°, Route Bc

5. Conclusions

In this study, biodegradable Mg-Zn-Zr alloy billets were processed using ECAP utilizing two ECAP dies with internal angles of 90° and 120°. At a temperature of 250 °C, several ECAP routes (A, Bc, and C) were employed, as well as varied passes (one pass, two passes, and four passes). The influence of ECAP conditions on microstructural development, corrosion behavior, tensile characteristics, and Vicker's microhardness was thoroughly examined. To optimize the ECAP processing parameters of an Mg-Zn-Zr alloy, RSM, ANOVA, GA, and RSM-GA were used. The following conclusions could be drawn:

1. The predicted results were very close to the actual experimental results with a narrow slight deviation.
2. The obtained regression models are adequate and could be useful to predict the optimization of ECAP parameters.
3. Route Bc is the most effective route in grain refinement
4. ECAP processing through four passes of route Bc displayed a more homogenous distribution of the ultrafine grains
5. For the multiple passes, the ECAP die angle has an insignificant effect on refining the grain size compared to the effect of the ECAP route type.
6. ECAP processing via 4Bc through the 90°-degree die revealed a better corrosion rate at 0.091mpy.
7. The 120°-die revealed higher corrosion resistance compared to the 90°-die.
8. 4Bc through the 120°-die resulted in enhancing the hardness by 86.5% relative to the AA counterpart.
9. 4Bc through the 90°-die revealed the best TS, while 2C through the 120°-die showed the best ductility at fracture.

Author Contributions: Conceptualization, A.I.A. and W.H.E.-G.; methodology, M.O.A. and A.B.; software, H.K. and S.E.-S.; validation, W.H.E.-G., M.O.A. and M.K.; formal analysis, A.I.A. and S.E.; investigation, M.O.A. and A.B.; resources, M.O.A.; data curation, H.K., S.E.-S., S.E. and M.K.; writing—original draft preparation, S.E.-S. and S.E.; writing—review and editing, W.H.E.-G. and M.K.; visualization, M.O.A.; supervision, A.I.A. and W.H.E.-G.; project administration, A.I.A., W.H.E.-G. and M.O.A.; All authors have read and agreed to the published version of the manuscript.

Funding: The Princess Nourah bint Abdulrahman University Researchers Supporting Project number PNURSP2022R230, Princess Nourah bint Abdulrahman University, Riyadh, Saudi Arabia.

Institutional Review Board Statement: Not applicable.

Informed Consent Statement: Not applicable.

Data Availability Statement: All the raw data supporting the conclusion of this paper were provided by the authors.

Acknowledgments: The authors acknowledge the Princess Nourah bint Abdulrahman University Researchers Supporting Project number PNURSP2022R230, Princess Nourah bint Abdulrahman University, Riyadh, Saudi Arabia.

Conflicts of Interest: The authors declare no conflict of interest.

Nomenclature

ECAP	Equal channel angular pressing
RSM	Response Surface Methodology
ANOVA	Analysis of Variance
HCP	Hexagonal close-packed
SPD	Severe plastic deformation
UFG	Ultra-Fine Grain
ε_{eq}	The equivalent strain

ϕ	ECAP die angle
Ψ	Outer corner angle
N	Number of passes
EBSD	Electron backscatter Diffraction
(SCE)	Saturated calomel electrode
(E_{corr})	Open-circuit potential
(EIS)	Electrochemical impedance spectroscopy
Hv	Vicker's microhardness test
GA	Genetic Algorithm
Y	Output response
f	The ECAP process independent input factors,
ε	The random error distributed about the response Y
YS	Ultimate tensile strength
TS	Tensile Strength
D	Ductility
x_1 and x_2	Dummy variables
R^2	Regression Coefficient

Appendix A

Table A1. Analysis of variance of grain size.

Source	DF	Grain Size (μm)			
		Sum of Squares	Mean Square	F-Value	*p*-Value
Model	7	0.1491	0.0213	78.74	<0.0001 [a]
A-No. of Passes	1	0.0511	0.0511	188.98	<0.0001 [a]
B-Die Angle	1	0.0029	0.0029	10.89	0.0109 [a]
C-Dummy x_1	1	0.0029	0.0029	10.55	0.0117 [a]
D-Dummy x_2	1	0.0008	0.0008	3.13	0.1147
AB	1	0.0063	0.0063	23.17	0.0013 [a]
AC	1	0.0057	0.0057	21.18	0.0018 [a]
AD	1	0.0025	0.0025	9.41	0.0154 [a]
Residual	8	0.0022	0.0003		
Lack of Fit	5	0.0021	0.0004	25.92	0.0113 [a]
Pure Error	3	0.0000	0.0000		
Cor Total	15	0.1513			
Fit Statistics		Std. Dev.	C.V. %	Adeq Precision	
		0.0164	4.76	29.8545	
		R^2	Adjusted R^2	Predicted R^2	
		0.9857	0.9732	0.9155	

[a] Within a 95% confidence interval, parameters referring to cells are significant.

Table A2. Analysis of variance of corrosion rate and corrosion resistance.

Source	DF	Corrosion Rate (mpy)				Corrosion Resistance ($\Omega \cdot cm^2$)				
		Sum of Squares	Mean Square	F-Value	*p*-Value	DF	Sum of Squares	Mean Square	F-Value	*p*-Value
Model	9	86.58	9.62	73.35	<0.0001 [a]	9	1.156×10^6	1.284×10^5	11.59	0.0037 [a]
A-No. of Passes	1	15.46	15.46	117.88	<0.0001	1	1.674×10^5	1.674×10^5	15.11	0.0081
B-Die Angle	1	0.0058	0.0058	0.0441	0.8405	1	88,102.62	88,102.62	7.95	0.0304
C-Dummy x1	1	0.0425	0.0425	0.3240	0.5898	1	1217.40	1217.40	0.1099	0.7516
D-Dummy x2	1	2.50	2.50	19.08	0.0047	1	5343.05	5343.05	0.4822	0.5134
AB	1	5.07	5.07	38.63	0.0008	1	1.212×10^5	1.212×10^5	10.94	0.0163
AD	1	26.95	26.95	205.47	<0.0001	1	17,219.21	17,219.21	1.55	0.2590
BD	1	0.0076	0.0076	0.0578	0.8180	1	12,575.49	12,575.49	1.13	0.3277
A^2	1	5.31	5.31	40.45	0.0007	1	2.714×10^5	2.714×10^5	24.49	0.0026

Table A2. *Cont.*

Source	DF	Corrosion Rate (mpy)				DF	Corrosion Resistance (Ω·cm²)			
		Sum of Squares	Mean Square	F-Value	*p*-Value		Sum of Squares	Mean Square	F-Value	*p*-Value
ABD	1	15.34	15.34	116.93	<0.0001	1	22,745.80	22,745.80	2.05	0.2019
Residual	6	0.7869	0.1311			6	66,483.81	11,080.64		
Lack of Fit	3	0.3265	0.1088	0.7093	0.6078 [an]	3	63,550.87	21,183.62	21.67	0.0155 [a]
Pure Error	3	0.4604	0.1535			3	2932.94	977.65		
Cor Total	15	87.36				15	1.222×10^6			
Fit Statistics		Std. Dev.	C.V. %	Adeq Precision			Std. Dev.	C.V. %	Adeq Precision	
		0.3621	9.87	34.1733			105.26	17.45	11.8548	
		R^2	Adjusted R^2	Predicted R^2			R^2	Adjusted R^2	Predicted R^2	
		0.9910	0.9775	0.9846			0.9456	0.8640	0.8798	

[a] Within a 95% confidence interval, parameters referring to filled cells are the significant, [an] are the insignificant terms.

Table A3. Analysis of variance of hardness at the center and edge.

Source	DF	Hardness at Center (HV)			
		Sum of Squares	Mean Square	F-Value	*p*-Value
Model	6	0.0004	0.0001	95.70	<0.0001 [a]
A-No. of Passes	1	0.0002	0.0002	372.73	<0.0001
B-Die Angle	1	0.0000	0.0000	46.84	<0.0001
C-Dummy X1	1	0.0000	0.0000	50.78	<0.0001
D-Dummy X2	1	3.150×10^{-6}	3.150×10^{-6}	5.03	0.0516
BC	1	0.0000	0.0000	49.59	<0.0001
A^2	1	0.0000	0.0000	19.34	0.0017
Residual	9	5.637×10^{-6}	6.263×10^{-7}		
Lack of Fit	6	5.480×10^{-6}	9.133×10^{-7}	17.42	0.0197 [a]
Pure Error	3	1.573×10^{-7}	5.243×10^{-8}		
Cor Total	15	0.0004			
Fit Statistics		Std. Dev.	C.V. %	Adeq Precision	
		0.0008	0.7072	24.5495	
		R^2	Adjusted R^2	Predicted R^2	
		0.9846	0.9743	0.9481	

Source	DF	Hardness at Edge (HV)			
		Sum of Squares	Mean Square	F-Value	*p*-Value
Model	7	0.0002	0.0000	63.99	<0.0001 [a]
A-No. of Passes	1	0.0001	0.0001	198.13	<0.0001
B-Die Angle	1	0.0000	0.0000	72.47	<0.0001
C-Dummy X1	1	9.004×10^{-6}	9.004×10^{-6}	25.34	0.0010
D-Dummy X2	1	5.651×10^{-6}	5.651×10^{-6}	15.91	0.0040
AB	1	5.436×10^{-6}	5.436×10^{-6}	15.30	0.0045
AD	1	2.090×10^{-6}	2.090×10^{-6}	5.88	0.0415
BC	1	0.0000	0.0000	64.43	<0.0001
Residual	8	2.842×10^{-6}	3.553×10^{-7}		
Lack of Fit	5	1.501×10^{-6}	3.001×10^{-7}	0.6711	0.6756 [an]
Pure Error	3	1.342×10^{-6}	4.472×10^{-7}		
Cor Total	15	0.0002			
Fit Statistics		Std. Dev.	C.V. %	Adeq Precision	
		0.0006	0.5560	26.6897	
		R^2	Adjusted R^2	Predicted R^2	
		0.9825	0.9671	0.9245	

[a] Within a 95% confidence interval, parameters referring to filled cells are significant, [an] are insignificant terms.

Table A4. Analysis of variance of YS and TS of ZK30.

Source	DF	YS (MPa)			
		Sum of Squares	Mean Square	F-Value	*p*-Value
Model	6	312.95	52.16	20.60	<0.0001 [a]
A-No. of Passes	1	35.12	35.12	13.87	0.0047
B-Die Angle	1	80.39	80.39	31.75	0.0003
C-Dummy x_1	1	4.06	4.06	1.60	0.2374
D-Dummy x_2	1	38.89	38.89	15.36	0.0035
AB	1	46.53	46.53	18.38	0.0020
A^2	1	10.25	10.25	4.05	0.0751
Residual	9	22.79	2.53		
Lack of Fit	6	20.02	3.34	3.62	0.1591 [an]
Pure Error	3	2.77	0.9217		
Cor Total	15	335.74			
Fit Statistics		Std. Dev.	C.V. %	Adeq Precision	
		1.59	1.74	12.5037	
		R^2	Adjusted R^2	Predicted R^2	
		0.9321	0.8869	0.7426	

Source	DF	TS (MPa)			
		Sum of Squares	Mean Square	F-Value	*p*-Value
Model	6	2931.98	488.66	48.53	<0.0001 [a]
A-No. of Passes	1	873.21	873.21	86.72	<0.0001
B-Die Angle	1	953.57	953.57	94.70	<0.0001
C-Dummy x_1	1	206.71	206.71	20.53	0.0014
D-Dummy x_2	1	9.10	9.10	0.9035	0.3667
AB	1	384.40	384.40	38.18	0.0002
AD	1	34.03	34.03	3.38	0.0992
Residual	9	90.62	10.07		
Lack of Fit	6	69.12	11.52	1.61	0.3738 [an]
Pure Error	3	21.50	7.17		
Cor Total	15	3022.60			
Fit Statistics		Std. Dev.	C.V. %	Adeq Precision	
		3.17	0.9804	19.0151	
		R^2	Adjusted R^2	Predicted R^2	
		0.97	0.95	0.9067	

[a] Within a 95% confidence interval, parameters referring to filled cells are significant, [an] are insignificant term.

Table A5. Analysis of variance of D% percentage of ZK30.

Source	DF	D% (mm/mm)			
		Sum of Squares	Mean Square	F-Value	*p*-Value
Model	7	145.30	20.76	74.74	<0.0001 [a]
A-No. of Passes	1	13.56	13.56	48.84	0.0001
B-Die Angle	1	21.04	21.04	75.76	<0.0001
C-Dummy x_1	1	1.01	1.01	3.62	0.0934
D-Dummy x_2	1	20.52	20.52	73.88	<0.0001
AC	1	8.76	8.76	31.55	0.0005
AD	1	9.17	9.17	33.03	0.0004
A^2	1	4.37	4.37	15.73	0.0041
Residual	8	2.22	0.2777		
Lack of Fit	5	1.64	0.3273	1.68	0.3554 [an]
Pure Error	3	0.5850	0.1950		
Cor Total	15	147.52			
Fit Statistics		Std. Dev.	C.V. %	Adeq Precision	
		0.5270	1.57	29.4038	
		R^2	Adjusted R^2	Predicted R^2	
		0.9849	0.9718	0.9410	

[a] Within a 95% confidence interval, parameters referring to filled cells are significant, [an] are insignificant terms.

References

1. Kulekci, M.K. Magnesium and its alloys applications in automotive industry. *Int. J. Adv. Manuf. Technol.* **2008**, *39*, 851–865. [CrossRef]
2. Friedrich, H.; Schumann, S. Research for a 'new age of magnesium' in the automotive industry. *J. Mater. Process. Technol.* **2001**, *117*, 276–281. [CrossRef]
3. Li, N.; Zheng, Y. Novel Magnesium Alloys Developed for Biomedical Application: A Review. *J. Mater. Sci. Technol.* **2013**, *29*, 489–502. [CrossRef]
4. Chen, Y.; Xu, Z.; Smith, C.; Sankar, J. Recent advances on the development of magnesium alloys for biodegradable implants. *Acta Biomater.* **2014**, *10*, 4561–4573. [CrossRef] [PubMed]
5. Gu, X.-N.; Zheng, Y.-F. A review on magnesium alloys as biodegradable materials. *Front. Mater. Sci. China* **2010**, *4*, 111–115. [CrossRef]
6. Saris, N.-E.L.; Mervaala, E.; Karppanen, H.; Khawaja, J.A.; Lewenstam, A. Magnesium: An update on physiological, clinical and analytical aspects. *Clin. Chim. Acta* **2000**, *294*, 1–26. [CrossRef]
7. Okuma, T. Magnesium and bone strength. *Nutrition* **2001**, *17*, 679–680. [CrossRef]
8. Wolf, F.I.; Cittadini, A. Chemistry and biochemistry of magnesium. *Mol. Asp. Med.* **2003**, *24*, 3–9. [CrossRef]
9. Staiger, M.P.; Pietak, A.M.; Huadmai, J.; Dias, G. Magnesium and its alloys as orthopedic biomaterials: A review. *Biomaterials* **2006**, *27*, 1728–1734. [CrossRef]
10. Yin, Y.; Huang, Q.; Liang, L.; Hu, X.; Liu, T.; Weng, Y.; Long, T.; Liu, Y.; Li, Q.; Zhou, S.; et al. In vitro degradation behaviour and cytocompatibility of ZK30/bioactive glass composites fabricated by selective laser melting for biomedical applications. *J. Alloys Compd.* **2019**, *785*, 38–45. [CrossRef]
11. Lin, X.; Tan, L.; Zhang, Q.; Yang, K.; Hu, Z.; Qiu, J.; Cai, Y. The in vitro degradation process and biocompatibility of a ZK60 magnesium alloy with a forsterite-containing micro-arc oxidation coating. *Acta Biomater.* **2013**, *9*, 8631–8642. [CrossRef] [PubMed]
12. Yamasaki, Y.; Yoshida, Y.; Okazaki, M.; Shimazu, A.; Kubo, T.; Akagawa, Y.; Uchida, T. Action of FGMgCO3Ap-collagen composite in promoting bone formation. *Biomaterials* **2003**, *24*, 4913–4920. [CrossRef]
13. Zreiqat, H.; Howlett, C.R.; Zannettino, A.; Evans, P.; Schulze-Tanzil, G.; Knabe, C.; Shakibaei, M. Mechanisms of magnesium-stimulated adhesion of osteoblastic cells to commonly used orthopaedic implants. *J. Biomed. Mater. Res.* **2002**, *62*, 175–184. [CrossRef] [PubMed]
14. Kirkland, N.T. Magnesium biomaterials: Past, present and future. *Corros. Eng. Sci. Technol.* **2012**, *47*, 322–328. [CrossRef]
15. Kannan, M.B.; Raman, R.K.S. In vitro degradation and mechanical integrity of calcium-containing magnesium alloys in modified-simulated body fluid. *Biomaterials* **2008**, *29*, 2306–2314. [CrossRef]
16. Li, Z.; Song, G.-L.; Song, S. Effect of bicarbonate on biodegradation behaviour of pure magnesium in a simulated body fluid. *Electrochim. Acta* **2014**, *115*, 56–65. [CrossRef]
17. Song, G. Control of biodegradation of biocompatable magnesium alloys. *Corros. Sci.* **2007**, *49*, 1696–1701. [CrossRef]
18. Song, Y.W.; Shan, D.Y.; Han, E.H. Electrodeposition of hydroxyapatite coating on AZ91D magnesium alloy for biomaterial application. *Mater. Lett.* **2008**, *62*, 3276–3279. [CrossRef]

19. El-Garaihy, W.H.; Alateyah, A.I.; Alawad, M.O.; Aljohani, T.A. Improving the Corrosion Behavior and Mechanical Properties of Biodegradable Mg-Zn-Zr Alloys Through ECAP for Usage in Biomedical Applications. In *Magnesium Technology*; Springer: Cham, Switzerland, 2022; pp. 259–269. Available online: https://link.springer.com/chapter/10.1007/978-3-030-92533-8_45 (accessed on 5 February 2022).
20. Hornberger, H.; Virtanen, S.; Boccaccini, A.R. Biomedical coatings on magnesium alloys–A review. *Acta Biomater.* **2012**, *8*, 2442–2455. [CrossRef]
21. Sun, Y.; Zhang, B.; Wang, Y.; Geng, L.; Jiao, X. Preparation and characterization of a new biomedical Mg–Zn–Ca alloy. *Mater. Des.* **2012**, *34*, 58–64. [CrossRef]
22. Alateyah, A.I.; Aljohani, T.A.; Alawad, M.O.; Elkatatny, S.; El-Garaihy, W.H. Improving the Corrosion Behavior of Biodegradable AM60 Alloy through Plasma Electrolytic Oxidation. *Metals* **2021**, *11*, 953. [CrossRef]
23. Almenaif, O.; Alhumaydan, Y.; Alnafisah, M.; Aldhalaan, M.; Alateyah, A.I.; El-Garaihy, W.H. A Computational Investigation into the Effect of Equal Channel Angular Processing on the Mechanical Properties of Severely Deformed ZK 60 Alloy Validated by Experiments. *Am. J. Eng. Appl. Sci.* **2020**, *13*, 296–310. [CrossRef]
24. Zong, Y.; Yuan, G.; Zhang, X.; Mao, L.; Niu, J.; Ding, W. Comparison of biodegradable behaviors of AZ31 and Mg–Nd–Zn–Zr alloys in Hank's physiological solution. *Mater. Sci. Eng. B* **2012**, *177*, 395–401. [CrossRef]
25. Ng, W.F.; Wong, M.H.; Cheng, F.T. Stearic acid coating on magnesium for enhancing corrosion resistance in Hanks' solution. *Surf. Coat. Technol.* **2010**, *204*, 1823–1830. [CrossRef]
26. Hermawan, H.; Dubé, D.; Mantovani, D. Developments in metallic biodegradable stents. *Acta Biomater.* **2010**, *6*, 1693–1697. [CrossRef]
27. Gunde, P.; Hänzi, A.C.; Sologubenko, A.S.; Uggowitzer, P.J. High-strength magnesium alloys for degradable implant applications. *Mater. Sci. Eng. A* **2011**, *528*, 1047–1054. [CrossRef]
28. Alateyah, A.I.; Alawad, M.O.; Aljohani, T.A.; El-Garaihy, W.H. Effect of ECAP Route Type on the Microstructural Evolution, Crystallographic Texture, Electrochemical Behavior and Mechanical Properties of ZK30 Biodegradable Magnesium Alloy. *Materials* **2022**, *15*, 6088. [CrossRef]
29. Alateyah, A.I.; Alawad, M.O.; Aljohani, T.A.; El-Garaihy, W.H. Influence of Ultrafine-Grained Microstructure and Texture Evolution of ECAPed ZK30 Magnesium Alloy on the Corrosion Behavior in Different Corrosive Agents. *Materials* **2022**, *15*, 5515. [CrossRef] [PubMed]
30. Wang, Y.; Guan, S.; Zeng, X.; Ding, W. Effects of RE on the microstructure and mechanical properties of Mg–8Zn–4Al magnesium alloy. *Mater. Sci. Eng. A* **2006**, *416*, 109–118. [CrossRef]
31. Witte, F.; Hort, N.; Vogt, C.; Cohen, S.; Kainer, K.U.; Willumeit, R.; Feyerabend, F. Degradable biomaterials based on magnesium corrosion. *Curr. Opin. Solid State Mater. Sci.* **2008**, *12*, 63–72. [CrossRef]
32. Gu, X.; Zheng, Y.; Cheng, Y.; Zhong, S.; Xi, T. In vitro corrosion and biocompatibility of binary magnesium alloys. *Biomaterials* **2009**, *30*, 484–498. [CrossRef] [PubMed]
33. Goodman, S.B.; Davidson, J.A.; Fornasier, V.L.; Mishra, A.K. Histological response to cylinders of a low modulus titanium alloy (Ti-13Nb-13Zr) and a wear resistant zirconium alloy (Zr-2.5Nb) implanted in the rabbit tibia. *J. Appl. Biomater.* **1993**, *4*, 331–339. [CrossRef]
34. Alateyah, A.I.; Aljohani, T.; Alawad, M.; El-Hafez, H.; Almutairi, A.; Alharbi, E.; Alhamada, R.; El-Garaihy, B.; El-Garaihy, W. Improved Corrosion Behavior of AZ31 Alloy through ECAP Processing. *Metals* **2021**, *11*, 363. [CrossRef]
35. Alateyah, A.I.; Ahmed, M.M.; Alawad, M.O.; Elkatatny, S.; Zedan, Y.; Nassef, A.; El-Garaihy, W. Effect of ECAP die angle on the strain homogeneity, microstructural evolution, crystallographic texture and mechanical properties of pure magnesium: Numerical simulation and experimental approach. *J. Mater. Res. Technol.* **2022**, *17*, 1491–1511. [CrossRef]
36. Yamashita, A.; Horita, Z.; Langdon, T.G. Improving the mechanical properties of magnesium and a magnesium alloy through severe plastic deformation. *Mater. Sci. Eng. A* **2001**, *300*, 142–147. [CrossRef]
37. El-Garaihy, W.H.; Fouad, D.M.; Salem, H.G. Multi-channel Spiral Twist Extrusion (MCSTE): A Novel Severe Plastic Deformation Technique for Grain Refinement. *Met. Mater. Trans. A* **2018**, *49*, 2854–2864. [CrossRef]
38. Fouad, D.M.; Moataz, A.; El-Garaihy, W.H.; Salem, H.G. Numerical and experimental analysis of multi-channel spiral twist extrusion processing of AA5083. *Mater. Sci. Eng. A* **2019**, *764*, 138216. [CrossRef]
39. Fouad, D.M.; El-Garaihy, W.H.; Ahmed, M.M.Z.; Seleman, M.M.E.; Salem, H.G. Influence of multi-channel spiral twist ex-trusion (MCSTE) processing on structural evolution, crystallographic texture and mechanical properties of AA1100. *Mater. Sci. Eng. A* **2018**, *737*, 166–175. [CrossRef]
40. Fouad, D.M.; El-Garaihy, W.H.; Ahmed, M.M.Z.; Albaijan, I.; Seleman, M.M.E.; Salem, H.G. Grain Structure Evolution and Mechanical Properties of Multi-Channel Spiral Twist Extruded AA5083. *Metals* **2021**, *11*, 1276. [CrossRef]
41. Alvarez-Lopez, M.; Pereda, M.D.; Del Valle, J.A.; Fernandez-Lorenzo, M.; Garcia-Alonso, M.C.; Ruano, O.A.; Escudero, M.L. Corrosion behaviour of AZ31 magnesium alloy with different grain sizes in simulated biological fluids. *Acta Biomater.* **2010**, *6*, 1763–1771. [CrossRef]
42. El-Shenawy, M.; Ahmed, M.M.Z.; Nassef, A.; El-Hadek, M.; Alzahrani, B.; Zedan, Y.; El-Garaihy, W.H. Effect of ECAP on the Plastic Strain Homogeneity, Microstructural Evolution, Crystallographic Texture and Mechanical Properties of AA2xxx Aluminum Alloy. *Metals* **2021**, *11*, 938. [CrossRef]

43. Gao, J.H.; Guan, S.K.; Ren, Z.W.; Sun, Y.F.; Zhu, S.J.; Wang, B. Homogeneous corrosion of high-pressure torsion treated Mg–Zn–Ca alloy in simulated body fluid. *Mater. Lett.* **2011**, *65*, 691–693. [CrossRef]
44. El-Garaihy, W.H.; Rassoul, E.A.; Alateyah, A.; Alaskari, A.M.; Oraby, S. Data Manipulation Approach and Parameters Interrelationships of the High-Pressure Torsion for AA6061-15%SiCp Composite. *SAE Int. J. Mater. Manuf.* **2018**, *11*, 167–182. [CrossRef]
45. El-Garaihy, W.; Rassoul, E.S.M.; Salem, H.G. Consolidation of High Performance AA6061 and AA6061-SiCp Composite Processed by High Pressure Torsion. *Mater. Sci. Forum* **2014**, *783–786*, 2623–2628. [CrossRef]
46. Alateyah, A.I.; Alharbi, M.; El-Hafez, H.M.; El-Garaihy, W.H. The Effect of Equal-Channel Angular Pressing Processing on Microstructural Evolution, Hardness Homogeneity, and Mechanical Properties of Pure Aluminum. *SAE Int. J. Mater. Manuf.* **2020**, *14*, 113–125. [CrossRef]
47. Ge, Q.; Dellasega, D.; Demir, A.G.; Vedani, M. The processing of ultrafine-grained Mg tubes for biodegradable stents. *Acta Biomater.* **2013**, *9*, 8604–8610. [CrossRef] [PubMed]
48. Wu, W.J.; Chen, W.; Zhang, L.; Chen, X.; Wang, H.; Wang, W.; Zhang, W. Improvement of tension/compression asymmetry for high-performance ZK61 magnesium alloy rod via tailoring deformation parameters: Upsetting-extrusion temperature and upsetting ratio. *Mater. Sci. Eng. A* **2021**, *823*, 141767. [CrossRef]
49. Savaedi, Z.; Mirzadeh, H.; Aghdam, R.M.; Mahmudi, R. Effect of grain size on the mechanical properties and bio-corrosion resistance of pure magnesium. *J. Mater. Res. Technol.* **2022**, *19*, 3100–3109. [CrossRef]
50. Němec, M.; Jäger, A.; Tesař, K.; Gärtnerová, V. Influence of alloying element Zn on the microstructural, mechanical and corrosion properties of binary Mg-Zn alloys after severe plastic deformation. *Mater. Charact.* **2017**, *134*, 69–75. [CrossRef]
51. Minarik, P.; Jablonska, E.; Kral, R.; Lipov, J.; Ruml, T.; Blawert, C.; Hadzima, B.; Chmelík, F. Effect of equal channel angular pressing on in vitro degradation of LAE442 magnesium alloy. *Mater. Sci. Eng. C* **2017**, *73*, 736–742. [CrossRef]
52. Li, X.; Jiang, J.; Zhao, Y.; Ma, A.; Wen, D.; Zhu, Y. Effect of equal-channel angular pressing and aging on corrosion behaviour of ZK60 Mg alloy. *Trans. Nonferrous Met. Soc. China* **2015**, *25*, 3909–3920. [CrossRef]
53. Valiev, R.Z.; Langdon, T.G. Principles of equal channel angular pressing as a processing tool for grain refinement. *Prog. Mater. Sci.* **2006**, *51*, 881–981. [CrossRef]
54. Sankuru, A.B.; Sunkara, H.; Sethuraman, S.; Gudimetla, K.; Ravisankar, B.; Babu, S.P.K. Effect of processing route on microstructure, mechanical and dry sliding wear behavior of commercially pure magnesium processed by ECAP with back pressure. *Trans Indian Inst. Met.* **2021**, *74*, 2659–2669. [CrossRef]
55. Ghaedi, M.; Azad, F.N.; Dashtian, K.; Hajati, S.; Goudarzi, A.; Soylak, M. Central composite design and genetic algorithm applied for the optimization of ultrasonic-assisted removal of malachite green by ZnO Nanorod-loaded activated carbon. *Spectrochim. Acta Part A Mol. Biomol. Spectrosc.* **2016**, *167*, 157–164. [CrossRef] [PubMed]
56. Daryadel, M. Study on Equal Channel Angular Pressing Process of AA7075 with Copper Casing by Finite Element-response Surface Couple Method. *Int. J. Eng.* **2020**, *33*, 2538–2548.
57. Alateyah, A.I.; El-Garaihy, W.H.; Alawad, M.O.; El Sanabary, S.; Elkatatny, S.; Dahish, H.A.; Kouta, H. The Effect of ECAP Processing Conditions on Microstructural Evolution and Mechanical Properties of Pure Magnesium—Experimental, Mathematical Empirical and Response Surface Approach. *Materials* **2022**, *15*, 5312. [CrossRef]
58. Saleh, B.; Jiang, J.; Xu, Q.; Fathi, R.; Ma, A.; Li, Y.; Wang, L. Statistical Analysis of Dry Sliding Wear Process Parameters for AZ91Alloy Processed by RD-ECAP Using Response Surface Methodology. *Met. Mater. Int.* **2021**, *27*, 2879–2897. [CrossRef]
59. Kilickap, E.; Huseyinoglu, M.; Yardimeden, A. Optimization of drilling parameters on surface roughness in drilling of AISI 1045 using response surface methodology and genetic algorithm. *Int. J. Adv. Manuf. Technol.* **2011**, *52*, 79–88. [CrossRef]
60. Dadrasi, A.; Fooladpanjeh, S.; Gharahbagh, A.A. Interactions between HA/GO/epoxy resin nanocomposites: Optimization, modeling and mechanical performance using central composite design and genetic algorithm. *J. Braz. Soc. Mech. Sci. Eng.* **2019**, *41*, 63. [CrossRef]
61. Hazir, E.; Ozcan, T. Response surface methodology integrated with desirability function and genetic algorithm approach for the optimization of CNC machining parameters. *Arab. J. Sci. Eng.* **2019**, *44*, 2795–2809. [CrossRef]
62. Elsanabary, S.; Kouta, H.K. Optimization of Inertia Friction Welding of Dissimilar Polymeric PA6-PVC Hollow Cylinders by Genetic Algorithm. *Port-Said Eng. Res. J.* **2021**, *25*, 91–100. [CrossRef]
63. Munusamy, T.D.; Chin, S.Y.; Khan, M.M.R. Photoreforming hydrogen production by carbon doped exfoliated g-C3N4: Optimization using design expert®software. *Mater. Today Proc.* **2021**, *57*, 1162–1168. [CrossRef]
64. Rezić, I. Prediction of the surface tension of surfactant mixtures for detergent formulation using Design Expert software. *Mon. Für Chem.-Chem. Mon.* **2011**, *142*, 1219–1225. [CrossRef]
65. Makki, A.A.; Stewart, R.A.; Panuwatwanich, K.; Beal, C. Development of a domestic water end use consumption forecasting model for South-East Queensland, Australia. In Proceedings of the 6th IWA Specialist Conference on Efficient Use and Management of Water, International Water Association, Dead Sea, Jordan, 29 Marth–2 April 2011.
66. Santhosh, A.J.; Tura, A.D.; Jiregna, I.T.; Gemechu, W.F.; Ashok, N.; Ponnusamy, M. Optimization of CNC turning parameters using face centred CCD approach in RSM and ANN-genetic algorithm for AISI 4340 alloy steel. *Results Eng.* **2021**, *11*, 100251. [CrossRef]
67. Gopal, M. Optimization of Machining Parameters on Temperature Rise in CNC Turning Process of Aluminium–6061 Using RSM and Genetic Algorithm. *Int. J. Mod. Manuf. Technol.* **2020**, *12*, 36–43.

68. Dritsa, V.; Rigas, F.; Doulia, D.; Avramides, E.J.; Hatzianestis, I. Optimization of culture conditions for the biodegradation of lindane by the polypore fungus Ganoderma australe. *Water Air Soil Pollut.* **2009**, *204*, 19–27. [CrossRef]
69. Asghar, A.; Raman, A.A.A.; Daud, W.M.A.W. A comparison of central composite design and Taguchi method for optimizing Fenton process. *Sci. World J.* **2014**, *2014*, 1–14. [CrossRef]
70. Aljohani, T.A.; Alawad, M.O.; Elkatatny, S.; Alateyah, A.I.; Bin Rubayan, M.T.; Alhajji, M.A.; AlBeladi, M.I.; Khoshnaw, F.; El-Garaihy, W.H. Electrochemical Behavior of SiC-Coated AA2014 Alloy through Plasma Electrolytic Oxidation. *Materials* **2022**, *15*, 3724. [CrossRef]
71. Peron, M.; Skaret, P.C.; Fabrizi, A.; Varone, A.; Montanari, R.; Roven, H.J.; Ferro, P.; Berto, F.; Torgersen, J. The effect of Equal Channel Angular Pressing on the stress corrosion cracking susceptibility of AZ31 alloy in simulated body fluid. *J. Mech. Behav. Biomed. Mater.* **2020**, *106*, 103724. [CrossRef]
72. Mostaed, E.; Vedani, M.; Hashempour, M.; Bestetti, M. Influence of ECAP process on mechanical and corrosion properties of pure Mg and ZK60 magnesium alloy for biodegradable stent applications. *Biomatter* **2014**, *4*, e28283. [CrossRef]
73. Alateyah, A.I.; Ahmed, M.M.Z.; Zedan, Y.; El-Hafez, H.A.; Alawad, M.O.; El-Garaihy, W.H. Experimental and Numerical Investigation of the ECAP Processed Copper: Microstructural Evolution, Crystallographic Texture and Hardness Homogeneity. *Metals* **2021**, *11*, 607. [CrossRef]
74. Lei, W.; Zhang, H. Analysis of microstructural evolution and compressive properties for pure Mg after room-temperature ECAP. *Mater. Lett.* **2020**, *271*, 127781. [CrossRef]
75. Mostaed, E.; Hashempour, M.; Fabrizi, A.; Dellasega, D.; Bestetti, M.; Bonollo, F.; Vedani, M. Microstructure, texture evolution, mechanical properties and corrosion behavior of ECAP processed ZK60 magnesium alloy for biodegradable applications. *J. Mech. Behav. Biomed. Mater.* **2014**, *37*, 307–322. [CrossRef] [PubMed]
76. Tolaminejad, B.; Dehghani, K. Microstructural characterization and mechanical properties of nanostructured AA1070 aluminum after equal channel angular extrusion. *Mater. Des.* **2012**, *34*, 285–292. [CrossRef]
77. Yuan, Y.; Ma, A.; Gou, X.; Jiang, J.; Arhin, G.; Song, D.; Liu, H. Effect of heat treatment and deformation temperature on the mechanical properties of ECAP processed ZK60 magnesium alloy. *Mater. Sci. Eng. A* **2016**, *677*, 125–132. [CrossRef]
78. Huang, S.-J.; Chiu, C.; Chou, T.-Y.; Rabkin, E. Effect of equal channel angular pressing (ECAP) on hydrogen storage properties of commercial magnesium alloy AZ61. *Int. J. Hydrog. Energy* **2018**, *43*, 4371–4380. [CrossRef]
79. Minárik, P.; Zimina, M.; Čížek, J.; Stráska, J.; Krajňák, T.; Cieslar, M.; Vlasák, T.; Bohlen, J.; Kurz, G.; Letzig, D. Increased structural stability in twin-roll cast AZ31 magnesium alloy processed by equal channel angular pressing. *Mater. Charact.* **2019**, *153*, 199–207. [CrossRef]

materials

MDPI

Article

Minimizing the Forces in the Single Point Incremental Forming Process of Polymeric Materials Using Taguchi Design of Experiments and Analysis of Variance

Nicolae Rosca [1], Tomasz Trzepieciński [2] and Valentin Oleksik [1,*]

1 Faculty of Engineering, "Lucian Blaga", University of Sibiu, Victoriei Bd. 10, 550024 Sibiu, Romania
2 Department of Manufacturing and Production Engineering, Faculty of Mechanical Engineering and Aeronautics, Rzeszow University of Technology, al. Powst. Warszawy 8, 35-959 Rzeszów, Poland
* Correspondence: valentin.oleksik@ulbsibiu.ro

Abstract: The aim of the present paper is that of conducting a study on the basis of which the optimal parameters for the manufacturing of polymer parts by means of the single point incremental forming process can be chosen in such a way that the process forces have minimum values. Two polymeric materials with a 3 mm thickness, polyamide and polyethylene, were chosen for the analysis. The other input parameters that were considered were: the punch diameter, the step on vertical direction and the wall angle. The Taguchi method was chosen for the design of experiments. Each of the input parameters, except for the material, were varied on three levels—for the punch diameter: 6 mm, 8 mm and 10 mm; for the step on vertical direction: 0.5 mm, 0.75 mm and 1 mm; and for the wall angle: 50°, 55° and 60°. Forces were measured in the three directions of the coordinate axes and the results were analyzed based on the signal-to-noise ratio and an analysis of variance with the aim of minimizing the values of the forces. Considering the input parameters analyzed, it was concluded that the forces are most influenced by the material, followed by the punch diameter, the step on vertical direction and the wall angle.

Keywords: polymer incremental forming; polyamide; polyethylene; statistical analysis; force minimizing

Citation: Rosca, N.; Trzepieciński, T.; Oleksik, V. Minimizing the Forces in the Single Point Incremental Forming Process of Polymeric Materials Using Taguchi Design of Experiments and Analysis of Variance. *Materials* **2022**, *15*, 6453. https://doi.org/10.3390/ma15186453

Academic Editor: Béla Iván

Received: 24 August 2022
Accepted: 14 September 2022
Published: 17 September 2022

1. Introduction

The single point incremental forming process has gained a reputation among forming processes over the last 20–30 years due to its advantages over other forming processes: the use of simple, modular dies which are much cheaper than the ones used in deep-drawing, the possibility of using computer numerical control (CNC) machine tools [1] or industrial robots [2] instead of presses, the possibility of obtaining parts with complex geometries even when limited to simple dies and, last but not least, the ability to obtain drastically reduced forming forces due to the progressive deformation of the material. However, the single point incremental forming process has its limitations as well, the most significant of which being the long manufacturing time, which leads to this process only being used for parts manufactured in small batches or one-off productions. Initially, the most commonly manufactured materials were metallic materials (steel, aluminum, copper, titanium and their alloys [3]), but more recently, this forming process has also found its application in polymeric materials. The first research on single point incremental forming (SPIF) of polymers was published in the early 2000s, and a first state-of-the-art synthesis in this field was published by Martins and his co-authors in 2009 [4]. Two other recent review papers have studied the latest advances in the incremental forming of polymer parts, ranging from room temperature and high temperature processing to single point or two point incremental forming [5,6]. Incremental forming has been used to manufacture parts from various polymeric materials: polyvinylchloride sheets [7], polycarbonate sheets [8], polypropylene sheets [9] and even shape memory polymer foam [10]. Special attention has been paid to the manufacturing of polymer composite parts. Thus, Ambrogio et al. [11] presented a study on the SPIF processing of a composite material made of a polyamide matrix in which short glass fibers were melted. The amount of short glass fibers in the volume of the composite material was 12%. The study focused on determining the accuracy of the deformed parts, as well as on a microscopic analysis of the material following the forming process. Another study investigated the possibility of making SPIF-processed cranial implants from a reinforced material made of polymethyl methacrylate (PMMA) [12]. Most papers related to the incremental forming of composite materials focused on polyamides with different reinforcing elements, such as: nanocomposites with montmorillonite (MMT) filler clay [13] and glass fibers in amounts of up to 30 wt.% [14]. Emani et al. presented a paper in which they analyzed methods of improving the SPIF formability of glass-fiber-reinforced polyamide composite materials [15]. They concluded that the SPIF formability of these materials increased with increasing forming temperature and decreasing fraction of glass fibers from the total volume of the workpiece. Another material manufactured with SPIF, in order to increase the rigidity, was paperboard with different fiber orientations (0°, 45° and 90°) and different moisture content (4.5%, 10% and 20%) [16]. The applications of polymer parts manufactured by SPIF are diverse and range from manufacturing parts for the aerospace industry [17], given the weight reduction while maintaining sufficient strength, to applications related to implant manufacturing [18], due to the fact that many of them are biocompatible materials. Centeno et al. [19] reported on the methodology for producing custom cranial implants, starting with computerized tomography and ending with the SPIF fabrication of polymeric materials, specifically a polyvinylchloride sheet and two polycarbonate sheets. In the study, the method of eliminating springback was also presented for the development of the punch trajectory, in order to achieve high accuracy. Another study analyzed the formability of polymeric materials during SPIF, both at room temperature and at elevated temperatures [20]. Experimental analysis of the failure modes of polycarbonate sheets was investigated by Rosa-Sainz et al. [7]. Experimental research related to plastic strains and failure modes of polyvinylchloride sheets [21] and polypropylene sheets [22] has contributed to understanding the deformation mechanism, as well as to estimating the formability of castor polymeric materials. Another review article summarized the formability of different polymer matrix composites [23]. The incremental forming process can be performed either in single point or with support from the opposite side of the punch. Most of the publications in the field of polymeric materials deal with the SPIF process, but there are also papers addressing the two point incremental forming [24,25]. In the case of parts made of polymeric materials and manufactured by incremental forming, the main defects that occurred were those caused by springback [26], wrinkle defects [27] and twist defects [28]. While springback and wrinkle defects are also common to the SPIF processing of other types of materials, the twist defects are more specific to polymeric materials that have a lower rigidity and twist during the forming around the vertical axis of the part. Yang and Chan investigated the twist phenomena occurrence mechanism in the SPIF of polyvinylchloride sheets [28]. They calculated the twist angle based on the law of conservation of energy. In order to increase the accuracy of thermoplastic parts processed by SPIF, Maas et al. proposed the overbending of the workpiece, with the aim of reducing the springback [29]. One phenomenon that occurs in the SPIF manufacturing of polymeric materials is that of self-heating due to the frictional forces between the punch and the polymer sheet. Formisano et al. studied this phenomenon and attempted to control it by the rotational speed of the punch, in such a way that the phenomenon improved the formability of the process [30]. A comparative analysis of the behavior of polymeric materials during incremental forming at elevated and room temperatures was performed by Kulkarni and Mocko [31]. Hot incremental forming of glass-fiber-reinforced polymers [32] or biocomposite materials developed from thermoplastic matrix [33,34] has been reviewed in recently published papers. A synthesis of the hot incremental forming process of polycarbonate sheets was published by Sridhar and Rajenthirakumar [35], addressing issues related to geometrical accuracy, thickness distribution and fracture formation. A number of published studies deal with the finite element method applied for the numerical simulation of the SPIF processing of polymeric materials [36,37]. These studies present different models: a thermomechanical coupled model [36], and a viscoplasticity based on overstress model [37]. Other studies not only perform numerical simulations, but also validate them by experimental investigations [38,39]. Another concern related to polymeric materials but also to the SPIF process is that of forming sandwich materials (steel/polymer/steel) [40]. The tools for the SPIF processing of polymeric materials can be made by classical cutting processes or by new, innovative methods such as rapid prototyping. A new method to produce the tools needed for the incremental forming process of polymeric materials, named direct rapid tooling, was presented by Afonso et al. [41]. Over the years, the SPIF process parameters for the manufacturing of polymeric materials have been analyzed in a number of papers. For instance, the influence of the punch diameter, wall-angle of the part, step on vertical direction and feed rate on the springback, thickness reduction and surface roughness for polypropylene sheets was studied by Jain et al. [42]. Another study examined the influence of the wall-angle of the part, step on vertical direction and feed rate on the same output parameters, but for polyvinylchloride sheets [43]. Thangavel et al. [44] studied the influence of the roller ball diameter, step on vertical direction, feed rate, initial sheet thickness and spindle speed during single point incremental forming with rolling friction for four polymeric materials: high density polyethylene, polycarbonate, polyvinylchloride and polypropylene. Formability (maximum depth before failure) and thickness distribution were chosen as response functions. Davarpanah et al. evaluated the effect of the vertical step and spindle speed on the failure modes and microstructural properties of polymeric materials formed with SPIF [45]. A first paper analyzing the forces during the SPIF process of polymeric materials is that of Bagudanch et al. [46]. They studied the influence of the vertical step, spindle speed, feed rate, punch diameter and initial sheet thickness on the maximum forming force during the SPIF of polyvinylchloride sheets. In addition to the maximum force, the maximum depth before failure and surface roughness were also chosen as output parameters. Considering the spindle feed, they also analyzed the self-heating process caused by the frictional forces occurring in the incremental forming process. Another work allows the prediction of forces in SPIF, based on the contact area and friction coefficient [47]. The analytical results were experimentally verified for two materials, polycarbonate and polyvinylchloride sheets. A recent paper presents the friction stir incremental forming process for polyoxymethylene sheets [48]. This study also presents the influence of the punch diameter, step on vertical direction, spindle speed and feed rate on the maximum force, temperature obtained as a result of the self-heating process and maximum depth before failure.

The present paper studies the single point incremental forming process applied to polymeric materials. As is well known, the process was developed as a result of economic necessity, as conventional processes led to high costs that could not be recovered in the cases of small series and one-off productions. Moreover, the paper is dedicated to the application of this technology to polymeric materials, which are low-density materials, increasingly used in the automotive industry, medical device industry, electrotechnics industry and so on, having been less studied than metallic materials. The impact of the results obtained in this study should not only be considered from a technical, engineering point of view, but from a much greater one, leading to economic or even environmental benefits, as confirmed by a recently published study [49]. This is also confirmed by previous studies analyzing the parameters influencing energy consumption in polymer processing or another study, also focusing on energy consumption, concerning the possibility of reusing waste through incremental forming [50]. Two other recent studies analyzed the energy consumption, carbon emissions and product cost in the context of the sustainability of the incremental forming process [51,52]. The environmental impact of the incremental forming process has also been analyzed from the life cycle assessment perspective [53]. Although the present study did not intend to analyze the incremental forming process with regard to these aspects, it nevertheless makes an essential contribution in this direction, by identifying ways of reducing the forces in the process and thus the amount of energy consumed.

All reviewed papers only study the force on vertical direction (on Oz direction) but not the other two horizontal components (on Ox and Oy direction), which are also very important in terms of the stress applied to the machine used in SPIF. Moreover, there is a lack of knowledge concerning the way in which the SPIF process parameters (step size, punch diameter and wall angle) affect the horizontal components of the forming force. This matter is of particular importance when forming polymeric materials that exhibit a formability that is different to that of metallic sheets commonly tested in incremental sheet forming. A forming tool mounted in a gripper of the shelf-mounted robot was used in the investigations. Achieving the lowest possible forming forces is also important in terms of minimizing the elastic deformation of the robot arms, which determine the forming accuracy. As can be seen from the above, although there are papers dealing with the issue of force variation during the SPIF process and the influence of various process parameters on the maximum force, the purpose of the present paper is, by means of the Taguchi design of experiments and an analysis of variance (ANOVA), to establish the influence of certain essential parameters, such as the punch diameter, step on vertical direction, wall angle and material type, on all forces in the process. The aim of the study is therefore to find the optimal variants of the input parameters that lead to the minimization of the forces in the process. Moreover, for the two chosen materials, specifically polyamide and polyethylene, the regression relations for the three categories of forces were also obtained. To achieve the purpose of the present research, experiments were carried out to form the two polymeric sheets (polyamide and polyethylene) by using the shelf-mounted Kuka Kr210-2 robot. The input parameters considered were the step size, the punch diameter and the wall angle. The Taguchi method was used to find the minimum number of experiments to be performed within the permissible limit of factors and levels. These SPIF parameters were varied on three levels—for the punch diameter: 6 mm, 8 mm and 10 mm; for the step on vertical direction: 0.5 mm, 0.75 mm and 1 mm; and for the wall angle: 50°, 55° and 60°. The PCB 261A13 force sensor was used to measure the two horizontal components of the forming force that act in the workpiece plane and the axial component in the direction of the forming tool. The results were analyzed based on the signal-to-noise (S/N) ratio and an analysis of variance. In order to obtain the mathematical relationships between the forming force components in SPIF and the input parameters, a regression analysis was performed.

Apart from the materials studied in the paper, namely polyamide, polyethylene and polytetrafluoroethylene, the behavior of other polymeric materials regarding the incremental forming process at room temperature, such as polyoxymethylene and polyvinylchloride, was also initially studied. Due to the fact that these materials developed fractures in the initial stages of the incremental forming process, it was decided not to include the respective materials in the present study.

2. Materials and Methods

The experimental research in the paper is structured in two directions: research quantifying the formability of polymeric materials and research strictly related to the incremental forming process of these materials. The research quantifying the formability of the materials is performed by uniaxial tensile testing of polymeric materials.

2.1. Uniaxial Tensile Testing of Polymeric Materials

Tensile testing always provides clues regarding the plastic deformation behavior of a material. Thus, in order to identify which polymeric materials are best subjected to deformation by incremental process, three polymeric materials were chosen: polyamide 6.6 (PA), high density polyethylene (HDPE) and polytetrafluoroethylene (PTFE).

An Instron 5587 tensile testing machine (Norwood, MA, USA) was used for the tests. The drive system of the Instron 5587 machine is electromechanical and the movable crossbar is guided by means of a ball screw guide system. The main features of the testing machine are: maximum load cell capacity of 300 kN, maximum test speed of 500 mm/min, load cell with a linearity of ±0.25% and repeatability of ±0.25% for readings taken at 0.4–100% of the maximum load cell capacity. The machine is connected to a computer via an Ethernet interface. Calibration and testing operations are controlled via the Bluehill 2.0 software package which comes with the machine and is installed on the computer. The software package allows the selection of the specimen type (standard ISO or ASTM types), specimen size, test speed, as well as the means by which specimen breakage is detected. The Bluehill 2.0 software package also monitors the testing, generates measurement reports and performs statistical processing of the data.

A 3 mm thickness was used for all three polymer materials. The calibrated area of the specimens had an initial width of 20 mm and length of 120 mm. To perform the tests sets of 5 specimens were made for each material. There are several methods by which the specimens can be cut: mechanical die cutting, laser cutting and waterjet cutting. Due to the specificity of the materials used, waterjet cutting was chosen as cutting method, on the one hand because laser cutting would not be viable, as the material is flammable, and on the other hand because cutting on a die would produce a significant burr due to the high elongation present in these types of materials.

After making the specimens, the settings for the actual test were defined in the testing machine's built-in program, Bluehill. A test speed of 20 mm/min was chosen, in accordance with current standards, and the maximum elongation was set to 700%, so as not to exceed the machine's limits. The following output data were chosen: the longitudinal modulus of elasticity (E) in megapascals (MPa), yield stress (Ys) in MPa, the maximum stress in MPa, and the strain corresponding to the maximum stress in %.

2.2. Experimental Layout for the Study of the Behavior of Polymeric Materials during Single Point Incremental Forming

All research strictly related to the incremental forming process of polymeric materials was carried out using the Kuka Kr210-2 robot (Figure 1) (Augsburg, Germany). In fact, three options were available to achieve the kinematics required to produce parts by incremental forming: the use of a numerically controlled machine, of dedicated equipment for incremental forming or of an industrial robot. In the absence of dedicated equipment, the only options left were a CNC machine or a robot. Both options provided their own advantages, as well as disadvantages.

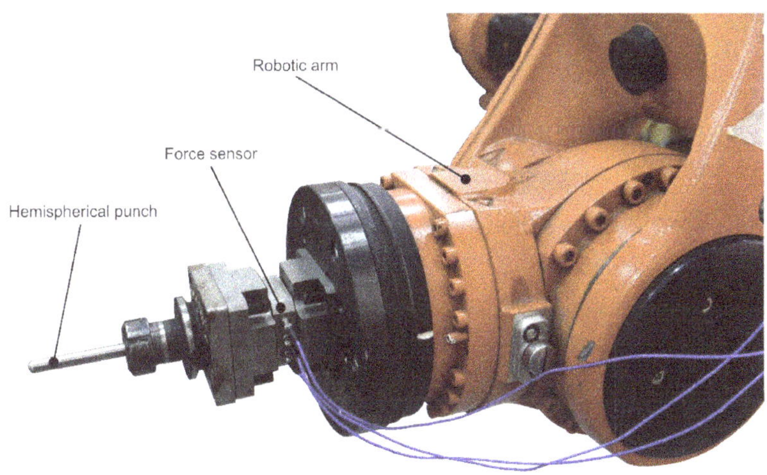

Figure 1. The PCB 261A13 force sensor mounted on the robotic arm of Kuka Kr210-2.

The CNC machine has higher rigidity but does not produce high forces in the x and y directions as it is not designed for plastic deformation processes. Moreover, one of the aims of the study is to measure the strains during the deformation process, which would not be possible due to the fact that the optical cameras of the measuring system cannot be positioned under the CNC machine table. The main disadvantage of robots is their lack of rigidity, but the vertical positioning of the semi-finished product would allow for the positioning of optical cameras. This problem can be significant in the case of hard materials with high tensile strength. However, lack of stiffness and hence precision is not an impediment in the case of polymeric materials, as they have a lower tensile strength and consequently develop lower deformation forces.

The Kuka KR210-2 robot was therefore chosen, a 6-axis robot with a maximum payload of 2100 N and a positioning repeatability of ± 0.06 mm. The total weight of the robot is 1445 kg, and the total work volume is 97 m^3. The robot is equipped with a KR C2 controller, 2005 edition.

The kinematics of the robot, required for the incremental forming processing of polymeric materials, are based on the desired model of the part, designed in Catia and exported to SprutCam, a program dedicated to generating by numerically controlled programs. A single type of trajectory was developed, a spiral trajectory for the production of truncated cone type parts. The deformation speed was of 400 mm/min.

The blank clamping system consists of a vertical frame and a clamping plate between which the semi-finished product is rigidly fixed. The free area of the blank is of 200 mm. The workpieces are in the form of square plates, 3 mm thick with sides of 250 mm.

To measure the forces in the incremental forming process of polymeric materials, a measuring system consisting of the PCB261A13 sensor (PCB piezotronics, Depew, NY, USA), the CMD600 signal amplifier (HBM, Darmstadt, Germany) and the Quantum X MX840B acquisition system (HBM, Darmstadt, Germany) was used. The force sensor allows force acquisition, and the amplifier increases the amplitude of the signal up to the range of voltages required for acquisition with the Quantum X system.

The PCB 261A13 force sensor produced by Piezotronics is specially designed to measure dynamic and quasi-static forces in three perpendicular directions simultaneously [54]. The sensor uses a set of quartz crystals arranged in a predefined pattern to optimize the measurement accuracy. The sensor features a classic three-directional piezoelectric sensor, clamped between two high-precision plates. Measurements in the z-axis direction are proportional to the compression, tensile or impact forces, while those in the x- and y-axis directions are proportional to the shear forces applied to the sensor.

In order to bring the signal acquired from the force sensor into the ± 10 V voltage range, the CMD600 amplifier from HBM was used [55]. Three such amplifiers were used for the experimental research, one for each channel (for each of the Fz, Fx and Fy forces). This amplifier model is dedicated to piezoelectric sensors and allows the use of sensors with a detection range of 50 pC to 600,000 pC.

The Quantum X MX840B acquisition system is an acquisition system developed by HBM and features 8 individually configurable and galvanically isolated measurement channels [56]. The system allows individual sample rates of up to 40 kS/s per channel and has a built-in active low pass filter (Bessel, Butterworth). It is equipped with a 24-bit Analog to Digital converter on each channel. The Catman software, provided with the acquisition system, was used for the acquisition of the electrical signal.

In order to evaluate the formability during the single point incremental forming process, the uniaxial tensile test was used for the three materials, namely polyamide, polyethylene and polytetrafluoroethylene. They were uniaxially stressed with constant speed until fracture occurred, in order to primarily determine the maximum tensile stress and the tensile strain at maximum tensile stress. Different behaviors were observed in the tensile tests of the three materials, both in terms of the values obtained and in terms of the conventional stress–strain curves.

3. Results and Discussion

3.1. Tensile Tests

Five sets of specimens were made for each of the three materials (polyamide, polyethylene and polytetrafluoroethylene), which were then subjected to tensile tests. The uniaxial tensile test was conducted in accordance with ISO 527-1:2019 which covers the general principles for testing plastic materials [57]. In addition, the shape of the specimen and the testing speed were chosen in accordance with ISO 527-3:2018 which defines this data for polymer sheets and films [58]. After the tensile tests were performed, the data obtained was statistically processed using the Minitab software. The data was therefore studied under two aspects, namely the elimination of outliers (outliers test) and the normal data distribution (normality test). Two tests were actually applied: the Anderson–Darling test to check for the normal distribution of the data and the Grubbs test to remove outliers. The mean value and the standard deviation (St. dev) were initially calculated.

The Anderson–Darling test is a statistical test used to determine whether or not a data set follows a particular probability distribution, for example, the normal distribution, considered in this thesis. The test involves calculating the Anderson–Darling indicator (AD value).

The two hypotheses for the Anderson–Darling test for normal distribution are the following:

Hypothesis 0 (H0). *The data follow the normal distribution (null hypothesis).*

Hypothesis 1 (H1). *The data do not follow the normal distribution (alternative hypothesis).*

The null hypothesis assumes that the data are normally distributed, while the alternative hypothesis assumes that the data are non normal. In most cases, a p-value is determined for the Anderson–Darling test (p-value for AD test), which can be used to determine whether the test is significant or not. For low p-values (e.g., ≤ 0.05), the data is concluded to not follow the normal distribution. The p-value depends on the significance level chosen, although the vast majority of cases use the value of 0.05, corresponding to a probability of 95%.

The Grubbs test is a test used to identify and eliminate outliers. It is employed to detect outliers in a data set by testing for one outlier at a time. Any detected outliers are deleted from the data set and the test is repeated until no more outliers are detected.

The basic hypothesis underlying the Grubbs test is that, outliers aside, the data are normally distributed. The null hypothesis is that there are no outliers in the data set. It is worth mentioning that the Grubbs method only works when testing the most extreme value in the sample. A p-value (p-value for Grubbs test) is calculated for the Grubbs test as well. Normally, an indicator of the Grubbs test (G value) is calculated for all values, and the smallest p-value corresponding to the largest G-value is estimated. This p-value represents the chance of observing a point so far away from the others if the data were all sampled from a Gaussian distribution.

Based on the previous considerations, the measured values and statistical processing of the two tests (Anderson–Darling and Grubbs for the three types of materials) are presented in Tables 1–3. The conventional characteristic stress–strain curves for all three tested materials are shown in Figure 2a–c. Due to the particularities of plastic flow in the cases of polyamide and polytetrafluoroethylene, it was not possible to determine the Yield stress, which is entirely accurate, as the observation of the conventional tensile graphs shows that these materials have practically no "Yield stress", but a flow area that is not exactly defined.

Table 1. Results of the tensile test for polyamide specimens.

Specimen Number	Young Modulus (MPa)	Yield Stress (MPa)	Maximum Tensile Stress (MPa)	Tensile Strain at Maximum Tensile Stress (%)
1.	1820.545	-	40.989	110.833
2.	1869.364	-	42.112	150.000
3.	1839.354	-	41.824	119.999
4.	1710.215	-	41.013	120.500
5.	1896.895	-	41.781	131.333
Mean	1827.274	-	41.544	126.533
St. dev	71.611	-	0.512	14.994
AD value	0.335		0.436	0.307
p-value for AD test	0.334	-	0.166	0.401
G value	1.630	-	1.110	1.570
p-value for Grubbs test	0.150	-	1.000	0.261

Table 2. Results of the tensile test for polyethylene specimens.

Specimen Number	Young Modulus (MPa)	Yield Stress (MPa)	Maximum Tensile Stress (MPa)	Tensile Strain at Maximum Tensile Stress (%)
1.	1044.393	12.699	24.339	502.167
2.	1068.163	12.692	24.735	476.667
3.	914.544	12.565	24.683	461.667
4.	961.303	12.613	25.241	538.833
5.	1035.568	12.574	24.651	504.167
Mean	1004.794	12.629	24.730	496.700
St. dev	64.357	0.064	0.325	29.536
AD value	0.330	0.389	0.373	0.219
p-value for AD test	0.345	0.231	0.258	0.673
G value	1.400	1.110	1.570	1.430
p-value for Grubbs test	0.583	1.000	0.247	0.530

Table 3. Results of the tensile test for polytetrafluoroethylene specimens.

Specimen Number	Young Modulus (MPa)	Yield Stress (MPa)	Maximum Tensile Stress (MPa)	Tensile Strain at Maximum Tensile Stress (%)
1.	472.717	-	23.796	386.499
2.	535.948	-	22.532	277.833
3.	471.926	-	21.166	312.500
4.	485.780	-	23.671	385.500
5.	424.333	-	22.959	358.833
Mean	478.141	-	22.825	344.233
St. dev	39.885	-	1.062	47.745
AD value	0.299		0.294	0.328
p-value for AD test	0.425	-	0.439	0.349
G value	1.450	-	1.560	1.390
p-value for Grubbs test	0.482	-	0.267	0.609

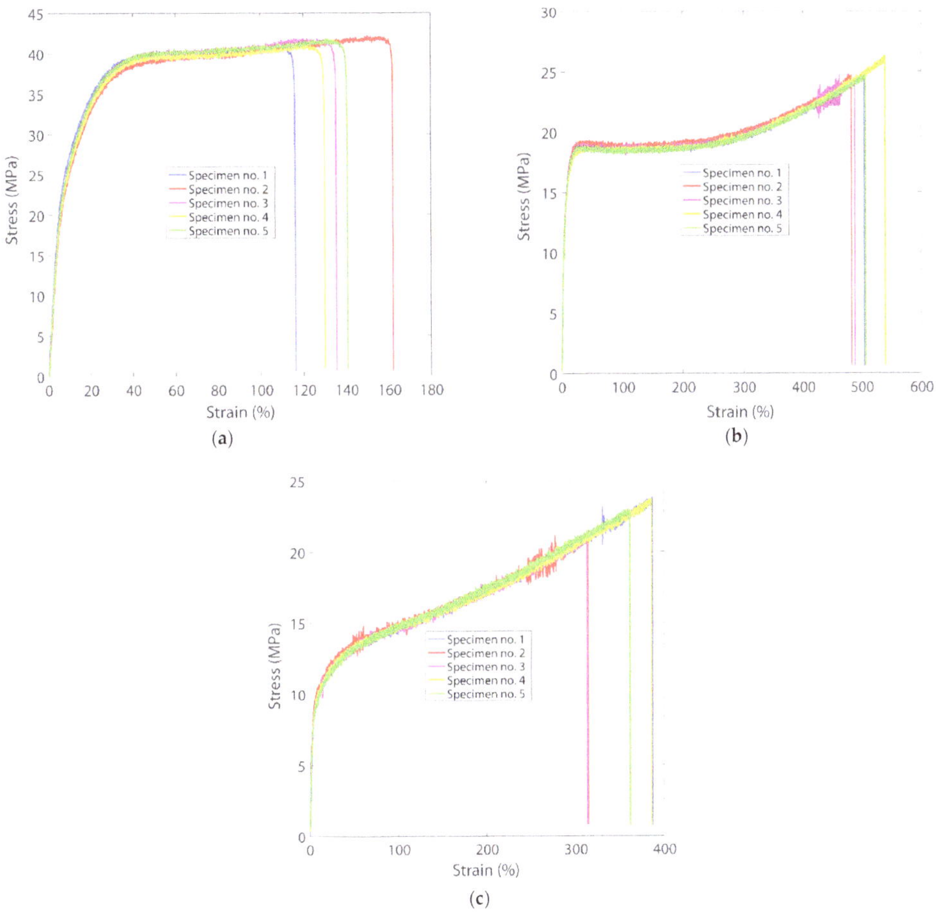

Figure 2. Stress–strain curves for polyamide (**a**), polyethylene (**b**) and polytetrafluorethylene (**c**).

The analysis of the results and statistical processing of the tensile tests of the four types of polymeric materials found no outliers for any of the measurements and a normal distribution of results for all obtained data. This is evidenced by the fact that the *p*-values are significantly greater than 0.05 in all cases.

The material with the lowest value of the elastic modulus is polytetrafluoroethylene (with an average value of 478.14 MPa), which indicates the fact that polytetrafluoroethylene has a much more noticeable elastic behavior. The Yield stress, which is only significant for polyethylene, averages 12.63 MPa, about 50% of the maximum value of the tensile stress in the material. The material with the highest value of the maximum tensile stress is polyamide with 41.54 MPa, followed by polyethylene and polytetrafluoroethylene with 24.73 MPa and 22.83 MPa, respectively. The material with the highest elongation, estimated on the basis of the tensile strain at maximum stress, is polyethylene with 496.7%, followed by polytetrafluoroethylene with 344.23% and polyamide with 126.53%.

Based on the observations presented above, the research on the behavior during SPIF was continued for all three polymeric materials previously discussed.

3.2. Force Determination during Incremental Forming of Polymeric Materials

One of the most important output parameters for the incremental forming process is the force developed in the process. The force during the incremental forming process, in contrast to conventional forming processes (e.g., deep drawing), consists of three components that vary significantly throughout the process. Therefore, the force measurement in the incremental forming of polymeric materials made necessary the use of the force sensor together with the conditioning module and the data acquisition system.

Based on the literature review, the main factors influencing the forces in the incremental forming process of polymeric materials were selected. These factors are presented in Table 4.

Table 4. The main influence factors and their level of variation.

Input Factors					Output Factors
Code	Type	Symbol	Unit	Variation Domain	y_1, y_2, y_3
x_1	Polymeric material	MAT	-	PA, PE, PTFE	Fz, Fx, Fy (kN)
x_2	Step	s	(mm)	0.5–1	Fz, Fx, Fy (kN)
x_3	Punch diameter	Dp	(mm)	6–10	Fz, Fx, Fy (kN)
x_4	Wall angle	α	(°)	50–60	Fz, Fx, Fy (kN)
x_5	Initial thickness	g	(mm)	0.5–5	Fz, Fx, Fy (kN)
x_6	Punch feed	v_s	(mm/min)	100–600	Fz, Fx, Fy (kN)
x_7	Punch angular speed	ω	(rot/min)	50–180	Fz, Fx, Fy (kN)

Of the seven input factors presented in Table 4, only the first four, namely the material type, the punch diameter, the step on vertical direction and the wall angle were selected for analysis for a number of reasons: two or more identical initial thicknesses were not available for all materials, the use of an industrial robot did not allow for a rotational movement of the punch and the value used for the punch feed was the maximum value, in order to reduce the processing time.

Previous to designing the experiments, given that the first four parameters presented in Table 4 were to be considered, several preliminary experiments were carried out, attempting to identify the limits of the process in terms of formability. For this reason, three sets of experiments were performed for all three types of materials. The first set had the following configuration: wall angle $\alpha = 60°$, punch diameter Dp = 10 mm and step s = 0.75 mm. The second set had the following input parameters: wall angle $\alpha = 55°$, punch diameter Dp = 6 mm and step s = 0.75 mm, while the third one had: wall angle $\alpha = 50°$, punch diameter Dp = 8 mm and step s = 0.75 mm. In fact, these experiments were conducted to observe whether, at different angles of the part and different diameters of the punch, all materials allow deformation without breaking. The step was set constant at an average value of 0.75 mm. These preliminary tests proved useful as they found that polytetrafluoroethylene broke in all 3 variations of the experiment. Figure 3a,b, show images of the part taken before and immediately after failure for the third set of experiments (wall angle $\alpha = 50°$, punch diameter Dp = 8 mm and step s = 0.75 mm).

A significant decrease in formability was observed for polytetrafluoroethylene, leading to the conclusion that this material is not suitable for incremental forming at room temperature. Possible future research could address the hot deformation of this material. As a result, it was decided to eliminate polytetrafluoroethylene from the experimental investigations and only keep polyamide and polyethylene.

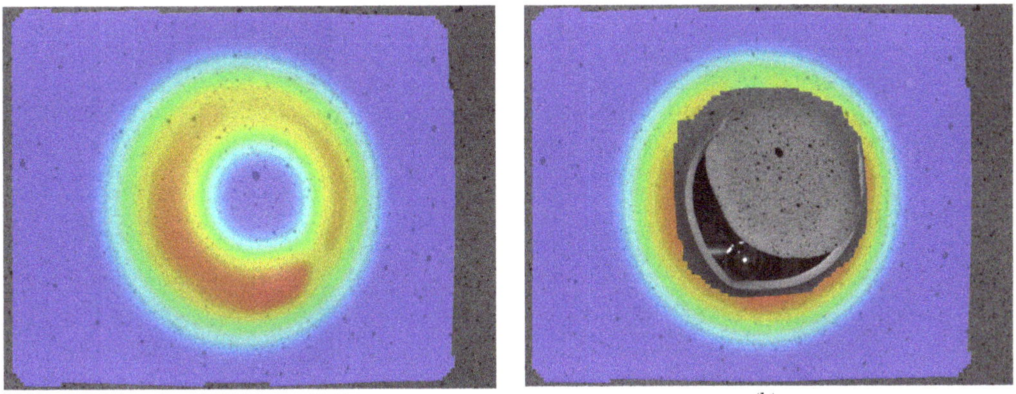

(**a**) (**b**)

Figure 3. The images of the part manufactured by polytetrafluorethylene with wall angle $\alpha = 50°$, punch diameter Dp = 8 mm and step s = 0.75 mm taken before (**a**) and immediately after (**b**) failure.

3.2.1. Analysis of the Signal-to-Noise Ratio for the Forces during the Incremental Forming of Polymeric Materials

For the design of experiments, there are software packages that automatically perform the experimental plan using: a general full-factorial experiment, a fractional experiment, a surface response methodology or the Taguchi methodology. The Taguchi method is derived from the factorial methodology and throughout time it has proven its efficiency over other methods in the design of experiments. For the study presented below the Taguchi design of experiment was chosen because of its most significant advantage over other methods, i.e., obtaining robust results with a small number of experiments.

Before outlining the experimental investigations using the Taguchi orthogonal arrays, it is worth mentioning the fact that in the general case of forming processes (not only in the particular case of incremental forming), the minimization of forces is always desired. This is because, in the case of forming processes, the machine is chosen according to the required force. All the more so in the present case, where an industrial robot which does not produce a force as great as that of a press or of a CNC machine, nor has a very high rigidity was used in the experiments, it is important to be able to reduce the forces in the process.

Therefore, the aim of the experimental investigation is to decrease the maximum force values found in the peaks of the variation graphs. The design of experiments based on the Taguchi method is used to analyze the effect of the input processing parameters on the forces during the incremental forming processing of polymeric materials. Four parameters were selected from the input parameters presented in Table 4, of which two parameters are related to the part to be produced and the other two are technological parameters, related to the operating regime used. The parameters concerning the part are the material of the part (MAT) and the angle of the part produced (α), whereas the technological parameters are the diameter of the punch (Dp) and the step on vertical direction (s).

The material of the part introduces two levels of variation, with polyamide and polyethylene having been selected as materials, as mentioned previously, while the other parameters are defined in three levels of variation, namely 50°, 55° and 60° for the angle of the part, 6 mm, 8 mm and 10 mm for the diameter of the punch and 0.5 mm, 0.75 mm and 1 mm for the step on vertical direction. The values chosen for the step may raise an issue. In the case of incremental forming of metallic materials, the step chosen has smaller values (0.1 mm, 0.25 mm) since a larger step would lead to "waves" on the inner surface of the part. Given the fact that polymeric materials have a very strong springback ability and that a small step size would lead to particularly long processing times, the step size values presented above were chosen in order to increase the productivity of the process.

To study the contribution of each input parameter on the values of the force during the incremental forming of polymeric materials, the Taguchi orthogonal array L18($2^1 \times 3^3$) with mixed levels (one parameter has two levels of variation and three parameters have three levels of variation) is selected. Table 5 contains the design of experiments, with the factors and levels of variation, as well as their codifications. To reduce the margin of error, all eighteen experiments were repeated twice. The forces in the process were measured along all three directions of the coordinate axes.

Table 5. The design of experiments using Taguchi orthogonal array L18 ($2^1 \times 3^3$).

Case	Mat. Code	α Code	Dp Code	s Code	Mat.	α (°)	Dp (mm)	s (mm)
F1	−1	−1	−1	−1	PA	50	6	0.50
F2	−1	−1	0	0	PA	50	8	0.75
F3	−1	−1	1	1	PA	50	10	1.00
F4	−1	0	−1	0	PA	55	6	0.75
F5	−1	0	0	1	PA	55	8	1.00
F6	−1	0	1	−1	PA	55	10	0.50
F7	−1	1	−1	1	PA	60	6	1.00
F8	−1	1	0	−1	PA	60	8	0.50
F9	−1	1	1	0	PA	60	10	0.75
F10	1	−1	−1	−1	PE	50	6	0.50
F11	1	−1	0	0	PE	50	8	0.75
F12	1	−1	1	1	PE	50	10	1.00
F13	1	0	−1	0	PE	55	6	0.75
F14	1	0	0	1	PE	55	8	1.00
F15	1	0	1	−1	PE	55	10	0.50
F16	1	1	−1	1	PE	60	6	1.00
F17	1	1	0	−1	PE	60	8	0.50
F18	1	1	1	0	PE	60	10	0.75

Depending on the purpose of the experiments, different signal-to-noise (S/N) ratios can be selected as follows: "smaller is better", "larger is better" and "nominal is better". For this study, as already mentioned, the "smaller is better" option was chosen for all forces present in the process.

The numerical results of the measurements, namely the mean value, standard deviation, S/N ratio and variation coefficient for all three types of forces are presented in Tables 6–8.

The force variation graphs for the F2, F9, F11 and F18 cases, two for polyamide (F2 and F9) and two for polyethylene (F11 and F18), obtained from the design using the Taguchi orthogonal array for the three directions of the coordinate axis system, are shown in Figures 4–15.

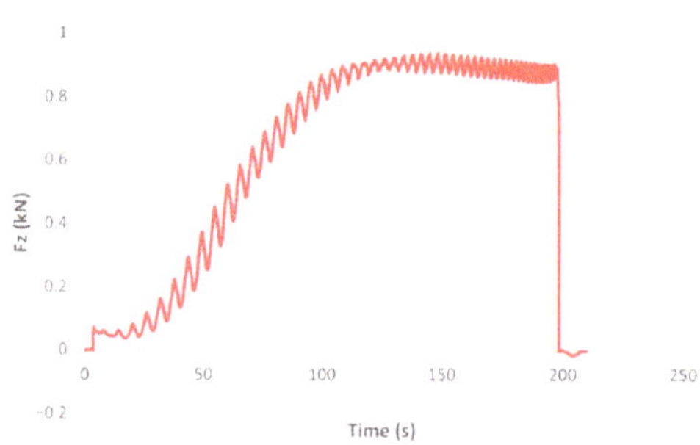

Figure 4. The Fz force variation for the parts produced from PA, with Dp = 8 mm, s = 0.75 and α = 50° (case F2).

Table 6. The results and the statistical processing for the Fz forces.

Cod	Mat.	α (°)	Dp (mm)	p (mm)	Fz_1 (kN)	Fz_2 (kN)	Fz Mean (kN)	Standard Deviation	S/N Ratio	Coefficient of Variation
F1	PA	50	6	0.5000	0.6857	0.6806	0.6831	0.0036	3.3099	0.0053
F2	PA	50	8	0.7500	0.9381	0.9348	0.9365	0.0023	0.5703	0.0025
F3	PA	50	10	1.0000	1.2168	1.2178	1.2173	0.0007	−1.7081	0.0006
F4	PA	55	6	0.7500	0.7742	0.7751	0.7746	0.0006	2.2181	0.0008
F5	PA	55	8	1.0000	1.0643	1.0717	1.0680	0.0053	−0.5715	0.0049
F6	PA	55	10	0.5000	0.9321	0.9569	0.9445	0.0176	0.4949	0.0186
F7	PA	60	6	1.0000	0.8565	0.8586	0.8575	0.0015	1.3351	0.0017
F8	PA	60	8	0.5000	0.8276	0.8060	0.8168	0.0153	1.7570	0.0187
F9	PA	60	10	0.7500	1.0995	1.0955	1.0975	0.0028	−0.8081	0.0026
F10	PE	50	6	0.5000	0.4518	0.4225	0.4371	0.0208	7.1825	0.0475
F11	PE	50	8	0.7500	0.6598	0.6554	0.6576	0.0031	3.6409	0.0048
F12	PE	50	10	1.0000	0.9084	0.9459	0.9272	0.0265	0.6552	0.0286
F13	PE	55	6	0.7500	0.4901	0.4937	0.4919	0.0026	6.1623	0.0053
F14	PE	55	8	1.0000	0.6612	0.6645	0.6628	0.0023	3.5718	0.0035
F15	PE	55	10	0.5000	0.6514	0.7126	0.6820	0.0433	3.3155	0.0634
F16	PE	60	6	1.0000	0.5424	0.4969	0.5197	0.0322	5.6772	0.0620
F17	PE	60	8	0.5000	0.5641	0.5700	0.5670	0.0042	4.9278	0.0074
F18	PE	60	10	0.7500	0.7226	0.7391	0.7308	0.0117	2.7235	0.0160

Table 7. The results and the statistical processing for the Fx forces.

Cod	Mat.	α (°)	Dp (mm)	p (mm)	Fx_1 (kN)	Fx_2 (kN)	Fx Mean (kN)	Standard Deviation	S/N Ratio	Coefficient of Variation
F1	PA	50	6	0.5000	0.3238	0.3183	0.3211	0.0039	9.8674	0.0121
F2	PA	50	8	0.7500	0.4523	0.4496	0.4509	0.0019	6.9181	0.0043
F3	PA	50	10	1.0000	0.5627	0.6189	0.5908	0.0398	4.5613	0.0673
F4	PA	55	6	0.7500	0.4002	0.3974	0.3988	0.0020	7.9853	0.0050
F5	PA	55	8	1.0000	0.5265	0.5313	0.5289	0.0034	5.5321	0.0065
F6	PA	55	10	0.5000	0.4767	0.4667	0.4717	0.0071	6.5262	0.0150
F7	PA	60	6	1.0000	0.4811	0.4836	0.4823	0.0018	6.3332	0.0037
F8	PA	60	8	0.5000	0.3735	0.4258	0.3996	0.0370	7.9479	0.0925
F9	PA	60	10	0.7500	0.5143	0.5915	0.5529	0.0546	5.1263	0.0987
F10	PE	50	6	0.5000	0.2065	0.2067	0.2066	0.0001	13.6975	0.0007
F11	PE	50	8	0.7500	0.3216	0.3287	0.3252	0.0050	9.7578	0.0154
F12	PE	50	10	1.0000	0.4589	0.4524	0.4556	0.0045	6.8273	0.0100
F13	PE	55	6	0.7500	0.2712	0.2736	0.2724	0.0017	11.2965	0.0063
F14	PE	55	8	1.0000	0.3869	0.3830	0.3850	0.0028	8.2911	0.0072
F15	PE	55	10	0.5000	0.3491	0.3500	0.3496	0.0006	9.1298	0.0017
F16	PE	60	6	1.0000	0.3438	0.3444	0.3441	0.0004	9.2655	0.0013
F17	PE	60	8	0.5000	0.2844	0.2822	0.2833	0.0015	10.9544	0.0054
F18	PE	60	10	0.7500	0.4108	0.4073	0.4091	0.0025	7.7641	0.0061

Table 8. The results and the statistical processing for the Fy forces.

Cod	Mat.	α (°)	Dp (mm)	p (mm)	Fy_1 (kN)	Fy_2 (kN)	Fy Mean (kN)	Standard Deviation	S/N Ratio	Coefficient of Variation
F1	PA	50	6	0.5000	0.3592	0.3089	0.3340	0.0356	9.4993	0.1064
F2	PA	50	8	0.7500	0.4844	0.4263	0.4553	0.0411	6.8159	0.0903
F3	PA	50	10	1.0000	0.5311	0.6373	0.5842	0.0751	4.6330	0.1286
F4	PA	55	6	0.7500	0.3935	0.4084	0.4009	0.0106	7.9370	0.0264
F5	PA	55	8	1.0000	0.5151	0.5408	0.5280	0.0182	5.5454	0.0345
F6	PA	55	10	0.5000	0.4646	0.4785	0.4715	0.0099	6.5285	0.0209
F7	PA	60	6	1.0000	0.4803	0.4832	0.4817	0.0020	6.3441	0.0042
F8	PA	60	8	0.5000	0.3782	0.4430	0.4106	0.0458	7.7050	0.1116
F9	PA	60	10	0.7500	0.4944	0.5997	0.5471	0.0745	5.1993	0.1361

Table 8. *Cont.*

Cod	Mat.	α (°)	Dp (mm)	p (mm)	Fy_1 (kN)	Fy_2 (kN)	Fy Mean (kN)	Standard Deviation	S/N Ratio	Coefficient of Variation
F10	PE	50	6	0.5000	0.2060	0.2062	0.2061	0.0002	13.7193	0.0008
F11	PE	50	8	0.7500	0.2877	0.3663	0.3270	0.0555	9.6468	0.1699
F12	PE	50	10	1.0000	0.4405	0.4737	0.4571	0.0235	6.7941	0.0514
F13	PE	55	6	0.7500	0.2715	0.2721	0.2718	0.0005	11.3140	0.0017
F14	PE	55	8	1.0000	0.3696	0.4127	0.3911	0.0305	8.1406	0.0780
F15	PE	55	10	0.5000	0.3491	0.3512	0.3501	0.0015	9.1154	0.0042
F16	PE	60	6	1.0000	0.3417	0.3461	0.3439	0.0031	9.2719	0.0091
F17	PE	60	8	0.5000	0.2816	0.2848	0.2832	0.0023	10.9581	0.0081
F18	PE	60	10	0.7500	0.4125	0.4083	0.4104	0.0029	7.7351	0.0072

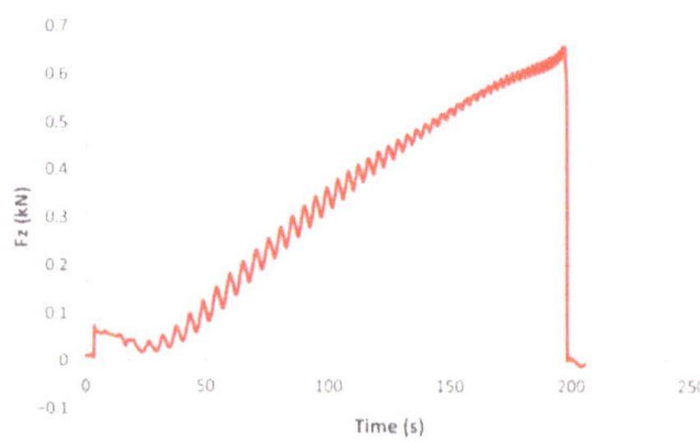

Figure 5. The Fz force variation for the parts produced from PE, with Dp = 8 mm, s = 0.75 and α = 50° (case F11).

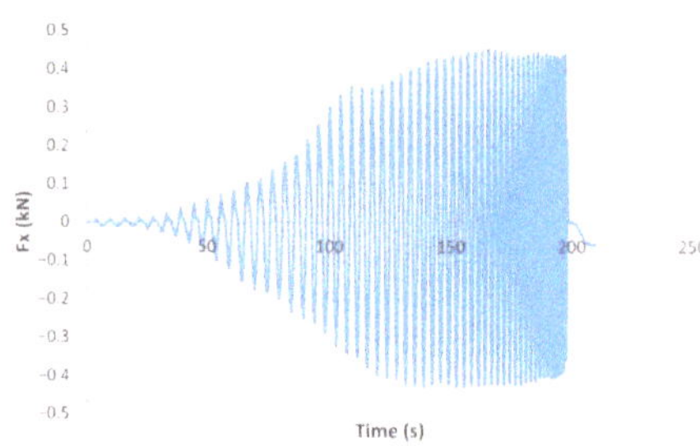

Figure 6. The Fx force variation for the parts produced from PA, with Dp = 8 mm, s = 0.75 and α = 50° (case F2).

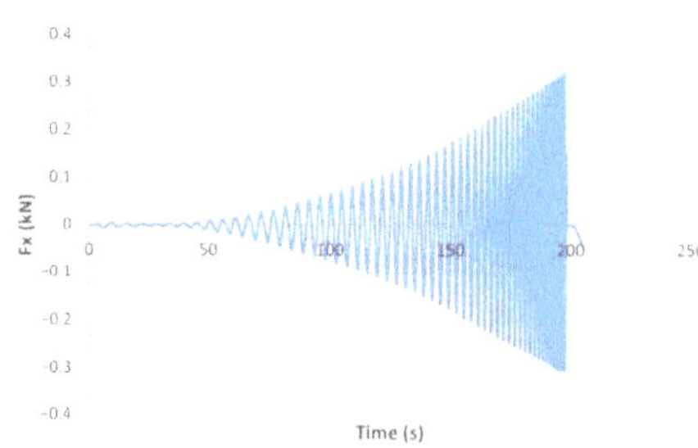

Figure 7. The Fx force variation for the parts produced from PE, with Dp = 8 mm, s = 0.75 and α = 50° (case F11).

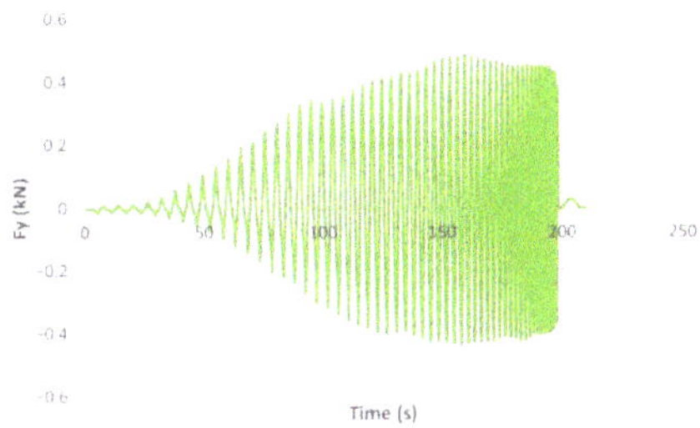

Figure 8. The Fy force variation for the parts produced from PA, with Dp = 8 mm, s = 0.75 and α = 50° (case F2).

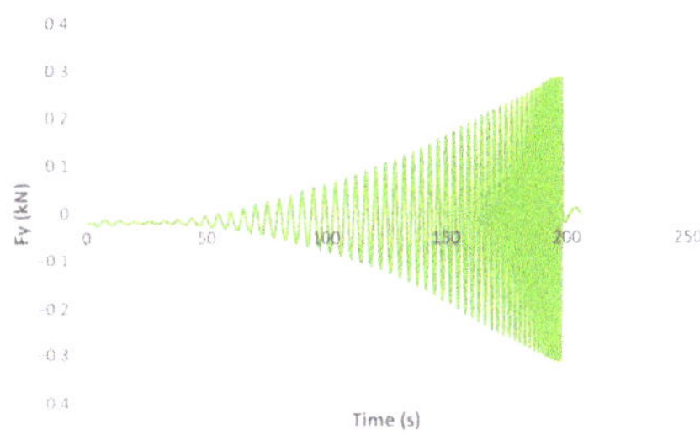

Figure 9. The Fy force variation for the parts produced from PE, with Dp = 8 mm, s = 0.75 and α = 50° (case F11).

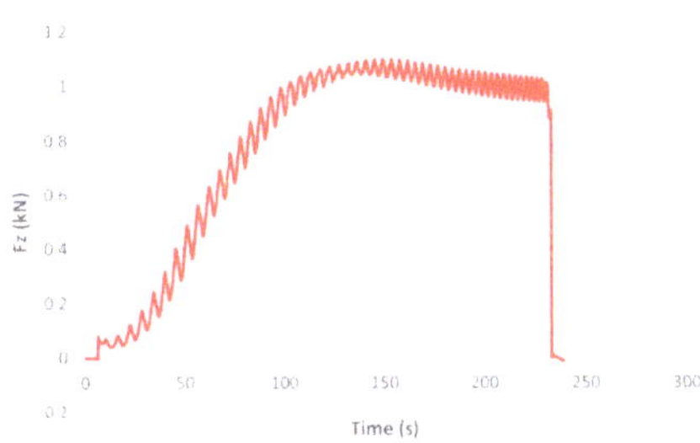

Figure 10. The Fz force variation for the parts produced from PA, with Dp = 10 mm, s = 0.75 and α = 60° (case F9).

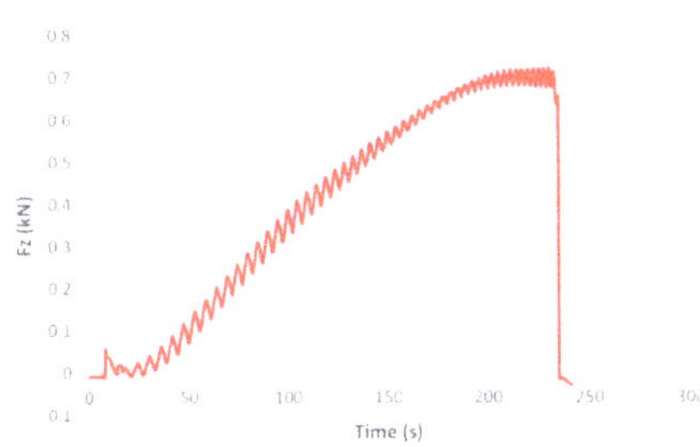

Figure 11. The Fz force variation for the parts produced from PE, with Dp = 10 mm, s = 0.75 and α = 60° (case F18).

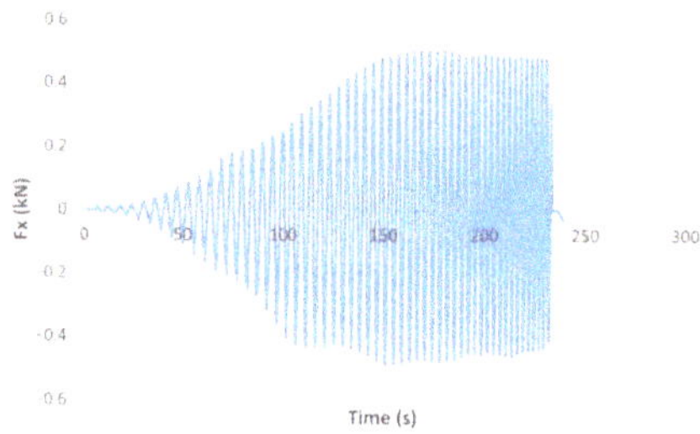

Figure 12. The Fx force variation for the parts produced from PA, with Dp = 10 mm, s = 0.75 and α = 60° (case F9).

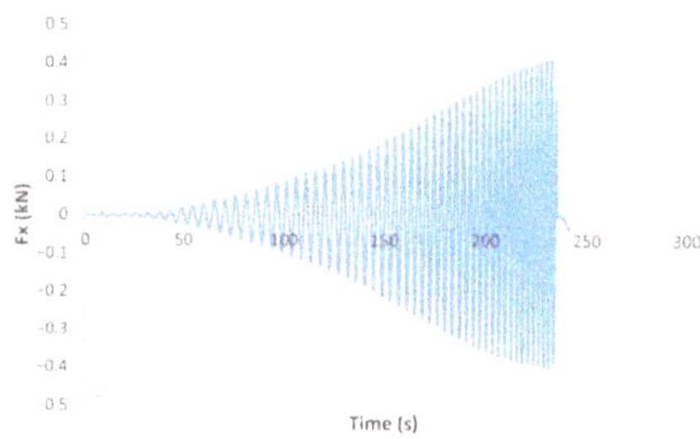

Figure 13. The Fx force variation for the parts produced from PE, with Dp = 10 mm, s = 0.75 and α = 60° (case F18).

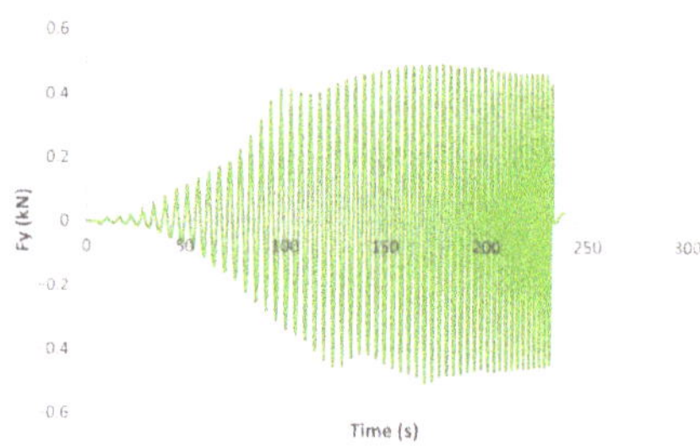

Figure 14. The Fy force variation for the parts produced from PA, with Dp = 10 mm, s = 0.75 and α = 60° (case F9).

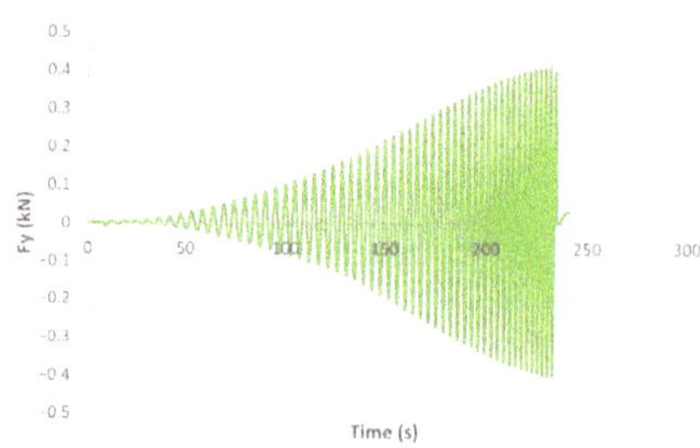

Figure 15. The Fy force variation for the parts produced from PE, with Dp = 10 mm, s = 0.75 and α = 60° (case F18).

The force in the Oz direction (Fz) increases for both materials, with local maxima and minima from the first contact of the punch with the blank, as can be seen in Figures 4, 5, 10 and 11. The local maxima occur in the areas where the punch penetrates the material by one step, while the minima occur in the moment prior to this moment. However, a slightly different behavior is observed for the two materials. In the case of polyamide, the force increases with a steeper slope, reaching a maximum value somewhere towards the middle of the working stroke and then remains relatively constant or even decreases slightly when using punches with smaller diameters. This is most noticeable when the punch has a 6 mm diameter (cases F1, F4 and F7). When it comes to polyethylene, the force increases with a less steep slope, slowly reaching its maximum value somewhere at the end of the working stroke, where it remains relatively constant. When using a 10 mm diameter punch (cases F12, F15 and F18), the force increases continuously all the way to the end.

Forces in the Ox and Oy directions (Fx) and (Fy) have a somewhat similar variation from each other, but different to that of the Fz force. Basically, the two forces increase and decrease around zero, reaching local maxima and minima with every circle the punch travels in the horizontal plane, as can be seen in Figures 6–9 and 12–15. The behavior of the two analyzed materials is slightly different here also. For polyamide, like for the Fz force, the local maxima of the Fx and Fy forces also increase to about half the stroke value and then stabilize until the end of the process. In the case of polyethylene, the stagnation towards the end of the working stroke is not so evident, with the values of the local maxima rising continuously until near the end. After the manufacturing process, the incrementally formed parts were CT scanned. The results of the scanning process for cases F9 and F18 are presented in Figure 16.

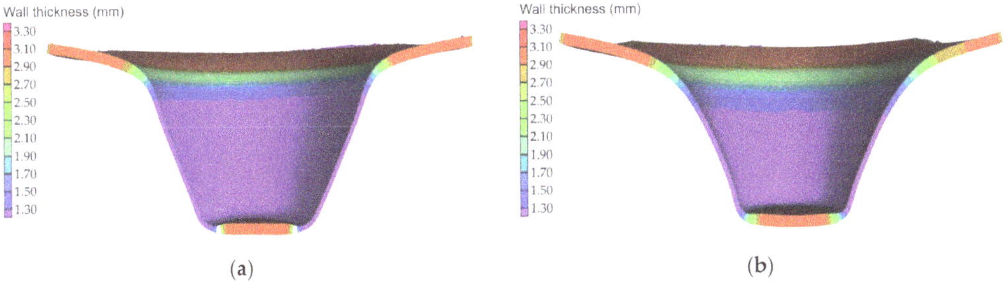

Figure 16. The CT scan of the parts obtained for cases F9—polyamide (**a**) and F18—polyethylene (**b**).

As can be seen in Figure 16, using the same trajectory for both materials, different geometries are obtained. The polyamide part results with a geometry similar to a frustum of cone (as it is desired), while the polyethylene part results with a concave geometry. For this reason, even though the parts have the same depth, the diameter of the section is smaller for polyethylene than for polyamide. It is for this reason that the variation of the forces follows a different pattern for the two polymeric materials.

The maximum values of the forces are always higher in the case of polyamide than in the case of polyethylene, for both the Fz force and the Fx and Fy forces. Thus, the maximum value of the Fz force occurs in case F3 (1.2173 kN), while those of the Fx and Fy forces are also obtained in case F3, with 0.5842 kN for Fx and 0.5908 kN for Fy, respectively. The minimum values of the forces occur in case F10—polyethylene (Fz = 0.4371 kN, Fx = 0.2066 kN and Fy = 0.2061 kN, respectively). Two parts made of polyamide and polyethylene with a frustum of pyramid type trajectory were also made, with the following input parameters: Dp = 8 mm, s = 0.75 mm and $\alpha = 50^{\circ}$, corresponding to cases F2 and F11 in Table 8. The variation graphs of the six forces (three for each material) are shown in Figures 17–22.

The analysis of the graphs for the frustum of pyramid type parts leads to the same conclusions as those for the frustum of cone type parts, i.e., in the case of polyamide the force reaches its maximum value at about half of the working stroke, after which it stays constant or even decreases, as shown in Figure 17, while in the case of polyethylene it increases constantly until close to the end of the working stroke, as shown in Figure 18.

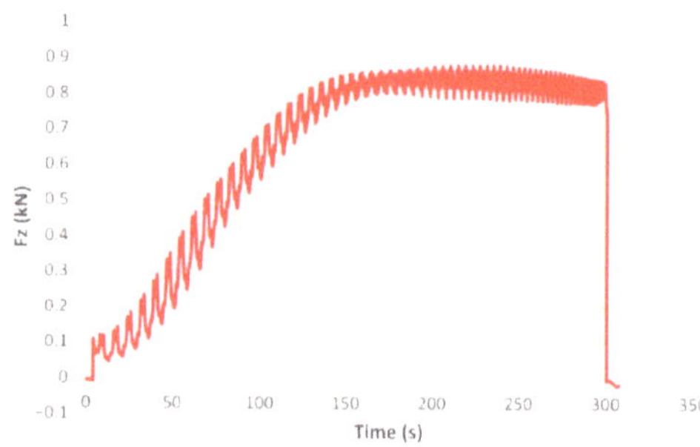

Figure 17. The Fz force variation for the frustum of pyramid parts produced from PA, with Dp = 8 mm, s = 0.75 and α = 50°.

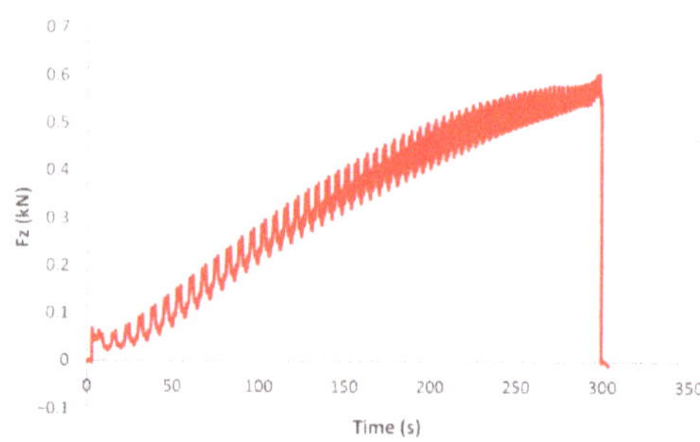

Figure 18. The Fz force variation for the frustum of pyramid parts produced from PE, with Dp = 8 mm, s = 0.75 and α = 50°.

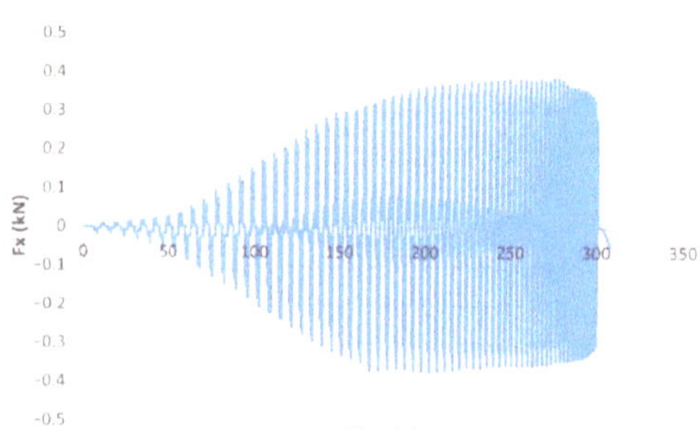

Figure 19. The Fx force variation for the frustum of pyramid parts produced from PA, with Dp = 8 mm, s = 0.75 and α = 50°.

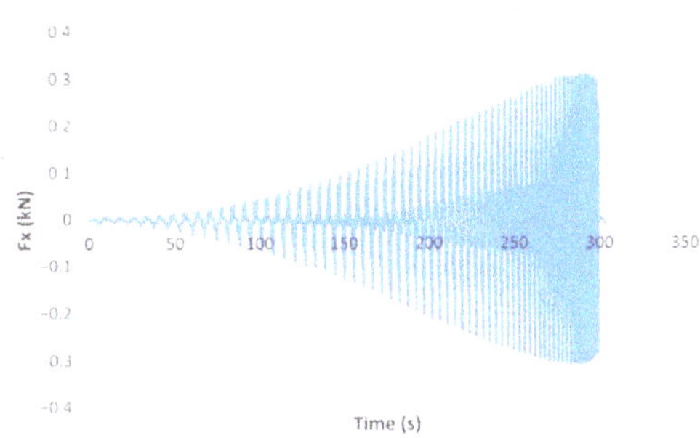

Figure 20. The Fx force variation for the frustum of pyramid parts produced from PE, with Dp = 8 mm, s = 0.75 and α = 50°.

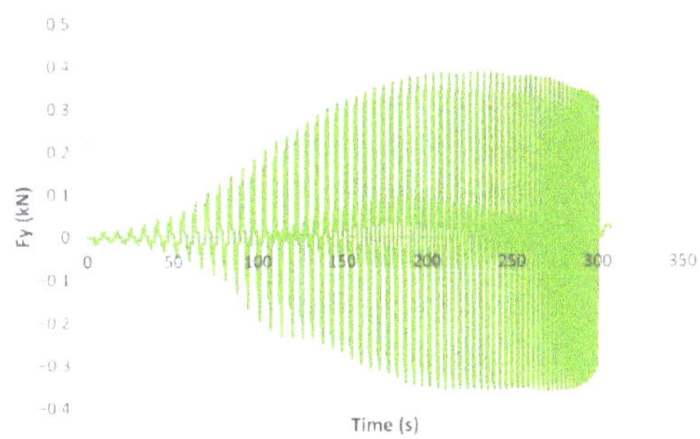

Figure 21. The Fy force variation for the frustum of pyramid parts produced from PA, with Dp = 8 mm, s = 0.75 and α = 50°.

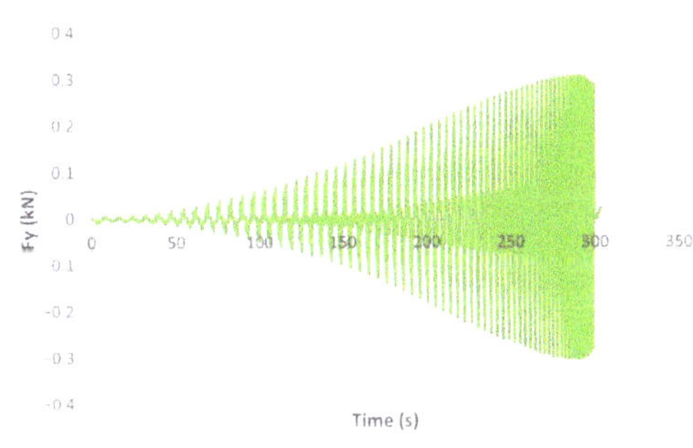

Figure 22. The Fy force variation for the frustum of pyramid parts produced from PE, with Dp = 8 mm, s = 0.75 and α = 50°.

The maximum Fz force values obtained for the frustum of pyramid type parts are lower for both polyamide (0.8810 kN) and polyethylene (0.6083 kN) than those obtained for the frustum of cone type parts (0.9365 kN—polyamide, 0.6576 kN—polyethylene). The maximum Fx force values obtained for the frustum of pyramid type parts are also lower for both polyamide (0.3876 kN), as shown in Figure 19, and polyethylene (0.3087 kN), as shown in Figure 20, compared to those obtained for the frustum of cone type parts (0.4509 kN—polyamide, presented in Figure 12, 0.3252 kN—polyethylene, presented in Figure 13). The same applies in the case of the Fy force, where the values for polyamide (0.3879 kN), presented in Figure 21, and polyethylene (0.3076 kN), presented in Figure 22, are lower than those obtained for the frustum of cone type parts (0.4553 kN—polyamide, as shown in Figure 14, 0.3270 kN—polyethylene, as shown in Figure 15).

The mean response values of the signal-to-noise ratios (S/N) for the Fz force (with the "smaller is better" condition) are shown in Table 9.

Table 9. Response Table for Signal-to-Noise Ratios for the Fz force (with the "smaller is better" condition).

Level	Polymeric Material	Wall Angle	Punch Diameter	Step
1	0.733	2.275	**4.314**	**3.498**
2	**4.271**	2.532	2.316	2.418
3		**2.602**	0.779	1.493
Delta	3.538	0.327	3.535	2.004
Rank	**1**	**4**	**2**	**3**

Table 9 provides the mean response values of the S/N ratios, which analyze the effect of the influencing factors (MAT, α, Dp and p) on the force in the z-direction (Fz). It shows the optimal levels (based on the S/N ratios) for which the control factors result in the lowest value of Fz. These are: level 2 for MAT (S/N = 4.271), level 3 for α (S/N = 2.602), level 1 for Dp (S/N = 4.314) and level 1 for s (S/N = 3.498). Delta and rank determine which of these four factors has the greatest impact on the response. Delta measures the magnitude of impact by subtracting the lowest mean response value of a factor from its highest value. Rank assigns an order number to the factors, from the highest to the lowest impact, based on their delta values. For Fz, the material has the highest impact (delta: S/N = 3.538 and rank = 1), but the punch diameter also has a very similar impact (delta: S/N = 3.535 and rank = 2), while α has the lowest impact (delta: S/N = 0.327 and rank = 4).

The mean response values of the signal-to-noise ratios (S/N) for the Fx force (with the "smaller is better" condition) are shown in Table 10.

Table 10. Response Table for Signal-to-Noise Ratios for the Fx force (with the "smaller is better" condition).

Level	Polymeric Material	Wall Angle	Punch Diameter	Step
1	6.566	**8.605**	**9.741**	**9.687**
2	**9.665**	8.127	8.234	8.141
3		7.899	6.656	6.802
Delta	3.099	0.706	3.085	2.885
Rank	**1**	4	2	3

Similar to the case presented for the Fz force, Table 10 indicates the optimal levels of the control factors for which the lowest value of Fx is obtained. Thus, the optimal values are obtained for MAT at level 2 (S/N = 9.665), α at level 1 (S/N = 8.605), Dp at level 1 (S/N = 9.741) and s at level 1 (S/N = 9.687). The material has the highest impact for Fx as well (delta: S/N = 3.099 and rank = 1), with the punch diameter again having a similar impact (delta: S/N = 3.085 and rank = 2), while α has the lowest impact (delta: S/N = 0.706 and rank = 4).

The mean response values of the signal-to-noise ratios (S/N) for the Fy force (with the "smaller is better" condition) are shown in Table 11.

Table 11. Response Table for Signal-to-Noise Ratios for the Fy force (with the "smaller is better" condition).

Level	Polymeric Material	Wall Angle	Punch Diameter	Step
1	6.575	**8.518**	**9.681**	**9.588**
2	**9.633**	8.097	8.135	8.108
3		7.869	6.668	6.788
Delta	3.058	0.649	3.013	2.799
Rank	**1**	**4**	**2**	**3**

An analysis of Table 11 leads to the optimal values resulting in the lowest value of Fy: MAT at level 2 (S/N = 9.633), α at level 1 (S/N = 8.518), Dp at level 1 (S/N = 9.681) and s at level 1 (S/N = 9.588). For Fy, as for Fx, the material has the highest impact (delta: S/N = 3.058 and rank = 1), followed by the punch diameter (delta: S/N = 3.013 and rank = 2), with α having the lowest impact (delta: S/N = 0.649 and rank = 4).

The graphical representation (the graph showing the main effects for the signal-to-noise (S/N) ratios) of the factor levels in Tables 9–11 for the three forces (Fz, Fx and Fy) is shown in Figures 23–25.

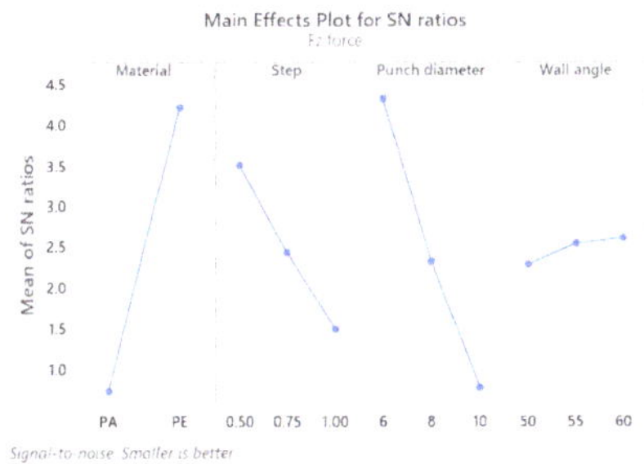

Figure 23. The main effects plot for S/N ratios for Fz force.

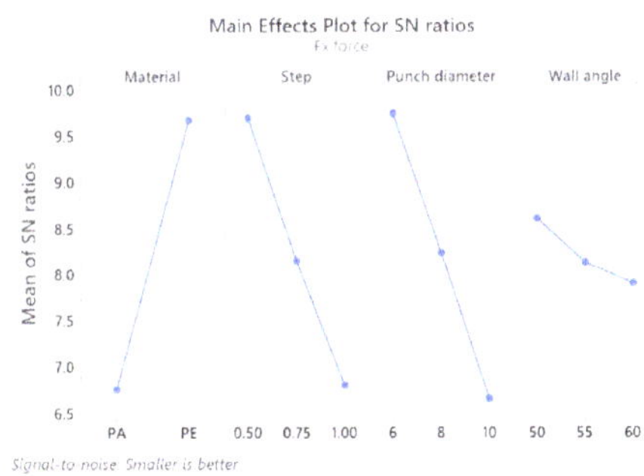

Figure 24. The main effects plot for S/N ratios for Fx force.

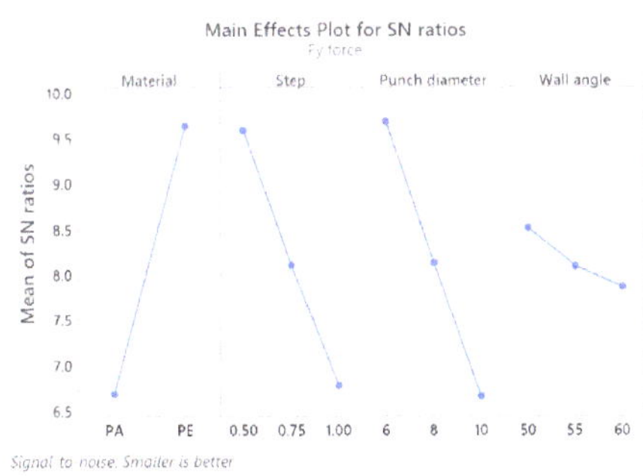

Figure 25. The main effects plot for S/N ratios for Fy force.

The graphs of the main effects show the optimal values of the control factors that lead to the lowest values of the forces. The main effect is also called the difference between means, caused by the different levels within a factor. If the graph line is horizontal, then there is no main effect for the levels of a factor, but a line with a greater slope or difference between vertical points will have a greater main effect.

For all three categories of forces, the main effect graphs for the material (MAT) and for the punch diameter (Dp) display graph lines with larger slopes, therefore having the largest main effect. The main goal of the experimental investigations is to find the control factor levels that minimize the impact of the noise factors on the responses. The optimal parameters for minimizing the forces can be easily determined from these graphs.

The best level for each control factor was found based on the highest S/N ratio in the levels of these input parameters (control factors). The range of the control factors throughout the experiments was mentioned earlier in this chapter. The analysis of Tables 9–11, as well as of the graphs showing the main effects, leads to the optimal values for which the Fz, Fx and Fy forces can be kept as low as possible. Thus, the optimal values for which the Fz force is minimal are the following: 6 mm for Dp, polyethylene for MAT, 0.5 mm for s and 60° for α. Minimum Fx and Fy forces are obtained with the following optimal values: Dp = 6 mm, MAT—polyethylene, s = 0.5 mm and α = 50°.

3.2.2. Analysis of Variance for the Forces during the Incremental Forming of Polymeric Materials

For further robustness of the results obtained by the Taguchi method based on the S/N ratios, an ANOVA statistical model is used. Unlike the Taguchi method, it calculates the contribution of each factor involved. In addition, ANOVA analysis also allows the assessment of interactions between different factors. For the present case, the interactions between the material and the technological factors, namely the diameter of the punch and the step, were chosen. The significance of the control factors in the ANOVA analysis is determined by comparing the F-values of each control factor. These tables were calculated with a 5% significance level and a 95% confidence level. If the *p*-value of the factor is greater than 0.05, then it will be considered a non-significant factor. Table 12 shows the ANOVA table for the Fz force.

Table 12. Analysis of Variance for Means (Fz force).

Source	Degree of Freedom	Adjusted Value of the Sum of the Squares	F Value	*p* Value	Contribution (%)
MAT	1	54.2855	382.8200	0.0000	51.0400
α	2	0.3556	1.2500	0.3510	0.3300
Dp	2	37.7085	132.9600	0.0000	35.4600
s	2	12.0801	42.5900	0.0000	11.3600
MAT*Dp	2	0.9885	3.4900	0.0990	0.9300
MAT*s	2	0.0848	0.3000	0.7520	0.0800
Residual error	6	0.8508			0.8000
Total	17	106.3540			

An analysis of the *p*-values in Table 12 shows that a *p*-value greater than 0.05 occurs in the case of the part angle and of two interactions (material * punch diameter and material * step). These factors are therefore non-significant for the Fz force. Table 13 shows the ANOVA table for the Fx force.

The *p*-values in Table 13 indicate a *p*-value greater than 0.05 for only two interactions (material * punch diameter and material * step). Therefore, all other factors are significant for the Fx force. Table 14 presents the ANOVA table for the Fy force.

Considering the *p*-values in Table 14, only two interactions (material * punch diameter and material * step) have a *p*-value greater than 0.05, as in the case of the Fx force.

The percentage contribution in the ANOVA tables shows the magnitude of the influence of the parameters. The contribution of the material is 51.04%, of the punch diameter is 35.46%, of the step is 11.36% and of the angle is only 0.33% for the value of the Fz force. Similarly, for the value of the Fx force, these percentages are 40.3% for the material, 30.21% for the punch diameter, 26.47% for the step and 1.65% for the angle. As for the value of the Fy force, the percentage contributions are 41.94% for the material, 29.31% for the punch diameter, 25.32% for the step and 1.4% for the angle. The most significant parameter for all three responses is the blank material. The percentage of error was considerably low, 0.8% for Fz, 0.59% for Fx and 0.57% for Fy, respectively.

Table 13. Analysis of Variance for Means (Fx force).

Source	Degree of Freedom	Adjusted Value of the Sum of the Squares	F Value	*p* Value	Contribution (%)
MAT	1	38.0950	407.5000	0.0000	40.3000
α	2	1.5591	8.3400	0.0190	1.6500
Dp	2	28.5578	152.7400	0.0000	30.2100
s	2	25.0207	133.8200	0.0000	26.4700
MAT*Dp	2	0.5526	2.9600	0.1280	0.5800
MAT*s	2	0.1841	0.9800	0.4270	0.1900
Residual error	6	0.5609			0.5900
Total	17	94.5302			

Table 14. Analysis of Variance for Means (Fy force).

Source	Degree of Freedom	Adjusted Value of the Sum of the Squares	F Value	*p* Value	Contribution (%)
MAT	1	38.9774	440.0500	0.0000	41.9400
α	2	1.3016	7.3500	0.0240	1.4000
Dp	2	27.2472	153.8100	0.0000	29.3100
s	2	23.5355	132.8600	0.0000	25.3200
MAT*Dp	2	0.8809	4.9700	0.0530	0.9500
MAT*s	2	0.4722	2.6700	0.1480	0.5100
Residual error	6	0.5315			0.5700
Total	17	92.9463			

3.2.3. Regression Analysis for the Forces during the Incremental Forming of Polymeric Materials

In order to obtain the mathematical relationships of the forces in the process for the input parameters, the two materials had to be separated and one relationship had to be obtained for each of them. Thus, this study resulted in finding relationships between the three response variables (forces Fz, Fx and Fy) and the three input variables (α, Dp and s). These are obtained by using the regression analysis. Three linear regression equations were modeled for each material. Several important parameters for evaluating regression models are R-sq (describes how much variation is present in the response), R-sq (adj) (is a modified R-sq that has been adjusted for the number of expressions), and R-sq (pred) (indicates how well the model predicts responses for new measurements). For the regression model to be of best performance, these values must be high.

Following the regression analysis performed with Minitab, the data given in Table 15 was obtained for the Fz force in polyamide.

Table 15. The coefficients of the regression analysis of the Fz force for polyamide.

Variable	Coefficient	SE Coefficient	T Value	*p* Value
Constant term	0.0740	0.1090	0.6700	0.5300
Wall angle (α)	−0.0022	0.0018	−1.2000	0.2830
Punch diameter (Dp)	0.0787	0.0045	17.4300	0.0000
Step (s)	0.4656	0.0361	12.9000	0.0000

As can be seen from the *p*-values obtained, the coefficients of the angle α and of the constant in the regression equation are not significant, having values greater than 0.05. These coefficients are consequently excluded, and the analysis is repeated. The regression equation is then obtained by Equation (1):

$$F_{z\_PA} = 0.07511 \cdot D_p + 0.4442 \cdot s \quad (1)$$

An excellent predictability of the proposed mathematical model was achieved (R-sq (pred) = 99.89%). A regression analysis was then performed for the Fz force in the case of polyethylene, the results of which are shown in Table 16.

Table 16. The coefficients of the regression analysis of the Fz force for polyethylene.

Variable	Coefficient	SE Coefficient	T Value	*p* Value
Constant term	0.2000	0.1740	1.1500	0.3020
Wall angle (α)	−0.0068	0.0029	−2.3800	0.0630
Punch diameter (Dp)	0.0742	0.0072	10.3600	0.0000
Step (s)	0.2823	0.0573	4.9200	0.0040

In this case also, the *p*-values indicate that the coefficients of the angle α and of the constant in the regression equation are not significant, since they are greater than 0.05. Therefore, again, these coefficients are eliminated, and the analysis is repeated, leading to the following regression equation, Equation (2):

$$F_{z\_PE} = 0.06062 \cdot D_p + 0.2004 \cdot s \quad (2)$$

Excellent predictability of the proposed mathematical model (R-sq (pred) = 99.09%) was also obtained in this case. Following the regression analysis performed with Minitab, the data presented in Table 17 for the Fx force in polyamide was obtained.

Table 17. The coefficients of the regression analysis of the Fx force for polyamide.

Variable	Coefficient	SE Coefficient	T Value	*p* Value
Constant term	−0.1460	0.0347	−4.2100	0.0080
Wall angle (α)	0.0024	0.0006	4.2000	0.0090
Punch diameter (Dp)	0.0344	0.0014	24.0600	0.0000
Step (s)	0.2731	0.0114	23.8500	0.0000

As can be seen from the *p*-values, all coefficients are significant, with values less than 0.05. This generates the following regression equation, Equation (3):

$$F_{x\_PA} = -0.1460 + 0.002402 \cdot \alpha + 0.03443 \cdot D_p + 0.2731 \cdot s \quad (3)$$

The value of R-sq(pred) is 98.51%, which means that again a good predictability of the regression model was obtained. A regression analysis was then conducted for the Fx force in polyethylene, with the results presented in Table 18.

Table 18. The coefficients of the regression analysis of the Fx force for polyethylene.

Variable	Coefficient	SE Coefficient	T Value	*p* Value
Constant term	−0.1867	0.0320	−5.8300	0.0020
Wall angle (α)	0.00164	0.0005	3.1000	0.0270
Punch diameter (Dp)	0.0326	0.0013	24.6600	0.0000
Step (s)	0.2302	0.0106	21.7700	0.0000

Similar to the Fx force in polyamide, it is observed that all coefficients are significant, with values less than 0.05. The following regression equation is then obtained by Equation (4):

$$F_{x\_PE} = -0.1867 + 0.001637 \cdot \alpha + 0.03260 \cdot D_p + 0.2302 \cdot s \quad (4)$$

In this case too, a good predictability of the proposed mathematical model is found (R-sq(pred) = 98.32%). The regression analysis performed with Minitab resulted in the data shown in Table 19 for the Fy force in polyamide.

Table 19. The coefficients of the regression analysis of the Fy force for polyamide.

Variable	Coefficient	SE Coefficient	T Value	*p* Value
Constant term	−0.0987	0.0224	−4.4000	0.0070
Wall angle (α)	0.0022	0.0004	5.9200	0.0020
Punch diameter (Dp)	0.0322	0.0009	34.7300	0.0000
Step (s)	0.2518	0.0074	33.9700	0.0000

The *p*-values show that all coefficients are significant. Therefore, the regression equation is as follows:

$$F_{y\_PA} = -0.0987 + 0.002193 \cdot \alpha + 0.032176 \cdot D_p + 0.2518 \cdot s \quad (5)$$

The value obtained for R-sq(pred) is 99.15%, again indicating a good predictability of the regression model. The regression analysis for the Fy force in polyethylene leads to the results shown in Table 20.

Table 20. The coefficients of the regression analysis of the Fy force for polyethylene.

Variable	Coefficient	SE Coefficient	T Value	*p* Value
Constant term	−0.1891	0.0238	−7.9600	0.0010
Wall angle (α)	0.0015	0.0004	4.0200	0.0100
Punch diameter (Dp)	0.0330	0.0010	33.6200	0.0000
Step (s)	0.2351	0.0079	29.9500	0.0000

Again, all coefficients are significant, having values less than 0.05. The regression equation obtained is:

$$F_{y\_PE} = -0.1891 + 0.001577 \cdot \alpha + 0.032989 \cdot D_p + 0.23512 \cdot s \quad (6)$$

The value of R-sq(pred) is 99.07%, which means that this model also has a good predictability.

Both the Taguchi method and the analysis of variance led to the same conclusion: the most important influence on the maximum value of the force belongs to the material, followed by the diameter of the punch, the vertical step and finally the wall angle of the part for all three types of forces. In fact, the analysis of variance shows that the influence of the wall angle is 0.33% for the Fz force, 1.65% for the Fx force and 1.4% for the Fy force, and it can therefore be concluded that the angle of the part has almost no impact on the values of the three types of forces.

Knowledge of how the analyzed input parameters influence the force variation in the single point incremental forming process, and more importantly knowing the optimal values that lead to minimizing the values of these forces, is of great importance for the academic-scientific community of sciences and engineering. This is because it allows us to choose the optimal forming machine for the process, or in case the machine cannot be changed, it helps us to choose the optimal values of the input parameters in order to minimize the forces in the process, as well as the afferent energy consumption. As is well known, in material forming processes, unlike machining processes, the machine (press) is chosen on the basis of the force required in the process and not on the basis of the power available, which obviously leads to solving an important problem in this process. The most significant difference between the Taguchi method and the other methods consists of the manner in which noise is taken into account in an experiment. While the other statistical methods attempt to eliminate the noise, the Taguchi method uses a composite performance indicator, the signal to noise (S/N) ratio, which is used to identify the optimal combinations of input parameters that are most insensitive to noise.

The optimum values leading to minimum forces in the incremental forming of polymeric materials must be correlated with minimizing the defects of the polymeric parts. For this reason, this study needs further updating with a follow-up considering the influence of the input parameters (polymeric materials, wall angle, punch diameter and vertical step) with respect to the springback, wrinkles and twist defects.

4. Conclusions

The main objective of the paper was to determine the optimal values of the input parameters for which the forces during the single point incremental forming of polymeric materials are minimal. Based on the study on the signal-to-noise ratio of the Taguchi method and on the analysis of variance, it was found that the main forming force, perpendicular to the initial plane of the table, decreases the smaller the punch diameter, the smaller the value of the step and the larger the value of the wall angle of the part, while the forces on the directions of the initial plane of the table decrease the smaller the punch diameter, the smaller the value of the step and the smaller the value of the wall angle of the part. Another observation shows that the forces obtained in the case of polyethylene are always lower than in the case of polyamide when manufactured under the same forming conditions (the same punch diameter, step and wall angle).

Author Contributions: Conceptualization, N.R., T.T. and V.O.; methodology, N.R., T.T. and V.O.; software, N.R. and V.O.; validation, N.R., T.T. and V.O.; formal analysis, N.R. and V.O.; investigation, N.R. and V.O.; resources, N.R. and V.O.; data curation, N.R. and T.T.; writing—original draft preparation, N.R., T.T. and V.O.; writing—review and editing, N.R. and T.T.; visualization, N.R. and V.O.; supervision, T.T. and V.O.; project administration, T.T. and V.O.; funding acquisition, V.O. All authors have read and agreed to the published version of the manuscript.

Funding: This research received no external funding.

Institutional Review Board Statement: Not applicable.

Informed Consent Statement: Not applicable.

Data Availability Statement: Not applicable.

Acknowledgments: This work is supported by the Ministry of Research, Innovation and Digitization through Program 1—Development of the national research-development system, Subprogram 1.2—Institutional performance—Projects for financing excellence in RDI, contract no. 28PFE/30 December 2021.

Conflicts of Interest: The authors declare no conflict of interest.

References

1. Koli, B.R.S.; Kumar, A.L. Experimental investigation of incremental sheet metal forming on polyvinylchloride (PVC) by using CNC vertical milling machine. *J. Chem. Pharm. Sci.* **2016**, *9*, 2956–2959.
2. Ostasevicius, V.; Eidukynas, D.; Grigaliunas, V.; Jurenas, V.; Paleviciute, I.; Gudauskis, M. Investigation of advanced robotized polymer sheet incremental forming process. *Sensors* **2021**, *21*, 7459. [CrossRef] [PubMed]
3. Szpunar, M.; Trzepieciński, T.; Ostrowski, R.; Zwolak, M. Research on forming parameters optimization of incremental sheet forming process for commercially pure titanium grade 2 sheets. *Arch. Metall. Mater.* **2022**, *67*, 1411–1418.
4. Martins, P.A.F.; Kwiatkowski, L.; Franzen, V.; Tekkaya, A.E.; Kleiner, M. Single point incremental forming of polymers. *CIRP Ann.-Manuf. Technol.* **2009**, *58*, 229–232. [CrossRef]
5. Zhu, H.; Ou, H.; Popov, A. Incremental sheet forming of thermoplastics: A review. *Int. J. Adv. Manuf. Technol.* **2020**, *111*, 565–587. [CrossRef]
6. Hassan, M.; Hussain, G.; Wei, H.; Qadeer, A.; AlKahtani, M. Progress on single-point incremental forming of polymers. *Int. J. Adv. Manuf. Technol.* **2021**, *114*, 1–26. [CrossRef]

7. Rosa-Sainz, A.; Centeno, G.; Silva, M.B.; Vallellano, C. Experimental failure analysis in polycarbonate sheet deformed by SPIF. *J. Manuf. Process.* **2021**, *64*, 1153–1168. [CrossRef]
8. Beşliu, I.; Tamaşag, I.; Slătineanu, L. An Experimental Study on Incremental Forming Process of Polycarbonate Sheets. *Macromol. Symp.* **2021**, *395*, 2000282. [CrossRef]
9. Kong, F.; Cha, G.; Yan, A.; Pu, Y. An improvement study on incremental forming resilience of PP plate. *Bol. Tec./Tech. Bull.* **2017**, *55*, 110–115.
10. Mohammadi, A.; Vanhove, H.; Attisano, M.; Ambrogio, G.; Duflou, J.R. Single point incremental forming of shape memory polymer foam. *MATEC Web Conf.* **2015**, *21*, 04007. [CrossRef]
11. Ambrogio, G.; Conte, R.; Gagliardi, F.; De Napoli, L.; Filice, L.; Russo, P. A new approach for forming polymeric composite structures. *Compos. Struct.* **2018**, *204*, 445–453. [CrossRef]
12. Clavijo-Chaparro, S.L.; Iturbe-Ek, J.; Lozano-Sánchez, L.M.; Sustaita, A.O.; Elías-Zúñiga, A. Plasticized and reinforced poly(methyl methacrylate) obtained by a dissolution-dispersion process for single point incremental forming: Enhanced formability towards the fabrication of cranial implants. *Polym. Test.* **2018**, *68*, 39–45. [CrossRef]
13. Borić, A.; Kalendová, A.; Urbanek, M.; Pepelnjak, T. Characterisation of polyamide (PA)12 nanocomposites withmontmorillonite (MMT) filler clay used for the incremental forming of sheets. *Polymers* **2019**, *11*, 1248. [CrossRef] [PubMed]
14. Bagheri, S.; Kami, A.; Shakouri, M. Single point incremental forming of polyamide/30 wt% short glass fiber composite. *J. Thermoplast. Compos. Mater.* 2022, *in press, first published online*. [CrossRef]
15. Emami, R.; Mirnia, M.J.; Elyasi, M.; Zolfaghari, A. An experimental investigation into single point incremental forming of glass fiber-reinforced polyamide sheet with different fiber orientations and volume fractions at elevated temperatures. *J. Thermoplast. Compos. Mater.* 2022, *in press, first published online*. [CrossRef]
16. Stein, P.; Franke, W.; von Elling, M.; Groche, P. Forming behavior of paperboard in single point incremental forming. *BioResources* **2019**, *14*, 1731–1764. [CrossRef]
17. Barimani-Varandi, A.; Nasrabadi, M.K.; Ravan, B.A.; Javadi, M. Rapid prototyping of aircraft canopy based on the incremental forming process. *J. Braz. Soc. Mech. Sci. Eng.* **2021**, *43*, 59. [CrossRef]
18. Diabb Zavala, J.M.; Leija Gutiérrez, H.M.; Segura-Cárdenas, E.; Mamidi, N.; Morales-Avalos, R.; Villela-Castrejón, J.; Elías-Zúñiga, A. Manufacture and mechanical properties of knee implants using SWCNTs/UHMWPE composites. *J. Mech. Behav. Biomed. Mater.* **2021**, *120*, 104554. [CrossRef]
19. Centeno, G.; Morales-Palma, D.; Gonzalez-Perez-Somarriba, B.; Bagudanch, I.; Egea-Guerrero, J.J.; Gonzalez-Perez, L.M.; García-Romeu, M.L.; Vallellano, C. A functional methodology on the manufacturing of customized polymeric cranial prostheses from CAT using SPIF. *Rapid Prototyp. J.* **2017**, *23*, 771–780. [CrossRef]
20. Durante, M.; Formisano, A.; Lambiase, F. Formability of polycarbonate sheets in single-point incremental forming. *Int. J. Adv. Manuf. Technol.* **2019**, *102*, 2049–2062. [CrossRef]
21. Alkas Yonan, S.; Silva, M.B.; Martins, P.A.F.; Tekkaya, A.E. Plastic flow and failure in single point incremental forming of PVC sheets. *Express Polym. Lett.* **2014**, *8*, 301–311. [CrossRef]
22. Zha, G.C.; Zhou, X.; Lu, C.K.; Zhao, W.; Sun, S.; Yan, A.X.; Li, Z.K. Experimental study on the limit test method of polypropylene sheet incremental forming. *Int. J. Adv. Manuf. Technol.* **2017**, *93*, 2369–2374. [CrossRef]
23. Liu, Z.; Cheng, K.; Peng, K. Exploring the deformation potential of composite materials processed by incremental sheet forming: A review. *Int. J. Adv. Manuf. Technol.* **2021**, *118*, 2099–2137. [CrossRef]
24. Davarpanah, M.A.; Zhang, Z.; Bansal, S.; Cao, J.; Malhotra, R. Preliminary investigations on Double Sided Incremental Forming of thermoplastics. *Manuf. Lett.* **2016**, *8*, 21–26. [CrossRef]
25. Esmaili, M.; Khalili, K. Two Point Incremental Forming of Metal–Polymer Three-Layer Sheets. *Iran. J. Sci. Technol. Trans. Mech. Eng.* **2021**, *45*, 181–196. [CrossRef]
26. Al-Ghamdi, K.A. Spring back analysis in incremental forming of polypropylene sheet: An experimental study. *J. Mech. Sci. Technol.* **2018**, *32*, 4859–4869. [CrossRef]
27. Feng, X.Y.; Zha, G.C.; Jiang, Z.D.; Yan, F.Y.; Mo, S.Y. Research on wrinkle defects of PE sheet based on incremental forming. *Suxing Gongcheng Xuebao/J. Plast. Eng.* **2021**, *28*, 61–68. [CrossRef]
28. Yang, Z.; Chen, F. Mechanism of twist in incremental sheet forming of thermoplastic polymer. *Mater. Des.* **2020**, *195*, 108997. [CrossRef]
29. Maaß, F.; Gies, S.; Tekkaya, A.E. Deformation characteristics of thermoplastics in single point incremental forming. *AIP Conf. Proc.* **2017**, *1896*, 050005.
30. Formisano, A.; Lambiase, F.; Durante, M. Polymer self-heating during incremental forming. *J. Manuf. Process.* **2020**, *58*, 1189–1199. [CrossRef]
31. Kulkarni, S.; Mocko, G.M. Comparison of Room Temperature and Heat Assisted Single Point Incremental Forming of Thermoplastics. Advances in Manufacturing Technology. *Adv. Manuf. Technol.* **2019**, *33*, 346–351.
32. Al-Obaidi, A.; Kunke, A.; Kräusel, V. Hot single-point incremental forming of glass-fiber-reinforced polymer (PA6GF47) supported by hot air. *J. Manuf. Process.* **2019**, *43*, 17–25. [CrossRef]
33. Hou, C.; Su, X.; Peng, X.; Wu, X.; Yang, D. Thermal-Assisted Single Point Incremental Forming of Jute Fabric Reinforced Poly(lactic acid) Biocomposites. *Fibers Polym.* **2020**, *21*, 2373–2379. [CrossRef]

34. Torres, S.; Ortega, R.; Acosta, P.; Calderón, E. Hot incremental forming of biocomposites developed from linen fibres and a thermoplastic matrix. *Stroj. Vestn./J. Mech. Eng.* **2021**, *67*, 123–132. [CrossRef]
35. Sridhar, R.; Rajenthirakumar, D. Polymer sheet hot incremental forming—An innovative polymer forming approach. *Polym. Polym. Compos.* **2016**, *24*, 447–454. [CrossRef]
36. García-Collado, A.; Medina-Sanchez, G.; Gupta, M.K.; Dorado-Vicente, R. Application of the finite element method to the incremental forming of polymer sheets: The thermomechanical coupled model and experimental validations. *Polymers* **2020**, *12*, 1715. [CrossRef] [PubMed]
37. Sy, L.V.; Hue, P.T.M. Implement a modified viscoplasticity based on overstress model into numerical simulation of the incremental sheet forming process. In *Proceedings of AETA 2016: Recent Advances in Electrical Engineering and Related Sciences*; Springer: Berlin/Heidelberg, Germany, 2017; Volume 415, pp. 23–33.
38. Ibrahim, A.K. A comparative study of simulation models on incremental forming process by using the enhancement tools. *Mater. Sci. Forum* **2019**, *957*, 93–102.
39. Rai, S.; Kumar, R.; Nirala, H.K.; Francis, K.; Agrawal, A. Experimental and simulation study of single point incremental forming of polycarbonate. In Proceedings of the International Manufacturing Science and Engineering Conference, MSEC2019-3026, Erie, PA, USA, 10–14 June 2019. [CrossRef]
40. Harhash, M.; Palkowski, H. Incremental sheet forming of steel/polymer/steel sandwich composites. *J. Mater. Res. Technol.* **2021**, *13*, 417–430. [CrossRef]
41. Afonso, D.; Pires, L.; de Sousa, R.A.; Torcato, R. Direct rapid tooling for polymer processing using sheet metal tools. *Procedia Manuf.* **2017**, *13*, 102–108. [CrossRef]
42. Jain, P.S.; Kagzi, S.A.; Patel, S.; Vasava, J. An analysis of the effect of various parameters on surface roughness, springback and thinning while performing single point incremental forming on polypropylene sheet. *Proc. Inst. Mech. Eng. Part E J. Process Mech. Eng.* 2021; *in press, first published online*. [CrossRef]
43. Patel, S.; Kagzi, S.; Jain, P. Investigation of Different Factors Influencing the Springback, Surface Roughness, and Thinning for Polyvinyl Chloride during Single Point Incremental Forming. *SAE Int. J. Mater. Manuf.* **2020**, *13*, 297–306. [CrossRef]
44. Thangavel, K.; Duraiswamy, R.; Nagarajan, S.; Ramasamy, S. Influence of roller ball tool in single point incremental forming of polymers. *Teh. Vjesn.* **2019**, *26*, 171–176. [CrossRef]
45. Davarpanah, M.A.; Mirkouei, A.; Yu, X.; Malhotra, R.; Pilla, S. Effects of incremental depth and tool rotation on failure modes and microstructural properties in Single Point Incremental Forming of polymers. *J. Mater. Process. Technol.* **2015**, *222*, 287–300. [CrossRef]
46. Bagudanch, I.; Garcia-Romeu, M.L.; Centeno, G.; Elías-Zúñiga, A.; Ciurana, J. Forming force and temperature effects on single point incremental forming of polyvinylchloride. *J. Mater. Process. Technol.* **2015**, *219*, 221–229. [CrossRef]
47. Medina-Sanchez, G.; Garcia-Collado, A.; Carou, D.; Dorado-Vicente, R. Force prediction for incremental forming of polymer sheets. *Materials* **2018**, *11*, 1597. [CrossRef]
48. Khalatbari, H.; Lazoglu, I. Friction stir incremental forming of polyoxymethylene: Process outputs, force and temperature. *Mater. Manuf. Process.* **2021**, *36*, 94–105. [CrossRef]
49. Kumar, S.P.; Elangovan, S.; Mohanraj, R.; Boopathi, S. Real-time applications and novel manufacturing strategies of incremental forming: An industrial perspective. *Mater. Today Proc.* **2021**, *46*, 8153–8164. [CrossRef]
50. Ingarao, G.; Zaheer, O.; Fratini, L. Manufacturing processes as material and energy efficiency strategies enablers: The case of Single Point Incremental Forming to reshape end-of-life metal components. *CIRP J. Manuf. Sci. Technol.* **2021**, *32*, 145–153. [CrossRef]
51. Riaz, A.A.; Hussain, G.; Iqbal, A.; Esat, V.; Alkahtani, M.; Khan, A.M.; Ullah, N.; Xiao, M.; Khan, S. Energy consumption, carbon emissions, product cost, and process time in incremental sheet forming process: A holistic review from sustainability perspective. *Proc. Inst. Mech. Eng. Part B J. Eng. Manuf.* 2022, *in press, first published online*. [CrossRef]
52. Torres, S.; Calderon, E.; Ortega, R.; Acosta, P. Sustainability in terms of energy consumption in processes of incremental forming and stamping. *Mater. Today Proc.* **2022**, *49*, 175–180. [CrossRef]
53. Cooper, D.R.; Gutowski, T.G. Prospective Environmental Analyses of Emerging Technology: A Critique, a Proposed Methodology, and a Case Study on Incremental Sheet Forming. *J. Ind. Ecol.* **2020**, *24*, 38–51. [CrossRef]
54. Component Charge Output Force Sensor. Model 261A13. Available online: https://www.pcbpiezotronics.fr/wp-content/uploads/261A13.pdf (accessed on 11 September 2022).
55. CMD 600 Digital Charge Amplifier. Available online: https://www.hbm.com/en/2659/paceline-cmd-digital-charge-amplifier (accessed on 11 September 2022).
56. QuantumX MX840B: 24-Bit Universal Measuring Amplifier. Available online: https://www.hbm.com/en/2129/quantumx-mx840b-8-channel-universal-amplifier (accessed on 11 September 2022).
57. *ISO 527-1:2019*; Plastics—Determination of Tensile Properties—Part 1: General Principles. ISO: Geneva, Switzerland, 2019.
58. *ISO 527-3:2018*; Plastics—Determination of Tensile Properties—Part 3: Test Conditions for Films and Sheets. ISO: Geneva, Switzerland, 2018.

materials

MDPI

Article

Plastic Anisotropy Effect on Variation of Mechanical and Structural Properties of VT23 Titanium Alloy Subjected to Impact-Oscillatory Loading

Mykola Chausov [1], Andrii Pylypenko [1], Pavlo Maruschak [2], Janette Brezinová [3,*], Jakub Brezina [3] and Ihor Konovalenko [2]

1 Department of Mechanics, National University of Life and Environmental Sciences of Ukraine, Heroiv Oborony Str. 15, 03041 Kyiv, Ukraine
2 Department of Industrial Automation, Ternopil National Ivan Puluj Technical University, Rus'ka Str. 56, 46001 Ternopil, Ukraine
3 Department of Technology, Materials and Computer Supported Production, Faculty of Mechanical Engineering, Technical University of Košice, Māsiarska 74, 04001 Košice, Slovakia
* Correspondence: janette.brezinova@tuke.sk; Tel.: +421-55-602-3512

Abstract: The main regularities in the impact of varying intensity impact-oscillatory loading on the variation of the mechanical and structural properties of the VT23 high-strength two-phase transverse-rolled sheet titanium alloy have been found. The intensity of the impulse introduction of energy into the alloy under the dynamic non-equilibrium process (DNP) was estimated by ε_{imp} (the increment of dynamic strain). The pulse intensity was found to change the shape of the static strain diagram with further tensioning, as compared to the initial state. This indicates the effect from the structure self-organization inherent in the VT23 titanium alloy upon the DNP. After the DNP (ε_{imp} = 1.44%), with further static deformation, the tensile diagram revealed yield sites up to 6.5% long. In most cases, the DNP was found to have a negative effect on the variation of the mechanical properties of the VT23 titanium alloy, especially if the latter was rolled in the transverse direction. The optimal DNP intensity is ε_{imp}~1.5%. In this case, the DNP can be used as an effective plasticization technology for the VT23 titanium alloy (regardless of the rolling direction) in the stamping of high-strength titanium alloys. Changes in the mechanical and structural condition of the VT23 titanium alloy subjected to the DNP were confirmed by the fractographic investigation of specimen fractures.

Keywords: titanium alloy; dynamic non-equilibrium process; mechanical properties; fracture

Citation: Chausov, M.; Pylypenko, A.; Maruschak, P.; Brezinová, J.; Brezina, J.; Konovalenko, I. Plastic Anisotropy Effect on Variation of Mechanical and Structural Properties of VT23 Titanium Alloy Subjected to Impact-Oscillatory Loading. *Materials* **2022**, *15*, 5718. https://doi.org/10.3390/ma15165718

Academic Editor: Elena Pereloma

Received: 26 June 2022
Accepted: 17 August 2022
Published: 19 August 2022

1. Introduction

Titanium alloys are widely used for various-purpose air transport and aerospace systems, special equipment, and deep-sea submarines [1–3]. Their high strength, corrosion resistance, and wide temperature range make it possible to significantly reduce the structure's weight and enhance the reliability of its operation.

Products made of titanium alloy can be subjected to dynamic loads over a wide range of strain rates. Therefore, many works are dedicated to changes in the mechanical properties of titanium alloys depending on the load rate and temperature [4–27]. Notably, the ability to self-organize inherent in the structure of high-strength titanium alloys upon the DNP has not been given proper attention so far [28]. The DNP is associated with the fast energy exchange between the system and the environment, and some features of the mechanical system's parameters, mainly their natural frequency. These differences (features) are mostly transient; they appear instantly due to a particular initial condition and vanish when the balance between the system and the environment is attained. The term "self-organization" has been generally accepted since the middle of the 20th century. It was introduced by the works of Haken, Ebeling, Nikolis and Prigogine to denote the main characteristic of a well-ordered structure formed in conditions of thermodynamic systems far from equilibrium. In Solid State Physics, it refers to the possibility of a well-ordered defect formed in a structure to promote the hydrodynamic plastic flow that reduces the dislocation slide. More specifically, it characterizes the hydrodynamic plastic flow rather than the dislocation slide [29]. Interestingly, such processes can be realized in materials of any statically indeterminate mechanical system when additional force energy is introduced into this system in a pulsed manner, for instance, due to the application of additional impulse loads. Given its great practical and scientific value, testing materials under the DNP requires further research.

Chausov et al. [30–32] propose the DNP realization method using impulse force loads under impact-oscillatory loading and tested it on materials of different classes. The basis for the method's development was the fact that strain rates of various materials, at which self-organizing processes are taking place, with new spatial dissipative structures being formed under impact-oscillatory loading, have their ranges. In particular, the Phantom v711 high-speed camera (maximum registration speed: 1,400,000 k/s) helped reveal two stages of the low- and high-speed processes responsible for the formation and development of a dissipative structure in materials under impact-oscillatory loading. The mean strain rate at the first stage of the process is 1–3 s^{-1} for aluminum alloys D16 and 2024-T3 [32] and 1–2 s^{-1} for stainless steel 12Kh17, respectively [33]. In the second stage, where the newly formed dissipative structure is likely to spread across the specimen's volume, a sharp increase in the strain rate is observed in local areas up to 50–60 s^{-1} for aluminum alloys D16 and 2024-T3 [32] and up to 6–10 s^{-1} for the 12Kh17 stainless steel [33]. That is, these processes primarily depend on the mechanical and structural properties of the material under study. We should also note that these processes occur at relatively low rates of additional impulse loads, when the mass transfer occurs in the absence of the energy dissipation in the classical sense (conversion of mechanical energy into heat). The oscillating load frequency, which occurs during the DNP of materials according to the proposed method, was found to be in the range of 1–2 kHz [32]. An experiment found that polycrystals have specific intensity ranges for introducing impulse energy into the material, which under further loading can show an increase or decrease in its ductility, toughness and static impact toughness. We should note that these effects may affect the strength of materials. Although most experiments were conducted at room temperature, many studies on titanium and aluminum alloys exposed specimens to temperatures of 77 K prior to the impulse introduction of energy into the specimen under the DNP and after this procedure [34,35]. The mechanical behavior patterns of materials after the DNP observed in the experiment indicate:

- The hazardous nature of additional impulse loads when applied to load-bearing structures during operation, as their mechanical properties can deteriorate significantly;
- The possibility of developing a technology for improving the mechanical properties of the source material using a simple operation of preliminary introduction of impulse energy in a given intensity range.

The authors performed many experiments to evaluate changes in the mechanical and structural parameters of high-strength two-phase titanium sheet alloys VT22, VT23, and VT23M under the DNP. However, most experiments were performed on specimens cut out of titanium sheets rolled in the longitudinal direction [36]. Given that all titanium alloys are anisotropic materials, it is advisable to study the DNP effect on specimens of titanium alloys rolled in the transverse direction.

This research aims to evaluate and generalize the anisotropy effect on the variation of the mechanical and structural properties ($\alpha + \beta$) of titanium alloy VT23 after the DNP.

2. Materials and Methods

The mechanical testing technique was implemented based on a modified ZD-100Pu (WPM, Leipzig, Germany) hydraulic installation for static testing and is described in detail in [30–32]. The proposed technique's main idea consists of high-speed tensioning of the material with the imposition of a high-frequency (1–2 kHz) oscillatory process, which corresponds to the natural frequency of the testing machine. Structurally, this is achieved by an inner contour introduced into the testing machine in addition to the outer contour (loaded frame of the testing machine). The inner contour is the simplest statically indeterminate structure in the form of three parallel elements loaded simultaneously—the central specimen and two satellites (brittle samples) of different cross-sections made of hardened steels 65 G or U8–U12. When this structure is tensioned, the satellites are destroyed, and the energy is introduced into the specimen under study in a pulsed manner. Satellites may get involved in the operation at any stage of preliminary static tension, making it possible to study the effect of the impulse introduction of energy on the degradation of mechanical properties by the material getting damaged in the process of static tension. By changing the initial diameter of the satellites, it is possible to control the intensity of the impulse introduction of force energy into the material.

Mechanical tests were performed on specimens (Figure 1) made of the VT23 sheet industrial titanium alloy with a thickness of 3 mm at a room temperature. The strain measurement base was 16 mm. In this case, a standard 0.5 accuracy class extensometer manufactured at the Antonov aircraft production plant, Ukraine, was used for strain measurements.

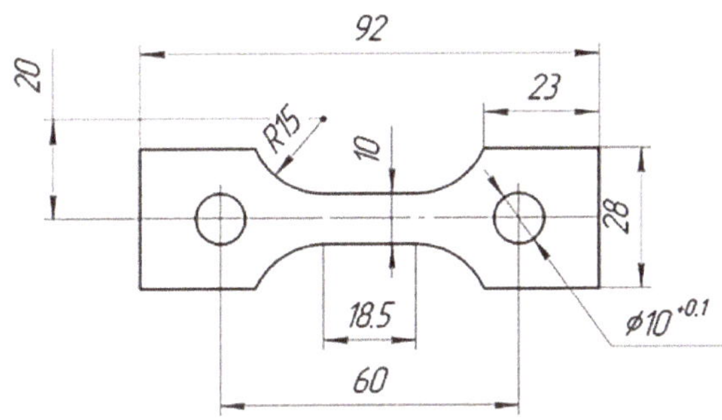

Figure 1. Test specimen (dimensions indicated in mm).

The value of ε_{imp} under the DNP was controlled by the optical method [32].

The initial mechanical properties of the alloy rolled in different directions are given in Table 1.

Table 1. The mechanical properties of titanium alloy VT23 depend on the rolling direction.

Rolling Direction	Mechanical Properties		
	σ_{ys}, MPa	σ_{us}, MPa	δ, %
longitudinal	1060–1080	1120–1180	16
transverse	960–980	1040–1080	15

The chemical composition of alloy VT23 is given in Table 2.

Table 2. Chemical composition of alloy VT23.

Fe	Cr	Mo	V	Ti	Al
0.5–0.8	1.0–1.4	1.8–2.5	4.3–5	86–89.3	4.4–6.3

The content of the α and β phases (in %) in the VT23 titanium alloy in the initial state was evaluated according to the data of radiographic studies [36]. The diffractometric study showed that the β phase occupies 43% of the mass in the VT23 titanium alloy, and the α phase occupies 57%. Characteristically, both phase components have a texture in the crystallographic direction of the specimen (002). This may be due to rolling or subjecting the specimens to another mechanical impact.

Fractographic studies of specimen fractures to detect morphological differences in the fracture micromechanisms were performed on the Zeiss EVO-40XVp (ZEISS International, Oberkochen, Germany) scanning electron microscope. An automated method for analyzing the shape and size of dimples of ductile tearing formed during static and impact fracture of titanium alloys VT23 was used. The method is based on the analysis of the image topology. The method contains the operations of smoothing the initial fractographic image; its convolution with a filter to identify topological ridges; thresholding with subsequent skeletonization to identify boundaries between dimples; clustering to isolate the connected areas that represent the sought objects—dimples. For each dimple, the following quantitative characteristics were calculated: dimple equivalent diameter *deq* and shape parameter *Kc*.

To determine to what degree the shape of dimples approaches the circle, calculated the coefficient of roundness for each *i*-th object. K_{ci} is the percentage of object pixels that fall in a circle with the same area, whose center is combined with the center of mass of the object [37]:

$$K_{ci} = \frac{\sum_{m=1}^{f_i} g\left(\vec{r}_m, d_i\right)}{f_i} \cdot 100\%, \; g\left(\vec{r}_m, d_i\right) = \begin{cases} 1, \text{ при } \left|\vec{r}_m\right| \le d_i/2 \\ 0, \text{ при } \left|\vec{r}_m\right| > d_i/2 \end{cases} \tag{1}$$

where f_i is the number of object pixels; d_i is the diameter of the equivalent circle; $g\left(\vec{r}_m, d_i\right)$ is the indicator function that shows whether the m-th pixel of the object falls within the boundary of the equivalent circle; $\vec{r}_m$ is the radius vector directed from the center of the equivalent circle $C_i(x_{ci}, y_{ci})$ to the m-th pixel of the object with coordinates (x_{cm}, y_{cm}) [38,39].

3. Results

3.1. Mechanical Properties

Figure 2 shows stress–strain diagrams of specimens from the VT23 titanium alloy loaded by static tensioning. The specimens are cut out of the sheet material and rolled in different directions. The effect of structural anisotropy on mechanical properties is revealed. It is noteworthy that in the longitudinal direction of rolling, the stress–strain diagram has almost no hardening area of the alloy. Immediately after attaining the tensile strength, a yield site occurs, after which the material is strengthened. In the transverse direction of rolling, the VT23 titanium alloy has a lower strength, but the diagram has a fairly long hardening area of the alloy, which indicates its strong load-bearing capacity.

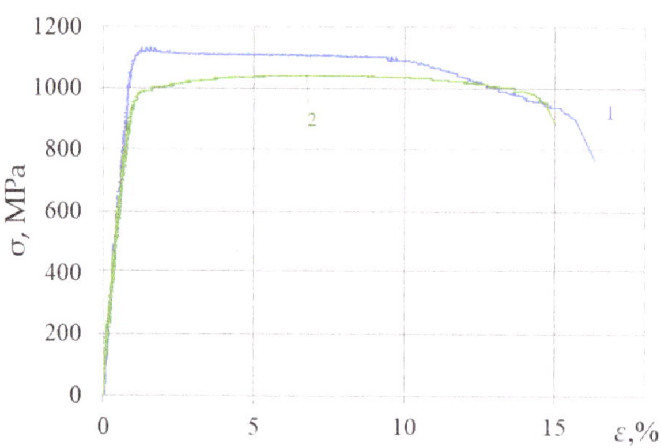

Figure 2. Static tensioning of specimens: 1—longitudinal rolling; 2—transverse rolling.

The alloy's total plasticity in different rolling directions is relatively uniform. Given these differences in the mechanical properties of the specimens rolled in different directions, we should expect significant differences in the variation of the mechanical and structural parameters after the DNP. For a detailed analysis of the DNP effect, the dynamic strain increments (ε_{imp}) of specimens in the process of impulse introduction of energy were set as (ε_{imp} = 0.8–5.5%). Given that at ε = 5.5%, the material's ultimate strength is nearly attained on the static tension diagram of the specimens, the maximum ε_{imp} was chosen. Previous DNP studies have shown that at such high stresses, the repeated static tensioning of titanium alloys has revealed deterioration of their mechanical properties.

Figure 3a–e give examples of the stress–strain diagrams obtained for specimens of the VT23 alloy rolled in the transverse direction after the DNP and repeated static tension. For comparison, all the graphs show a stress–strain diagram of the alloy in the initial state.

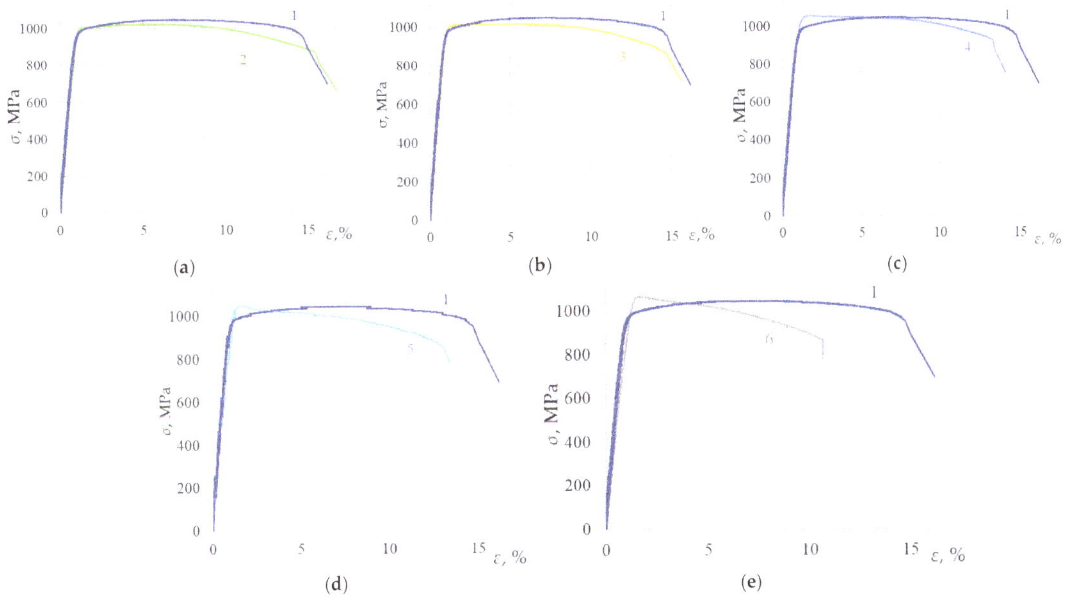

Figure 3. Stress–strain diagrams of the VT23 titanium alloy rolled in the transverse direction after the DNP and subsequent static tensioning, as compared to the static tensioning in the initial state (curve 1): (**a**) (2—ε_{imp} = 0.81%); (**b**) (3—ε_{imp} = 1.44%); (**c**) (4—ε_{imp} = 2.56%); (**d**) (5—ε_{imp} = 4.56%); (**e**) (6—ε_{imp} = 5.5%).

The analysis of the results shows that at ε_{imp} > 2.56%, the mechanical properties of the VT23 titanium alloy deteriorated significantly with repeated static tensioning. Most interesting effects were found at ε_{imp} = 1.44% when yield sites up to 6.5% long were formed on the stress–strain diagram with further static tensioning. At the same time, the yield strength of the alloy increased to 1000 MPa, that is, by 5–6%, compared to the initial state. The ultimate strength of the alloy decreased by approximately the same percentage, and the total ductility remained unchanged. The stress–strain diagram of the VT23 titanium alloy (rolled in the longitudinal and transverse directions) under the DNP (ε_{imp} = 1.44%) and subsequent static tension were compared Figure 4.

The research findings indicate that after the DNP of the same intensity, one can obtain stress–strain diagrams similar in shape for specimens of the VT23 titanium alloy cut in different directions of rolling. At the same time, they both have significant yield areas. This effect can be used for technological purposes.

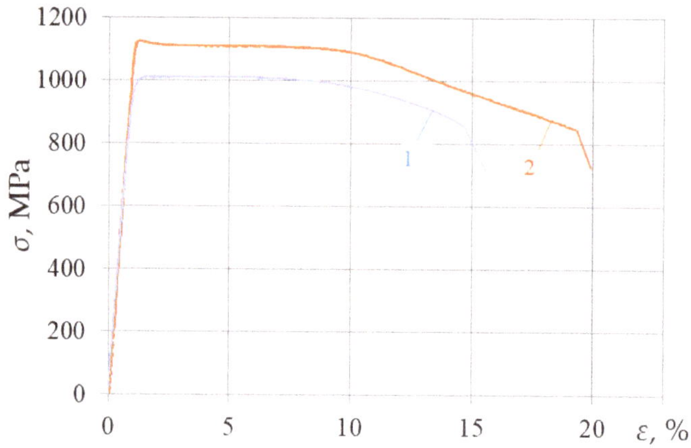

Figure 4. Stress–strain diagrams of the VT23 alloy after the DNP: 1—transverse direction of rolling, ε_{imp} = 1.44%; 2—longitudinal direction of rolling, ε_{imp} = 1.5% [36].

3.2. Fractographic Research Results

Characteristic changes in the mechanical properties of the VT23 titanium alloy at different DNP intensities (see Figure 4) are directly related to the characteristic structural changes, as evidenced by the results of fractographic research (Figure 5).

The research was carried out both in the central zones of the specimens, where the fracture occurred by the dimple separation mechanism, and in the lateral areas characterized by shear fracture mechanisms. Figure 5 presents their images. The DNP changed the fracture micromechanisms of the VT23 titanium alloy under study. Thus, at low ε_{imp}, static tensioning diagrams are characterized by a slight strengthening of the alloy on the ascending branch of the stress–strain diagram (see Figure 3a). This, in turn, causes an increase in the fibrous cup zone characterized by a significant number of small dimples (Figure 5a,b) in the central zone of the specimen separation. At the same time, apart from viscous chipping areas that occur in different directions, many dimples of different sizes formed on the lateral walls indicate a significant ductility of the alloy's fracture.

An increase in ε_{imp} (see Figure 5c–f) causes the formation of local micro separation areas in the central zones of the specimen separation. The relief formed by dimples of different sizes changes appreciably, as evidenced by the depth of separation dimples. All this indicates that the accumulated defects that merge into pores and microcracks dominate the descending branches of the stress–strain diagrams, which primarily affects their length (see Figure 3c,d). The flat chipping areas on the lateral walls of the specimens indicate a decrease in the fracture toughness of the alloy with an increase in the DNP intensity.

Above is the qualitative analysis of the fractographic research findings. For their quantitative analysis, the authors used the designated, highly efficient computer diagnostic system for the automated processing of the images obtained [37,40]. Figure 6 shows the fractographic images of the fracture microrelief formed on specimen surfaces, and their binarized mappings.

It is noteworthy that during the DNP followed by static tensioning, the central fracture zones of specimens made of the VT23 alloy are characterized by dimple microrelief under all loading conditions considered; however, its morphology is somewhat different due to the varying intensity of the impulse energy introduction. In particular, at ε_{imp} = 0.81%, 412 dimples were found within the area analyzed, the diameters of which were mostly up to 5 μm, with the maximum size reaching 8 μm. Of particular note is the shape coefficient parameter, the value of which (Kc = 70–80, according to the approximation) indicates the homogeneity of the dimple geometry and its closeness to the circular shape (see Figure 7a). These data indicate a relatively high ductility and crack resistance of the alloy in this structural state.

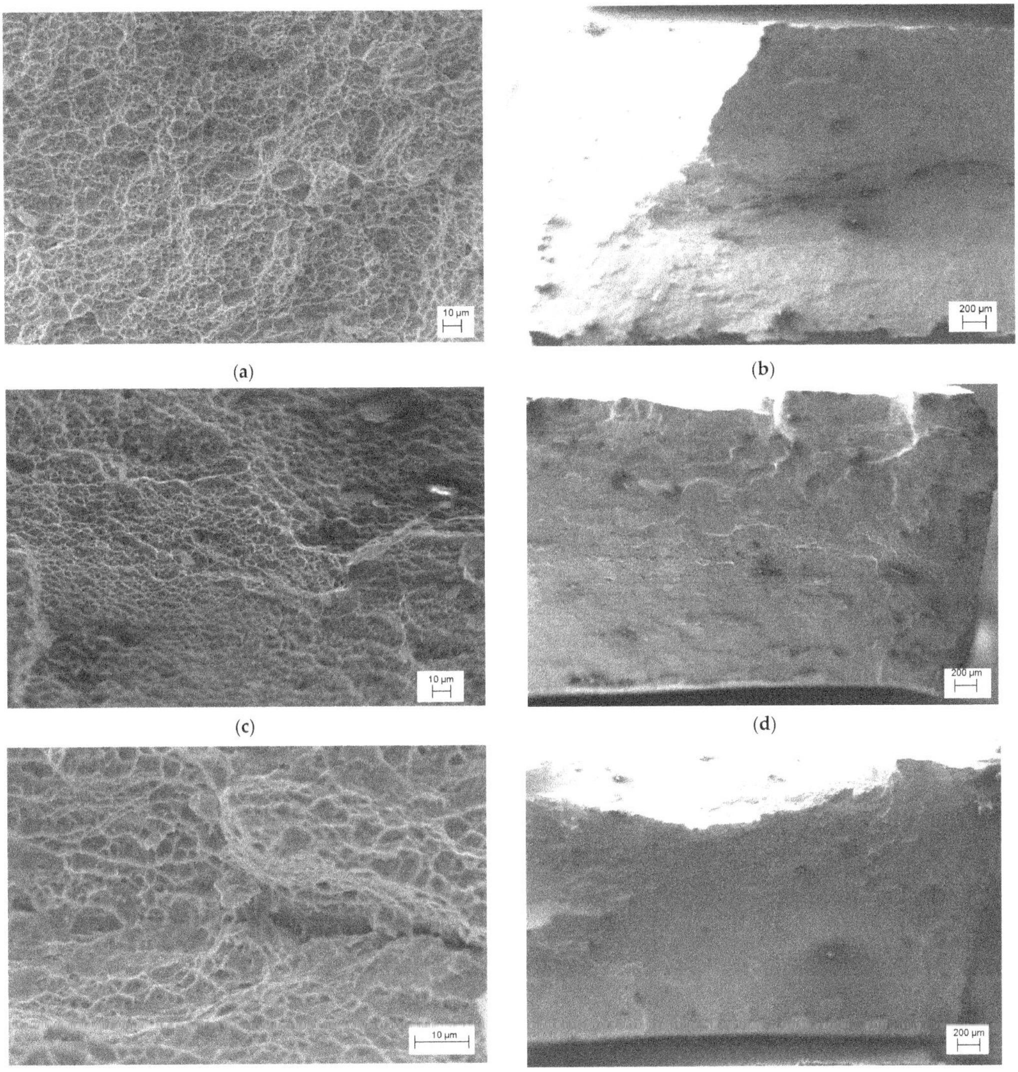

Figure 5. Fracture micromechanisms of specimens from titanium alloy VT23 rolled in the transverse direction after the DNP and subsequent static tensioning: (**a**,**b**)—ε_{imp} = 0.81%; (**c**,**d**)—ε_{imp} = 2.56%; (**e**,**f**)—ε_{imp} = 4.56%.

The fracture of specimens after the realization of the DNP (ε_{imp} = 2.56%) is also characterized by a dimple structure, which, however, differs significantly from the previous one. The dimples found on the surface do not have a clearly defined shape, as evidenced by the value of the shape coefficient (Kc = 25–85, according to the approximation). A significant scatter of size is inherent in the dimples, while their diameter may reach 16 μm. In addition, zones of elongation [41] were found around the dimples, indicating the alloy strain localization. The presence of such elements and a significantly larger area of membranes between the dimples indicate a decreased microplastic deformation and increased brittle component in the specimen fracture.

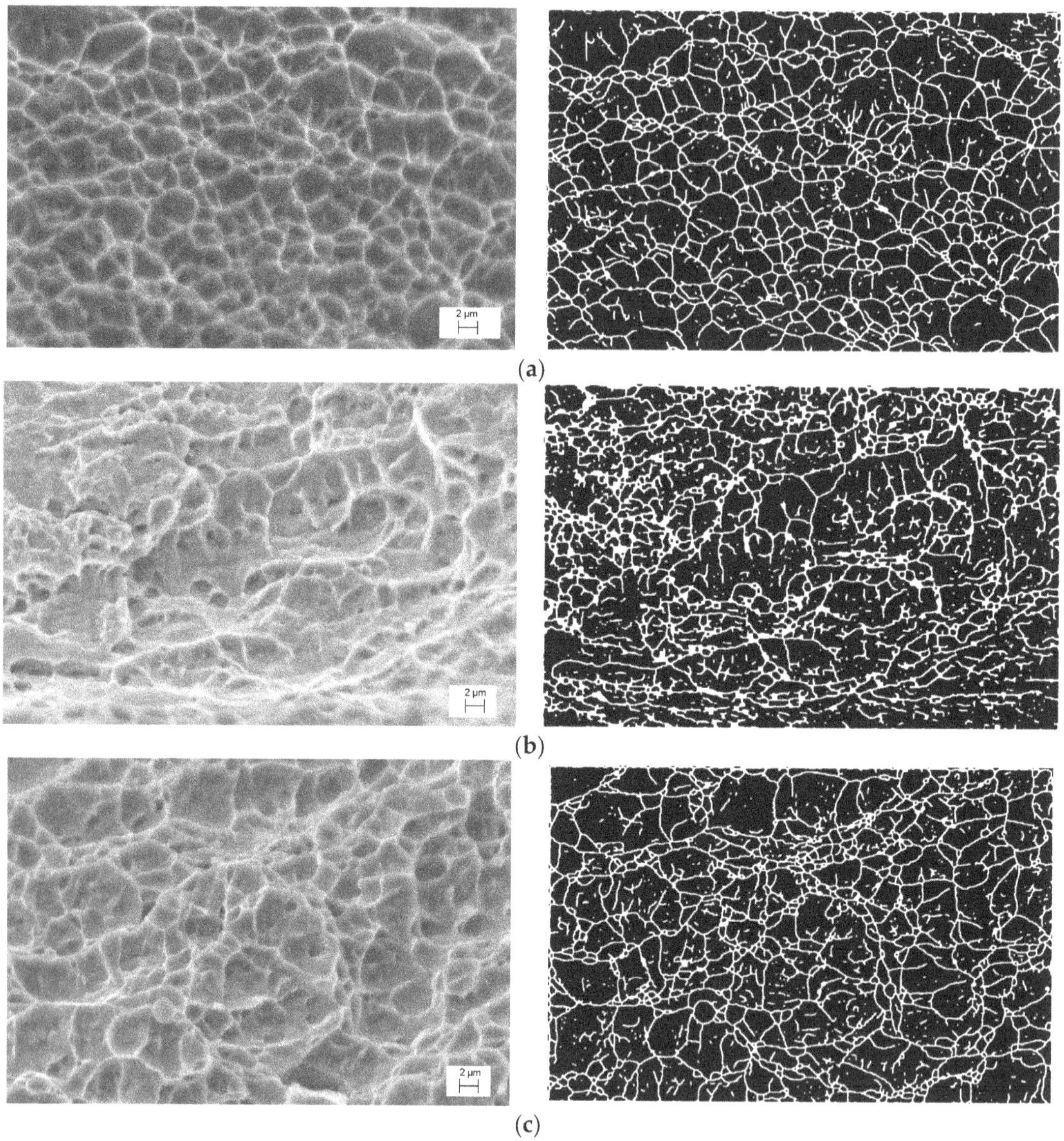

Figure 6. Fractograms of fracture surfaces of specimens made of the VT23 alloy rolled in the transverse direction, and their binarized mappings obtained after the impulse energy introduction under impact-oscillatory loading followed by static tensioning: (**a**)—ε_{imp} = 0.81%; (**b**)—ε_{imp} = 2.56%; (**c**)—ε_{imp} = 4.56%.

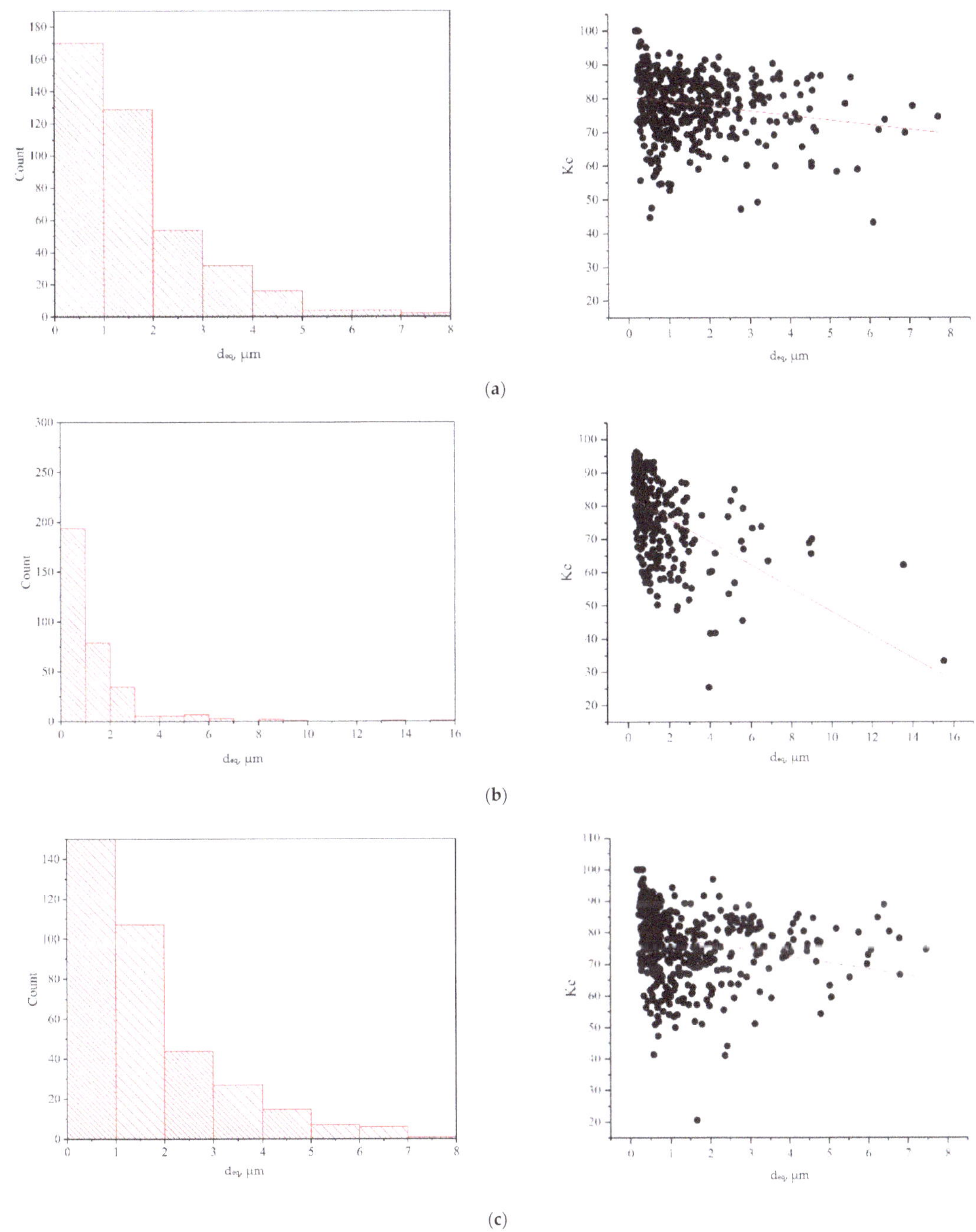

Figure 7. Histograms of dimple diameters and diagrams of dimple shape coefficients on fracture zones of specimens made of the VT23 alloy rolled in the transverse direction after the introduction of impulse energy due to impact-oscillatory loading followed by static tensioning: (**a**)—ε_{imp} = 0.81%; (**b**)—ε_{imp} = 2.56%; (**c**)—ε_{imp} = 4.56%.

Fracture of specimens after the DNP (ε_{imp} = 4.56%) results from the combined effect of the DNP and relaxation processes in the alloy in case of a significant intensity of introduction of impulse energy. On the one hand, the DNP activates deformation processes and fractions (fragments) the original β phase of the alloy, as shown earlier [42]. On the other hand, with a significant intensity of introduction of impulse energy into the alloy, relaxation processes occur in the alloy. Apart from the smaller structural elements formed (grains, subgrains, fragments), these relaxation processes are associated with the activation of rather intense dislocation redistributions, the consequence of which is the annihilation of intra-volume dislocations [42]. This caused the formation of the fracture surface similar to that shown in Figure 6a (the dimple diameter was mostly up to 5 μm, with the maximum size reaching 8 mm), however, with a different shape coefficient of the dimples (Kc = 65–80, according to the approximation). In addition, the number of dimples with sizes larger than 5 μm is different, and one can see that all the dimples have a flat bottom (the dimple diameter is much larger than its depth, compare Figure 6a,c). This indicates the presence of a brittle component in the micromechanism of fracture. Sharp boundaries between the dimples, which are clearly visible on the binarized image, also testify to an appreciable embrittlement of the material (see Figure 6c). Thus, fractographic research indirectly indicates significant structural changes in the alloy caused by the previous impulse introduction of energy of different intensities into the VT23 titanium alloy, which is associated with the self-organization of the alloy structure during the DNP. It should be noted that a similar procedure for evaluating the differences in the fracture relief of specimens made of the VT23 titanium alloy rolled in the longitudinal direction after the DNP of varying intensity was previously described in [43].

4. Discussion

To evaluate the effect of anisotropy on the variation of the mechanical properties of the alloy under the DNP, the authors used the previously obtained results for the VT23 titanium alloy rolled in the longitudinal direction [36] (Figure 8) and the test data obtained for specimens cut in the transverse rolling direction described in this article.

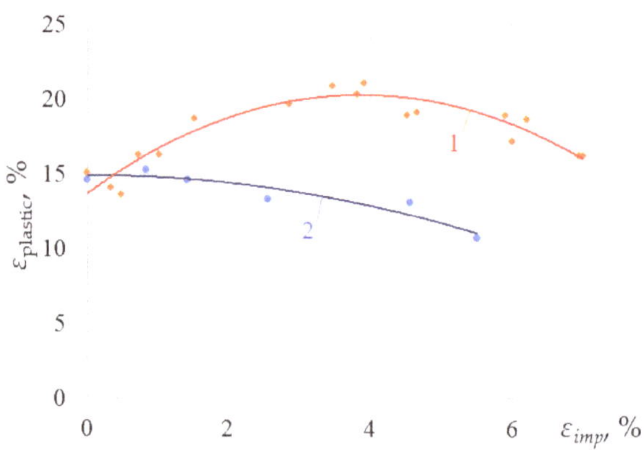

Figure 8. Dependence of the maximum strain of the VT23 alloy rolled in different directions on the increase in the dynamic strain increment ε_{imp} of specimens during the DNP: 1—longitudinal direction. [36]; 2—transverse direction.

Here, the abscissa shows the intensity of impulse energy introduced into the alloy, and the ordinate shows the maximum total deformation of the alloy after repeated static tension.

The transverse direction of rolling. In most cases, the previous DNP deteriorated the mechanical properties of the VT23 titanium alloy obtained by subsequent static tension. In addition, specimens cut in the transverse direction of rolling are more sensitive to this phenomenon than those cut in the longitudinal direction [44–47]. At the same time, a narrow ε_{imp} range was found, in which the ductility of the VT23 titanium alloy can be improved. This range is provided at ε_{imp}~1.5% (see Figure 4), when yield areas of significant length appear on the static stress–strain diagrams. An increase in ε_{imp} causes a deterioration in ductility and strength (see Figure 3).

The longitudinal direction of rolling. The maximum plastic properties were achieved at ε_{imp} = 4% while maintaining the strength properties for the specimens cut in the longitudinal direction of rolling. With DNP intensity ε_{imp}~1.5%, the total deformation of the VT23 titanium alloy increased by 18% (see Figures 2 and 4) due to the yield areas formed, while the strength properties also remained unchanged.

In the authors' opinion, the range of the impulse introduction of energy (ε_{imp}~1.5%) described is optimal in terms of improving the initial mechanical properties of the VT23 titanium alloy rolled in different directions, and can be used as a basis for developing the alloy plasticization technology [48,49].

5. Conclusions

The main regularities in the plastic anisotropy effect on the variation of the mechanical and structural properties of the VT23 high-strength two-phase titanium sheet alloy subjected to the DNP are established. For specimens of the VT23 titanium alloy rolled in the transverse direction, the DNP was found to be more harmful than for those rolled in the longitudinal direction. The strength and plastic properties of the VT23 titanium alloy deteriorated over nearly the entire DNP range (ε_{imp} = 0.8–5.5%) investigated. However, at ε_{imp}~1.5%, the positive effects of yield areas (up to 6.5% long) were revealed, practically without reducing the total plasticity of the source material. Notably, the impulse introduction of energy into the alloy rolled in the longitudinal direction led to the improved plastic properties of the alloy in a wide range of the ε_{imp} parameter variation (ε_{imp} = 1–7%). In particular, at ε_{imp} = 1.5%, the VT23 titanium alloy's ductility increased by 18% compared to the initial state. Moreover, specimens rolled in the longitudinal direction at a given DNP intensity were also characterized by yield areas up to 8% long.

Therefore, the optimal DNP range for the VT23 titanium alloy (ε_{imp}~1.5%) was found, in which the initial mechanical properties of the alloy rolled in different directions are improved simultaneously, making it possible to use this result for technological purposes. In particular, for developing the stamping technology for high-strength titanium alloys.

Author Contributions: The research was conceptualized by M.C., P.M. and A.P.; the experiments were conducted by A.P. and M.C.; and the results were analyzed and discussed by M.C., P.M., A.P., J.B. (Janette Brezinová), J.B. (Jakub Brezina) and I.K. The manuscript was written and reviewed by M.C., P.M., A.P., J.B. (Janette Brezinová), J.B. (Jakub Brezina) and I.K. All authors have read and agreed to the published version of the manuscript.

Funding: This research was funded by the Scientific Grant Agency, "Application of progressive technologies in restoration of functional surfaces of products" (1/0497/20); the Cultural and Educational Grant Agency KEGA 036TUKE-4/2021 "Distance form of education in the field of engineering technologies using modern IT tools for foreign students"; and the Slovak Research and Development Agency APVV-20-0303 "Innovative approaches to the restoration of functional surfaces by laser weld overlaying".

Institutional Review Board Statement: Not applicable.

Informed Consent Statement: Not applicable.

Data Availability Statement: The data presented in this study are available upon request from the corresponding author.

Acknowledgments: This contribution is the result of the project implementation: "Innovative approaches to the restoration of functional surfaces by laser weld overlaying" (APVV-20-0303), supported by the Slovak Research and Development Agency and supported by the Ministry of Education of Slovakia Foundation under grant projects VEGA No. 1/0497/20 "Application of progressive technologies in restoration of functional surfaces of products" and KEGA 036TUKE-4/2021 "Distance form of education in the field of engineering technologies using modern IT tools for foreign students". This support is highly appreciated by the authors.

Conflicts of Interest: The authors declare no conflict of interest.

References

1. Khanna, N.; Davim, J.P. Design-of-experiments application in machining titanium alloys for aerospace structural components. *Meas. J. Int. Meas. Confed.* **2015**, *61*, 280–290. [CrossRef]
2. Peters, M.; Kumpfert, J.; Ward, C.H.; Leyens, C. Titanium alloys for aerospace applications. *Adv. Eng. Mater.* **2003**, *5*, 419–427. [CrossRef]
3. Brewer, W.D.; Bird, R.K.; Wallace, T.A. Titanium alloys and processing for high speed aircraft. *Mater. Sci. Eng. A* **1998**, *243*, 299–304. [CrossRef]
4. Gladkovskii, S.V.; Volkov, V.P.; Salikhyanov, D.R.; Veselova, V.E.; Patselov, A.M. Rheological Behavior of a VT23 Alloy during Deformation in a Wide Temperature Range. *Russ. Metall.* **2020**, *2020*, 1147–1150. [CrossRef]
5. Kalashnikov, K.N.; Chumaevskii, A.V.; Kalashnikova, T.A.; Osipovich, K.S.; Kolubaev, E.A. Defect Formation in Titanium Alloy during Non-stationary Process of Local Metallurgy. *Russ. Phys. J.* **2020**, *63*, 962–967. [CrossRef]
6. Babaréko, A.A.; Égiz, I.V.; Khorev, A.I.; Martynova, M.M.; Samarin, E.B. Superplasticity of titanium alloys of different classes. *Met. Sci. Heat Treat.* **1995**, *37*, 251–256. [CrossRef]
7. Polyanskii, V.M.; Klyatskina, V.Y.; Silis, M.I. Evaluation of the brittleness of titanium alloys in terms of mechanical properties and fracture patterns. *Met. Sci. Heat Treat.* **2003**, *45*, 61–64. [CrossRef]
8. Khorev, A.T.; Babaréko, A.A.; Krasnozhon, A.I.; Betsofen, S.Y. Effect of structure, substructure, and crystallographic texture on the mechanical properties of titanium alloys. *Met. Sci. Heat Treat.* **1982**, *24*, 463–467. [CrossRef]
9. Zhao, Q.; Sun, Q.; Xin, S.; Chen, Y.; Wu, C.; Wang, H.; Xu, J.; Wan, M.; Zeng, W.; Zhao, Y. High-strength titanium alloys for aerospace engineering applications: A review on melting-forging process. *Mater. Sci. Eng. A* **2022**, *845*, 143260. [CrossRef]
10. Xin, C.; Wang, Q.; Ren, J.; Zhang, Y.; Wu, J.; Chen, J.; Zhang, L.; Sang, B.; Li, L. Plastic deformation mechanism and slip transmission behavior of commercially pure Ti during in situ tensile deformation. *Metals* **2022**, *12*, 721. [CrossRef]
11. Zhang, S.; Zeng, W.; Zhao, Q.; Ge, L.; Zhang, M. In situ SEM study of tensile deformation of a near-β titanium alloy. *Mater. Sci. Eng. A* **2017**, *708*, 574–581. [CrossRef]
12. Guan, B.; Xin, Y.; Huang, X.; Liu, C.; Wu, P.; Liu, Q. The mechanism for an orientation dependence of grain boundary strengthening in pure titanium. *Int. J. Plast.* **2022**, *153*, 103276. [CrossRef]
13. Xu, S.; Zhou, P.; Liu, G.; Xiao, D.; Gong, M.; Wang, J. Shock-induced two types of {101 2} sequential twinning in Titanium. *Acta Mater.* **2019**, *165*, 547–560. [CrossRef]
14. Wang, L.; Zheng, Z.; Phukan, H.; Kenesei, P.; Park, J.-S.; Lind, J.; Suter, R.; Bieler, T. Direct measurement of critical resolved shear stress of prismatic and basal slip in polycrystalline Ti using high energy X-ray diffraction microscopy. *Acta Mater.* **2017**, *132*, 598–610. [CrossRef]
15. Zhang, Z.; Zhang, P.; Li, L. Fatigue cracking at twin boundaries: Effects of crystallographic orientation and stacking fault energy. *Acta Mater.* **2012**, *60*, 3113–3127. [CrossRef]
16. Luster, J.; Morris, J.M. Compatibility of deformation in two-phase Ti-Al alloys: Dependence on microstructure and orientation relationships. *Metall. Mat. Trans. A* **1995**, *26*, 1745–1756. [CrossRef]
17. Wang, B.; Liu, H.; Zhang, Y.; Zhou, B.; Deng, L.; Wang, C.; Chen, J.; Zhang, Y. Effect of grain size on twinning behavior of pure titanium at room temperature. *Mater. Sci. Eng. A* **2021**, *827*, 142060. [CrossRef]
18. Emelianova, E.; Pisareva, M.; Zinovieva, O.; Romanova, V. The effect of texture sharpness on deformation-induced surface roughening in α-titanium. *Procedia Struct. Integr.* **2022**, *35*, 203–209. [CrossRef]
19. Won, J.W.; Park, K.-T.; Hong, S.-G.; Lee, C.S. Anisotropic yielding behavior of rolling textured high purity titanium. *Mater. Sci. Eng. A* **2015**, *637*, 215–221. [CrossRef]
20. Wu, X.; Kalidindi, S.; Necker, C.; Salem, A. Prediction of Crystallographic texture evolution and anisotropic stress–strain curves during large plastic strains in high purity α-titanium using a Taylor-type crystal plasticity model. *Acta Mater.* **2007**, *55*, 423–443. [CrossRef]
21. Sahu, V.K.; Yadava, M.; Chakraborty, P.; Gurao, N.P. Effect of notch severity and crystallographic texture on local deformation and damage in commercially pure titanium. *Int. J. Plast.* **2022**, *155*, 103318. [CrossRef]
22. Chouhan, D.K.; Biswas, S. Multiaxial plane-strain forging and rolling of biomedical grade titanium: Evolution of microstructure, texture, and mechanical properties. *Mater. Lett.* **2021**, *291*, 129540. [CrossRef]
23. Verleysen, P.; Peirs, J. Quasi-static and high strain rate fracture behaviour of Ti6Al4V. *Int. J. Impact Eng.* **2017**, *108*, 370–388. [CrossRef]
24. D'yakova, M.A.; L'vova, E.A. Phase transformations in dual-phase titanium alloys during mechanical testing at various temperatures. *Met. Sci. Heat Treat.* **1988**, *30*, 205–211. [CrossRef]
25. Revil-Baudard, B.; Cazacu, O.; Flater, P.; Kleiser, G. Plastic deformation of high-purity α-titanium: Model development and validation using the Taylor cylinder impact test. *Mech. Mater.* **2015**, *80*, 264–275. [CrossRef]

26. Głowacki, D.; Moćko, W.; Marczak, M.; Głowacka, A.; Kraśkiewicz, C. Energy absorbing properties analysis of layers structure of titanium alloy Ti6Al4V during dynamic impact loading tests. *Materials* **2021**, *14*, 7209. [CrossRef]
27. Cao, J.; Li, F.; Ma, W.; Li, D.; Wang, K.; Ren, J.; Nie, H.; Dang, W. Constitutive equation for describing true stress–strain curves over a large range of strains. *Philos. Mag. Lett.* **2020**, *100*, 476–485. [CrossRef]
28. Chausov, M.; Pylypenko, A.; Berezin, V.; Volyanska, K.; Maruschak, P.; Hutsaylyuk, V.; Markashova, L.; Nedoseka, S.; Menou, A. Influence of dynamic non-equilibrium processes on strength and plasticity of materials of transportation systems. *Transport* **2018**, *33*, 231–241. [CrossRef]
29. Zasimchuk, E.; Turchak, T.; Chausov, N. Hydrodynamic plastic flow in metal materials. *Results Mater.* **2020**, *226*, 100090. [CrossRef]
30. Chausov, N.G.; Voityuk, D.G.; Pilipenko, A.P.; Kuz'menko, A.M. Setup for testing materials with plotting complete stress–strain diagrams. *Strength Mater.* **2004**, *36*, 532–537. [CrossRef]
31. Chausov, N.G.; Pilipenko, A.P. Influence of dynamic overloading on fracture kinetics of metals at the final stages of deformation. *Mechanika* **2004**, *48*, 13–18.
32. Chausov, M.G.; Berezin, V.B.; Pylypenko, A.P.; Hutsaylyuk, V.B. Strain field evolution on the surface of aluminum sheet alloys exposed to specific impact with oscillation loading. *J. Strain. Anal.* **2014**, *50*, 61–62. [CrossRef]
33. Chausov, M.; Hutsaylyuk, V.; Sniezek, L.; Sobchak, A.; Pylypenko, A. Strain field evolution on the surface of stainless sheet steel 12Cr17 exprosed to a specific impact with oscillation loading. In Proceedings of the 11th International Conference on Intelligent Technologies in Logistics and Mechatronics Systems, Panevėžys, Lithuania, 28–29 April 2016; 47–52.
34. Turchak, T.; Chausov, N.; Hutsaylyuk, V.; Berezin, V. Influence of combined loading on microstructure and properties of aluminum alloy 2024-T3. *J. Mater. Eng Perform* **2013**, *22*, 3421–3429. [CrossRef]
35. Chausov, M.G.; Maruschak, P.O.; Hutsaylyuk, V.; Sniezek, L.; Pylypenko, A.P. Effect of complex combined loading mode on the fracture toughness of titanium alloys. *Vacuum* **2018**, *147*, 51–57. [CrossRef]
36. Chausov, M.; Brezinová, J.; Pylypenko, A.; Maruschak, P.; Titova, L.; Guzanová, A. Modification of mechanical properties of high-strength titanium alloys VT23 and VT23M due to impact-oscillatory loading. *Metals* **2019**, *9*, 80. [CrossRef]
37. Konovalenko, I.; Maruschak, P.; Brezinová, J.; Brezina, J. Morphological characteristics of dimples of ductile fracture of VT23M titanium alloy and identification of dimples on fractograms of different scale. *Materials* **2019**, *12*, 2051. [CrossRef]
38. Konovalenko, I.; Maruschak, P.; Prentkovskis, O.; Junevičius, R. Investigation of the rupture surface of the titanium alloy using convolutional neural networks. *Materials* **2018**, *11*, 2467. [CrossRef]
39. Konovalenko, I.; Maruschak, P.; Prentkovskis, O. Automated method for fractographic analysis of shape and size of dimples on fracture surface of high-strength titanium alloys. *Metals* **2018**, *8*, 161. [CrossRef]
40. Chausov, M.G.; Pylypenko, A.P.; Maruschak, P.O. *A Method for the Improvement of Plastic Properties of Sheet Two-Phase High-Strength Titanium Alloys Caused by Impact-Oscillatory Loading: Scientific and Methodical Recommendations for the Ukrainian Plants for the Design of Production of Agricultural Equipment*; FOP V.A. Palyanytsya: Ternopil, Ukraine, 2017; p. 48. Available online: http://elartu.tntu.edu.ua/handle/lib/22618 (accessed on 24 July 2022). (In Ukrainian)
41. Sun, L.; Xu, Z.; Peng, L.; Lai, X. Grain-size-dependent ductile-to-brittle fracture mechanism of titanium sheets. *Scr. Mater.* **2022**, *219*, 114877. [CrossRef]
42. Chausov, M.G.; Pylypenko, A.P.; Berezin, V.B.; Markashova, L.I.; Kushnariova, O.S.; Hutsaylyuk, V.B. Influence of impact-oscillatory loading upon the mechanical properties of the VT-22 titanium alloy sheet. *Mater. Eng. Perfom* **2016**, *25*, 3482–3492. [CrossRef]
43. Maruschak, P.O.; Chausov, M.G.; Konovalenko, I.V.; Yasnii, O.P.; Panin, S.V.; Vlasov, I.V. Effect of shock and vibration loading on the fracture mechanisms of a VT23 titanium alloy. *Strength Mater.* **2020**, *52*, 252–261. [CrossRef]
44. Khorev, A.I. High-strength titanium alloy VT23 and its applications in advanced welded and brazed structures. *Weld. Int.* **2010**, *24*, 276–281. [CrossRef]
45. Khorev, A.I. Alloying and heat treatment of structural ($\alpha + \beta$) titanium alloys of high and superhigh strength. *Russ. Engin. Res.* **2010**, *30*, 682–688. [CrossRef]
46. Vodolazskiy, F.V.; Illarionov, A.G.; Shirinkina, N.A. Microstructure, phase composition, physical and mechanical properties of titanium alloy VT23 hot-extruded tube. *Defect Diffus. Forum* **2021**, *410*, 324–329. [CrossRef]
47. Bryukhanov, A.A.; Ivanii, V.S.; Vasyliv, A.I.; Moroz, I.A. Recrystallization texture and anisotropy of the properties of cold rolled α titanium. *Met. Sci. Heat Treat.* **1977**, *19*, 500–502. [CrossRef]
48. Bryukhanov, A.A.; Gokhman, A.R. Orientation distribution function of texturized sheet stock in hexagonal alpha-titanium. *Russ. Metall. Met.* **1985**, *5*, 140–144.
49. Zakharchenko, I.G.; Gokhman, A.R.; Bryukhanov, A.A.; Mikhajlivskij, Y.G. Orientation dependence of the mechanical properties of textured Ti-3 Al-1.5 V alloy sheets. *Fiz. Met. I Metalloved.* **1993**, *76*, 164–169.

materials

Article

Frictional Characteristics of Deep-Drawing Quality Steel Sheets in the Flat Die Strip Drawing Test

Marek Szewczyk [1], Krzysztof Szwajka [1] and Tomasz Trzepieciński [2,*]

[1] Department of Integrated Design and Tribology Systems, Faculty of Mechanics and Technology, Rzeszow University of Technology, ul. Kwiatkowskiego 4, 37-450 Stalowa Wola, Poland; m.szewczyk@prz.edu.pl (M.S.); kszwajka@prz.edu.pl (K.S.)
[2] Department of Manufacturing and Production Engineering, Rzeszow University of Technology, al. Powst. Warszawy 8, 35-959 Rzeszow, Poland
* Correspondence: tomtrz@prz.edu.pl

Abstract: Friction is one of the most important technological phenomena and has a large influence on the flow characteristics of a deformed material. A strip drawing friction test was used to evaluate the friction characteristics of 0.8 mm thick DC04 steel sheets in a sheet forming operation. Friction tests were carried out using a specially designed friction simulator and uniaxial tensile tests were carried out to determine the mechanical properties of the specimens. In addition, measurements of the sheet surface topography were carried out to identify the tribological properties of the specimens. The friction tests were conducted under different pressure and lubrication conditions. A comparative analysis of the results of the friction tests revealed different changes in the surface topography of the test sheets which can be associated with specific friction mechanisms. It was found that the effectiveness of lubrication depends on the lubricant viscosity and nominal pressure. Increasing the nominal pressure intensifies the phenomenon of asperity flattening and reduces the volume of closed pockets of lubricant.

Keywords: coefficient of friction; deep drawing; sheet metal forming; steel sheet

Citation: Szewczyk, M.; Szwajka, K.; Trzepieciński, T. Frictional Characteristics of Deep-Drawing Quality Steel Sheets in the Flat Die Strip Drawing Test. *Materials* **2022**, *15*, 5236. https://doi.org/10.3390/ma15155236

Academic Editor: Carlos Garcia-Mateo

Received: 13 June 2022
Accepted: 26 July 2022
Published: 28 July 2022

1. Introduction

Friction and wear is an indispensable physical phenomenon accompanying the movement of two bodies in relation to each other. Holmberg and Erdemir [1] found that ~23% of world energy consumption comes from tribological contacts. It commonly assumed that 20% of this energy is used to overcome friction, and 3% is the cost of regenerating worn parts due to failure or wear. The phenomenon of friction occurring in the processes of sheet metal forming (SMF) differs significantly from the phenomena occurring at low loads and in the kinematic nodes of machines. Strong plastic deformation and high temperatures can intensify many tribological phenomena in the contact zone [2,3]. The amount of frictional resistance depends on the surface roughness of the contacting surfaces and on the contact pressures, the value of which may significantly exceed the yield stress of the deformed material [4]. Friction is a phenomenon that has a key impact on the flow of the sheet metal [5,6], the surface finish of the product [7], and also directly affects the wear of tools [8].

Friction in SMF is a complex function of the physical and mechanical properties of the tool and workpiece, process parameters, contact conditions (type of lubricant, oil viscosity) and surface topography of the workpiece and tool [9,10]. Deep drawing is one of the most important technologies, particularly in the engineering, automotive and aircraft industries. During SMF, the relative frictional movement of the tool against sheet metal changes the mechanical properties of the deformed material (e.g., hardness, directional topography and surface roughness) and requires specific properties of the tool surface (e.g., texture, protective coating, thermal-mechanical properties) [11,12]. In order to optimally design the SMF process and control tribological phenomena, it is necessary to understand the interactions between the tribological and surface engineering aspects [13,14]. In order to fully describe the role of friction in sheet metal forming processes, it is necessary to understand the mutual interactions of the tribological system [15], which consists of four main elements—the technological and mechanical parameters of the deformed material and the tool, the lubricant properties, the forming parameters and the characteristics of the tool material [16–18].

The strip drawing test is used to model frictional conditions in the flange area of the drawpiece. In this test, the surfaces of the countersamples representing the tool surface are pressed against the sheet metal strip, which is simultaneously drawn [19]. The sheet metal strip is pulled between non-rotating countersamples, which are flat [20] or have a cylindrical shape [19,21]. The possibility of direct measurement of the friction (tangential) and normal force allows high accuracy to be achieved when measuring the value of the coefficient of friction (COF). A change of friction conditions is produced by changing the sliding speed, normal pressure, shape and radius of the countersamples, surface roughness of the countersamples and lubrication conditions [22,23].

The strip drawing test is the most common method of testing the COF and wear of sheet metals. Severo et al. [24] tested the tribological behaviour of W-Ti-N coatings in strip drawing tests. For industrially deposited coatings, a lower COF was measured in dry friction tests for low loads. Tavares et al. [25] studied the friction of commercially coated forming tools. The strip drawing tests showed that the CrN/TiN (HV0.2 = 354 kgf·mm^2) and VN (HV0.2 = 229 kgf·mm^2) coatings presented lower and more stable COF values, leading to more favourable tribological conditions. Schell et al. [26] used the strip drawing test to study the influence of the wear performance and friction of different lubricants on a die surface. The results revealed that oil and solid lubricants show distinct behaviour characteristics in the high temperature forming of aluminium sheets. The results from the strip drawing tests conducted by Sutcliffe et al. [27] show a significant effect of the lubrication regime while the properties of the different die materials are less important. Trzepieciński and Fejkiel [19] tested the COF of pre-strained deep drawing quality steel sheets in the strip drawing test with cylindrical countersamples. They concluded that the COF of samples decreases with increasing nominal pressure of the countersamples under both dry and lubricated conditions (machine oil). Trzepieciński [22] tested 0.8 mm thick DC04 steel sheets between two non-rotating countersamples with radii of 10 and 200 mm in lubricated conditions using lubricants based on a combination of boric acid and bio-based oils. The high viscosity of heavy-draw oil (1157 mm^2·s^{-1}) suggested that for low pressures it creates conditions for the formation of a stable lubricant film completely separating the friction surfaces. In another paper, Bazan and Trzepieciński [2] investigated the influence of the surface parameters of the strip specimens, surface parameters of the countersamples and nominal pressure on the COF in the strip drawing friction test. In lubricated conditions with machine oil, the decrease in value of the kurtosis Sku parameter was greater than in dry friction conditions. The lubricated conditions also caused a smaller reduction in the skewness of the Ssk surface roughness parameter than for dry friction conditions.

The aim of this paper is to study of the effect of lubrication conditions and nominal pressure on the frictional properties of low-carbon DC04 steel sheets in the sheet metal–blankholder interface. Commonly available oils were used as lubricants. These oils are relatively cheap compared to professional deep-drawing lubricants used in sheet metal forming. Friction analysis using the strip drawing test is mainly carried out on cylindrical or rounded surface specimens. However, there are a limited number of papers on friction testing using the flat die strip drawing test. Flat countersamples are closer to a real-life model of the friction conditions of the contact properties in the sheet metal–blankholder interface. Therefore, in this article, it was decided to present the results of friction tests of DC04 low-carbon steel sheets with the use of a specially designed simulator which allows the COF to be determined when pulling a sheet strip between flat countersamples. Measurements of the sheet surface topography have been carried out to describe the tribological properties of the specimens. The friction tests have been conducted under different nominal pressure and lubrication conditions in order to find the effect of these parameters on lubrication efficiency and the change in the surface roughness parameters of the specimen surface under contact geometry reflecting friction in the blankholder zone in a conventional deep-drawing process.

2. Material and Methods

2.1. Test Material

Strip samples for the friction test, approximately 130 mm long and 20 mm wide, were cut along the rolling direction from DC04 (thickness g = 0.8 mm) sheet metal. DC04 (1.0338) is a grade of low-carbon cold-rolled deep-drawing plate mainly intended for use in the automotive industry. The quality requirements for the chemical composition of DC04 steel in accordance with EN 10130:2006 [28] are presented in Table 1.

Table 1. Chemical composition of DC04 in accordance with the requirements of EN 10130: 2006 (in wt.%).

C (Max.)	Mn (Max.)	P (Max.)	S (Max.)	Fe
0.08	0.4	0.03	0.03	balance

The values of the basic mechanical parameters of the test sheets (Table 2) were determined in a uniaxial tensile test in accordance with ISO 6892-1: 2020 [29]. The samples cut along the rolling direction of the sheet were subjected to uniaxial stretching. Based on the results of the tensile test conducted with three repetitions, the mean values of the following parameters were determined: yield stress $R_{p0.2}$, ultimate tensile strength R_m, strain hardening coefficient K, and strain hardening exponent n, elongation A_{50} and percentage reduction of area Z. The relation of the true plastic strain and yield stress of DC04 sheets (Figure 1) is exponential and described by the Hollomon equation $\sigma = K \cdot \varepsilon^n$.

Table 2. Basic mechanical parameters of DC04 steel sheet.

$R_{p0.2}$, MPa	R_m, Mpa	A_{50}, %	Z, %	K, Mpa	n
165.2	309.3	37.1	45.7	526.9	0.23

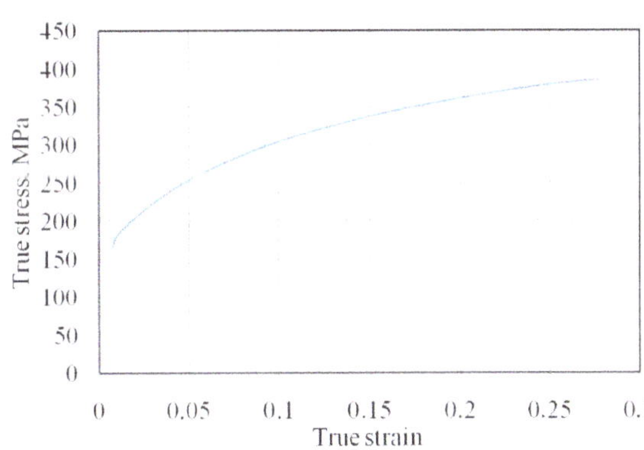

Figure 1. True stress vs. true strain curves for the DC04 steel sheet.

The hardness of the as-received specimens was tested using the Vickers hardness tester series Qness 60 EVO at a test force of 49.03 N. The average hardness was found to be 97 HV5. A view of the test point image is shown in Figure 2.

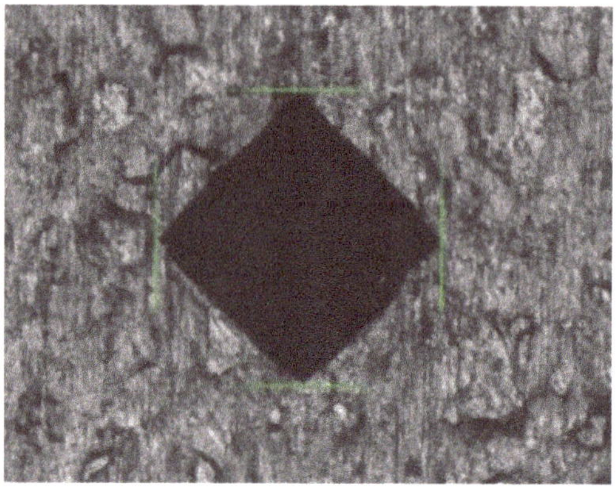

Figure 2. Test point image.

The topography of the DC04 sheet in the as-received state is shown in Figure 3a. The surface topography of the sheet before and after the friction process was measured with a Hommel-Etamic T8000RC stationary profilometer in accordance with the requirements of the ISO 25178 standard. The basic height parameters of the geometric structure of the surface were determined: arithmetical mean height Sa, skewness Ssk, kurtosis Sku, maximum profile peak height Sp, maximum profile valley depth Sv, maximum height Sz and root mean square deviation Sq. Surface roughness parameters of the as-received surface are as follows: Sa = 1.55 µm, Ssk = −0.0307, Sku = 2.68, Sp = 8.02 µm, Sv = 6.93 µm, Sz = 15.0 µm and Sa = 1.26 µm. The morphologies of the sheet surfaces subjected to friction were examined using a MIRA 3 TESCAN scanning electron microscope (SEM). The MIRA 3 Tescan microscope with a Schottky field emission gun was equipped with a four-quadrant BSE detector. The morphology of the as-received sheet surface is shown in Figure 3b. The surfaces of as-received sheet metals have a granular structure resulting from the sheet manufacturing process (sheet rolling).

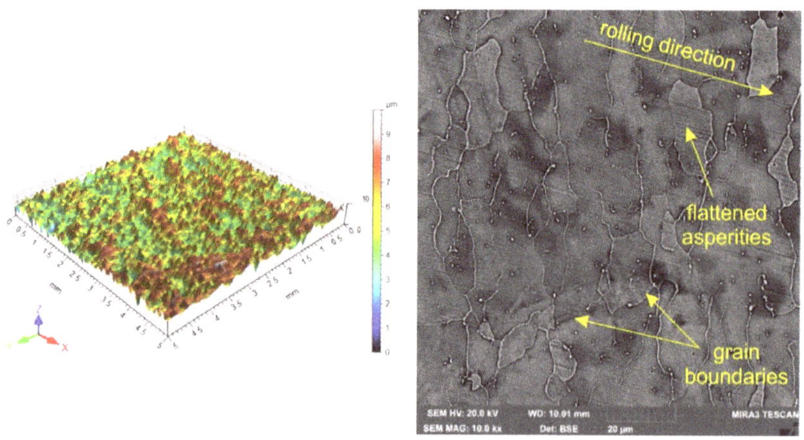

(a) (b)

Figure 3. (**a**) Topography of the DC04 sheet and (**b**) SEM micrograph of the in the as-received state.

2.2. The Friction Test Process

The strip drawing test consists in pulling a sheet metal strip between two countersamples with a flat working surface. The model and view of the specially designed tester are shown in Figure 4. The presence of frictional forces on the two contact surfaces is conducive to obtaining greater accuracy of the average value of the COF. The device has the advantage that it is a simple structure with an uncomplicated measurement of the force parameters during the test.

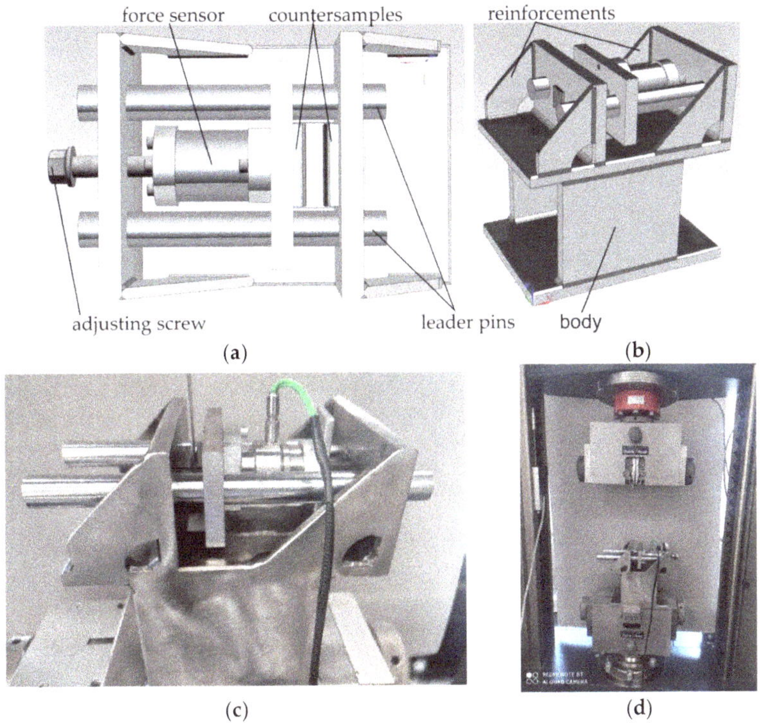

Figure 4. Geometric model of the friction tester in (**a**) top and (**b**) isometric views, (**c**) an actual photo and (**d**) friction tester mounted in Zwick/Roell Z100 testing machine.

The body of the apparatus was mounted in the lower holder of a Zwick/Roell Z100 testing machine. After the sample was placed between the countersamples, the upper free end of the sample was clamped in the upper holder of the testing machine. The technical parameters of the Zwick/Roell Z100 (Zwick/Roell, Ulm, Germany) testing machine are as follows: maximum testing force −100 kN, traverse speed between 0.0005 and 600 mm/min, accuracy of set rate 0.003% V_{nom}, force measurement accuracy with load cell-from 0.2 KN class 1 (from 1 kN class 0.5), resolution of crosshead travel 0.0083 μm/impuls and position accuracy ±2 μm.

The value of the clamping force of the countersamples was set in static conditions (before starting the sample movement) with the help of the adjusting screw (Figure 3a). The normal (clamp) force was measured with a Kistler type 9345B force sensor with a measuring range ±10 kN. Its high resolution allows measurement of the slightest dynamic changes in large forces and torques in non-rotating elements. LabVIEW 2016 software (National Instruments, Austin, Texas, USA) was used to record the clamping force of the countersamples. The value of the tangential (friction) force was measured using the measuring system of the Zwick/Roell Z100 testing machine. Values of both friction and normal forces were correlated in the LabVIEW program based on a Megatron Series SPR18 potentiometric linear transducer.

The samples in the form of sheet strips were cut from the same piece of metal sheet. The strips were tested in as-delivered surface condition [22,23,30]. In general, during friction tests related to sheet metal forming it is assumed [19,22,23,30–34] that samples cut from the same piece of metal sheet have the same surface roughness. These conditions correspond to industrial sheet metal forming, where the surface topography of the workpiece is not strictly the same when forming a series of products.

The sample was pulled through the countersample set with a constant travel speed of the top grip of the testing machine of 2 mm/s. The tests were carried out with the following nominal pressures p 3, 6, 9, 12 MPa. The values of these pressures are within the range of pressures prevailing in sheet metal forming [35,36]. The average value of the COF was determined according to dependence (1), disregarding the transient range $r_1 + r_2$ (Figure 4) related to the change of static friction conditions into kinematic friction.

$$\mu = \frac{F_T}{2F_N} \quad (1)$$

where: F_T—friction (pulling) force, F_N—normal (clamping) force.

In the first range (r_1 in Figure 5), plastic deformation of the highest peaks of the asperities occurs with the formation of large stresses in the material (Figure 6a). At the moment of starting the movement, the deformation of the summits of the asperities increases. At the same time, as a result of the strain hardening phenomenon, their mechanical strength is significantly increased. In such conditions, when a certain stress value is exceeded, the material cracks (Figure 6b). The cracking of the material releases some of the energy accumulated in the material, as a result of which the nominal pressure decreases (the broken particles fill the spaces in the valleys of the asperities (Figure 6c). As a result, the coefficient of friction decreases (r_2 in Figure 5). In the area of stabilised friction, all asperities of the sheet surface along the length of the contact path with the countersamples are subjected to the flattening mechanism. In the stabilised range r_3 of the COF (Figure 5), the shear of the summits of asperities occurs at the entrance to the countersample zone. The mean value of the COF was determined for the stabilised range r_3.

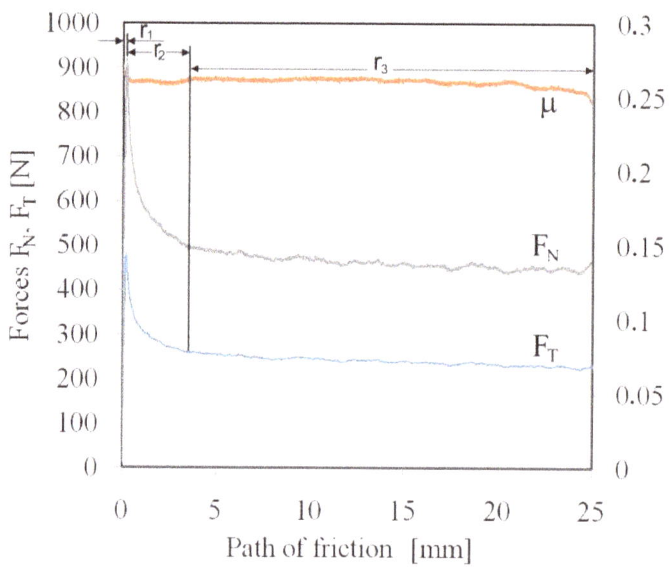

Figure 5. Variation of the friction force, normal force and coefficient of the friction during friction test (dry friction, P = 3 MPa).

Apart from specialist greases intended for plastic working, many greases and oils manufactured with the intention of being used in mechanical transmissions, engines and hydraulic systems are commonly used in sheet metal forming processes to reduce the value of frictional resistance [37]. The tests were carried out under dry friction conditions (in the as-received sheet metal state) and during lubrication with Castrol Axle EPX 80W-90 gear oil and Castrol EDGE 5W-30 engine oil. EPX 80W-90 is a multi-purpose axle oil which may be used in differentials, final drives and other applications in passenger cars and commercial vehicles. Castrol EDGE with Fluid Strength Technology TM 5W30 oil is a fully synthetic engine oil used in conditions requiring lower viscosity and stronger oils for longer oil change intervals. The criteria for selecting the oils were their general availability and relatively low cost. The viscosity grade of these oils is standardised according to the Society of Automotive Engineers (SAE) and many manufacturers produce oils with the same physico-chemical characteristics. The oils used are generally available and several times cheaper than professional oils intended for deep-drawing operations. Previous investigations on friction of titanium alloy sheets [38] confirmed the effectiveness of typical engine and gear oils in reducing the COF.

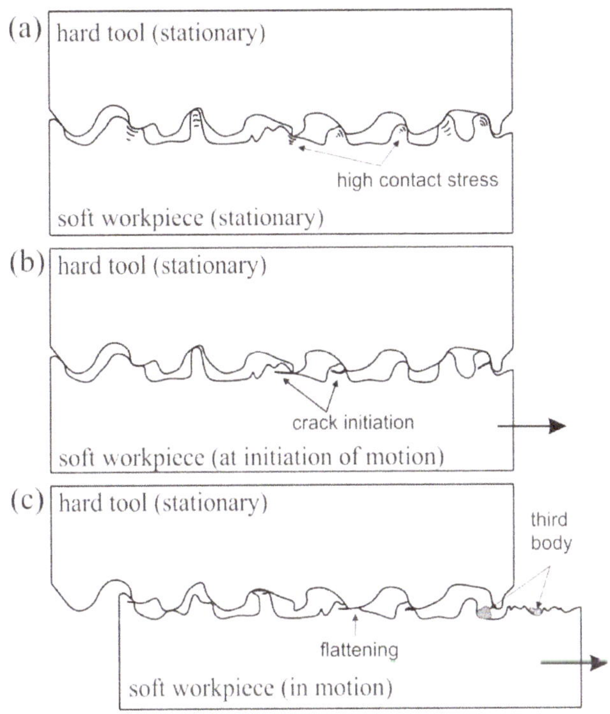

Figure 6. Character of the contact between the surfaces of the hard tool and the soft workpiece: (**a**) no movement, (**b**) initiation of motion and (**c**) in motion.

The lubrication procedure complies with the guidelines of Payen et al. [39]: apply oil with a soft brush, regularise the deposit with the brush emptied of excess oil (i.e., no drop forms at the end of hairs if weighting brush downward), let the sheet strip stand up for approximately 10 min so that the oil layer homogenises and the excess oil flows away by gravity. The drawing speed was 0.5 mm/s. The basic physical and chemical properties of the oils used in the tests are presented in Table 3.

Table 3. Basic physico-chemical properties of the oils.

Oil	Property	Test Method	Value
Castrol Axle EPX 80W-90	Density at 15 °C	ASTM D4052	0.896 g/mL
	Kinematic viscosity at 40 °C	ASTM D445	134 mm^2/s
	Viscosity index	ASTM D2270	101
	Pour point	ASTM D97	−33 °C
	Flash point	ASTM D92	222 °C
Castrol EDGE 5W-30	Density at 15 °C	ASTM D4052	0.851 g/ml
	Kinematic viscosity at 40 °C	ASTM D445	70 mm^2/s
	Viscosity index	ASTM D2270	169
	Pour point	ASTM D97	−42 °C
	Flash point	ASTM D93	202 °C

Countersamples in the form of inserts (Figure 7) with dimensions of 45 mm wide and 20 mm high incorporated in the construction of the device were made of 145Cr6 (1.2063) cold-worked tool steel. The 145Cr6 steel is recommended for use in forming tools, including stamping dies, punching dies, wire drawing dies, cutting plates and shear knives. The countersamples were fabricated to an overall flatness tolerance of 10 µm. The average hardness was found to be 215 HV10. Twelve measurements were used to calculate the average value of hardness. The topography of the countersamples and basic surface roughness parameters are shown in Figure 7b and Table 4, respectively.

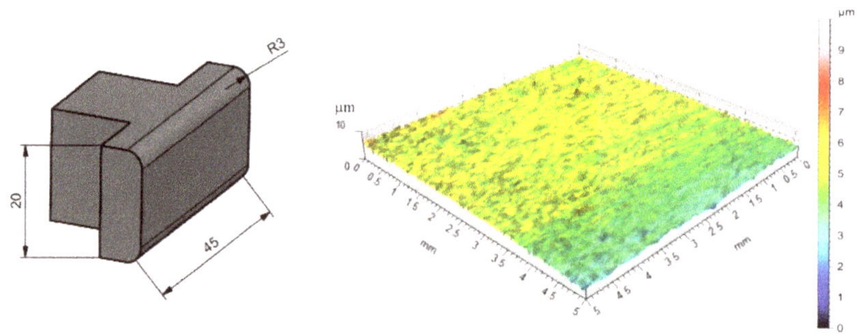

(a) (b)

Figure 7. (**a**) Shape and dimensions of the countersamples (units in mm) and (**b**) topography of the countersamples.

Table 4. Basic surface roughness parameters of countersamples.

Sa, µm	Sq, µm	Sku	Ssk	Sp, µm	Sv, µm	Sz, µm
0.636	0.810	3.76	−0.544	4.89	5.11	10.0

3. Results and Discussion

There was a trend of a decreasing value of the COF with increasing pressure which was related to the nominal contact surface area (Figure 8). This relationship applies to all the friction conditions analysed. This may be a result of the dependence of friction on the clamping (normal) force, where beyond a certain load the relationship between friction force and clamping force is nonlinear and the value of the COF is not constant. In other words, it changes with normal load. This relation was also observed by Kirkhorn et al. [40], Trzepiecinski et al. [19,41] and Vollertsen and Hu [42]. Friction occurring at high pressure values may significantly differ from the phenomena occurring at low loads and in the kinematic nodes of machines due to the large degree of plastic deformation that can intensify many frictional phenomena in the contact zone. When compared with conventional kinematic machine nodes, the material strength of one element of the friction pair (tool) is assumed to be much greater than the strength of a formed material that undergoes intentional plastic deformation in the sheet metal forming process. The initial topography of the sheet surface under the influence of large deformations is constantly evolving during SMF.

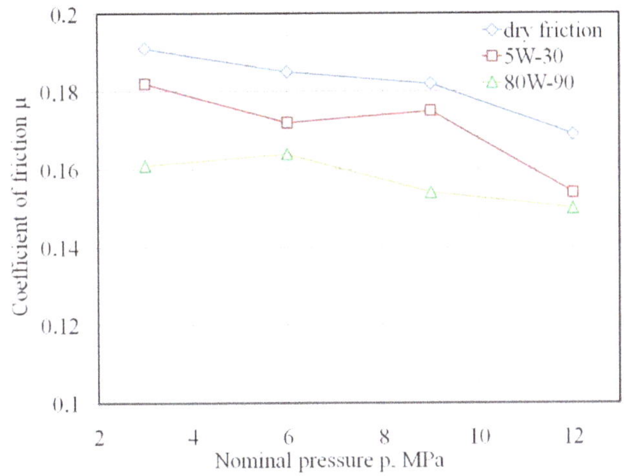

Figure 8. Effect of nominal pressure on the value of coefficient of friction.

As confirmed by Kirkhorn et al. [40], ten Thije et al. [43], Trzepieciński and Fejkiel [19] and Vollertsen and Hu [42], under such conditions the COF (as a coefficient of proportionality between the friction force and the normal force) has no constant value. In other words, it changes with normal load. It is obvious that the real contact area increases with increasing nominal load due to the intensification of the asperity flattening mechanism. However, determination of the real contact area in sheet metal forming is not fully understood and, due to the random nature of asperities, difficult to precisely determine experimentally. For this reason, the literature on friction testing in SMF is focused on the analysis of nominal pressure.

To determine the influence of the lubricant on the change of friction conditions, an effectiveness of lubrication coefficient W_L was introduced as the ratio between the COF determined for dry friction μ_{dry} and that for lubricated conditions μ_{oil} [19]:

$$W_L = \frac{\mu_{dry} - \mu_{oil}}{\mu_{dry}} \cdot 100\% \tag{2}$$

During the plastic working of metals, initially there is a small and effective contact area. In elastic contact the surfaces only stick to each other with the summits of the asperities. As the load is increased, there is an increase in the real contact area. The roughness asperities are plastically deformed until the contact surface created in this way is sufficient to transfer the load [44,45]. For smooth surfaces, adhesion is the dominant tribological phenomenon occurring during metallic contact. The asperities of the surface are flattened and sheared which increases the real contact area. A clear flattening of the surface asperities was observed for all analysed friction conditions (Figure 9a–c). It is obvious that the increase in pressure resulted in a more intensive flattening of the asperities. As the surface roughness increases, the influence of adhesion decreases, while the share of the shear mechanism of the sheet metal surface increases.

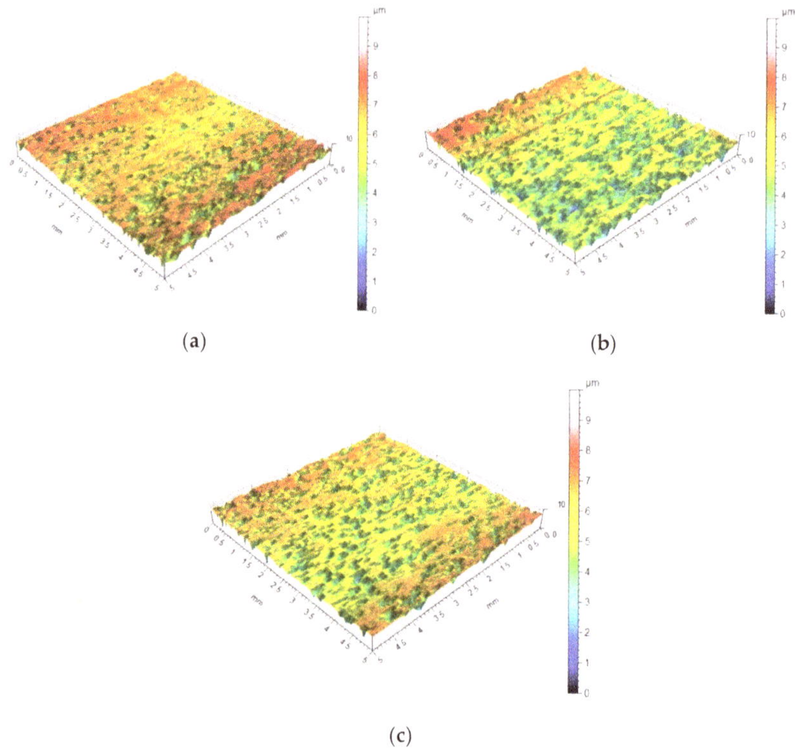

Figure 9. Surface topography of the specimens after a friction test at a nominal pressure of 6 MPa in (**a**) dry friction conditions (Sa = 0.876 µm, Ssk = −1.43, Sku = 5.56, Sp = 2.04 µm, Sv = 6.66 µm, Sz = 8.70 µm, Sa = 0.635 µm), and with (**b**) 5W-30 engine oil (Sa = 1.27 µm, Ssk = −0.698, Sku = 3.65, Sp = 3.47 µm, Sv = 5.38 µm, Sz = 8.85 µm, Sa = 0.965 µm) and (**c**) 80W-90 gear oil (Sa = 1.05 µm, Ssk = −1.36, Sku = 4.78, Sp = 2.46 µm, Sv = 5.88 µm, Sz = 8.34 µm, Sa = 0.782 µm).

The COF was in the range between 11.24% and 15.7% (Figure 10). With 5W-30 engine oil, which has a kinematic viscosity lower by about 50% compared to gear oil, there is a lower ability to reduce frictional resistance (3.84–8.87%). The effectiveness of lubrication depends on two basic factors: the viscosity of the lubricant and the value of the pressure acting on the surface, which determines the real area of contact. At low pressure, the dominant friction mechanism is the mechanical engagement of the summits of the asperities, and at the same time there are open pockets of lubricant which are not able to generate the appropriate pressure of lubricant separating the mechanical cooperation of the contact surfaces. Under these conditions, a lubricant that requires high viscosity plays a key role. With further increase in nominal pressure, the lubrication efficiency decreases (at 6 MPa for 80W-90 gear oil and at 9 MPa for 5W-30 engine oil). We can discuss a certain optimal pressure, which is a kind of balance between the mechanism of mechanical interaction of the summits of the asperities, the adhesion mechanism, and the volume of closed lubricant pockets generating sufficient hydrostatic pressure to create boundary lubrication. Such favourable conditions can be observed for a pressure of 9 MPa, for lubrication with 5W-30 engine oil. With a further increase in pressure, despite the favourable conditions for creating a "lubricant pad" separating the rubbing surfaces, the real contact area is so large that the mechanism for the flattening of the relatively soft sheet material by the tool material of much greater hardness is intensified. As a consequence, the value of the COF increases for lubrication with 5W-30 engine oil.

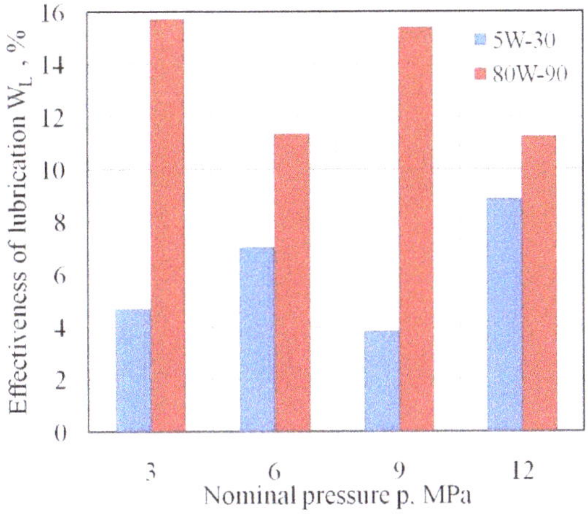

Figure 10. Effect of nominal pressure on the effectiveness of lubrication.

Figure 9 adds to the understanding of Figure 7, i.e., that the effectiveness of lubrication, W_L, increases as the nominal pressure increases, when the lubricant is less viscous (oil 5W-30). Conversely, W_L decreases as the nominal pressure increases, when the lubricant is more viscous (oil 80W-90) although the average value of W_L is still greater. There are oscillations in W_L, but they are probably attributable to experimental errors. If the experiment were repeated sufficient times such that the size of the error bars would be clear, then the oscillations would be wiped out, or highlighted.

The pressure–viscosity coefficient α refers to the relationship between the load placed on the oil film (pressure) and the thickness of the oil film (viscosity) at that load, when all other factors (pressure, sliding speed, geometry, temperature) are constant. The units of the pressure–viscosity coefficient α, are mm^2/N, which are units of area over force. They derive from the relation between viscosity and pressure, commonly expressed by the Barus equation $\eta_p = \eta_0 e^{\alpha p}$, where η_p is viscosity at pressure p (N/mm^2), η_0 is viscosity at atmospheric pressure (N/mm^2) and α is pressure–viscosity coefficient (mm^2/N). When a is multiplied by pressure with units N/mm^2, the units cancel giving the nondimensional exponent of the natural number e. An increase in pressure over a confined lubricant results in increasing the lubricant's viscosity. Such changes in viscosity under mechanical pressure are also known as piezo-viscous effects. This situation exists in certain regimes such as mixed lubrication. Accordingly, as pressure increases—which for a constant force implies a reduced contact area—so does the viscosity, though the relation is not necessarily linear. Additionally, in the case of granular media, it is known that a viscosity increase is accompanied by an increased friction coefficient.

Under dry friction conditions, the values of average roughness Sa and root mean square roughness Sq parameters decrease in the pressure range between 3 and 6 MPa (Figure 11a,b) as a result of the flattening mechanism. Increasing the pressure activates the ploughing mechanism involving the mechanical interaction of the highest summits of the tool surface with the flattened surface of the sheet metal. As a result, the COF increases slightly. Under lubricated conditions, the adhesion phenomenon in the areas of real contact practically does not occur, while the flattening and ploughing mechanisms are limited by the hydrodynamic interaction of the lubricating film accumulated in the valleys of the surface topography. When lubricating the sample surfaces with the oils investigated, there is a clear tendency to a reduction of the Sa and Sq parameters with increasing nominal pressure.

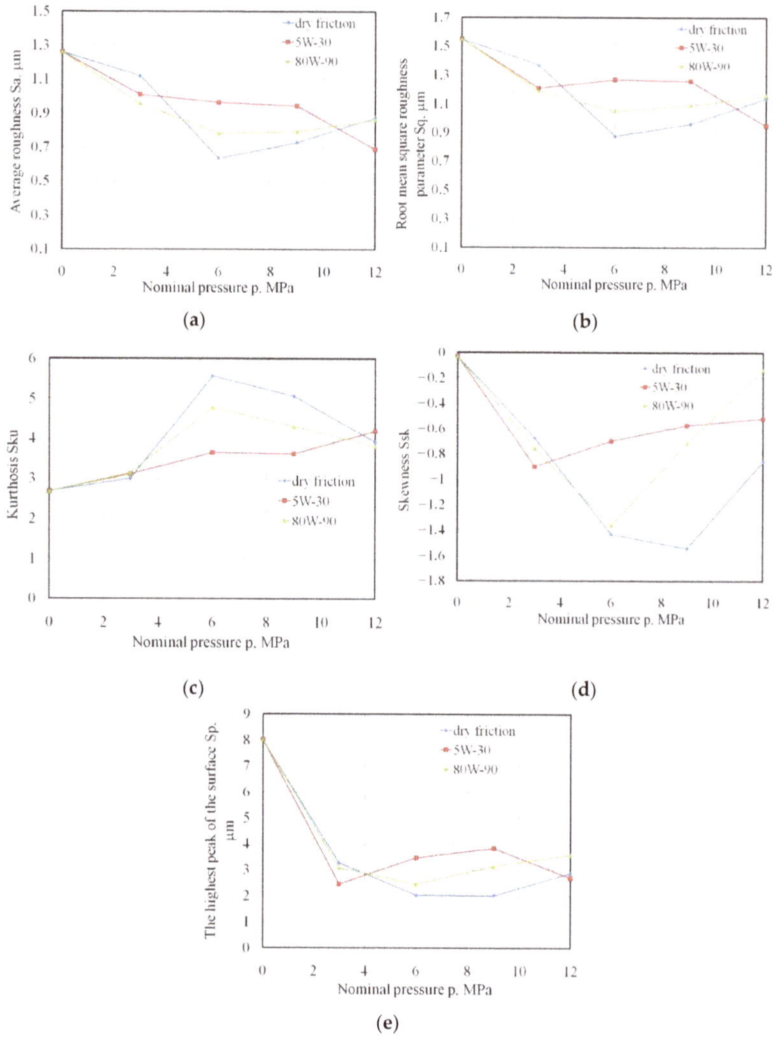

Figure 11. Effect of nominal pressure on variation of the (**a**) Sa, (**b**) Sq, (**c**) Sku, (**d**) Ssk, (**e**) Sp parameters.

For engine oil lubrication conditions, the surface slope coefficient, named kurtosis Sku, increases almost linearly with increasing nominal pressure (Figure 11c). In the remaining two friction conditions analysed, after a sharp increase in this parameter during a change in pressure from 3 to 6 MPa, kurtosis decreases.

A positive value of the surface asymmetry coefficient, named skewness Ssk, reduces the COF, while in the case of negative skewness, the friction is more intense. Hence, the most unfavourable friction conditions related to the skewness parameter occur at pressures of 3 MPa, 6 MPa and 9 MPa, for the lubrication conditions with 5W-30 engine oil, 80W-90 gear oil and lack of lubrication, respectively (Figure 11d).

The occurrence of the flattening phenomenon during the cooperation of the sheet surface with the countersample surface is evidenced by the reduction of the highest peak of the surface Sp by over 50% in relation to the as-received surface. There is also a certain correlation between the character of changes in the parameters Sq (Figure 11b) and Sp (Figure 11e).

During loading, the summits of the asperities deform elastically and plastically, changing the surface topography. In the case of elastic–plastic metals, the increase in pressure during the forced sliding movement causes an increase in the real contact surface, which may lead to a reduction in the volume of open pockets of lubricant [46]. Closed pockets of lubricant are separated from the outer edges of the material and store the lubricant in the closed volume of the valleys. Under load, the pressure of the lubricant increases. This forms a kind of hydrostatic cushion, which takes over part of the load [46].

SEM micrographs of the sheet surface after friction tests carried out under the nominal pressure P = 3 MPa and P = 12 MPa are shown in Figures 12 and 13, respectively. The summits of the surface were deformed creating flattened areas on the surface of the sheet (Figure 12). Increasing the nominal pressure increases the surface roughness of the sheet with clear traces of flattening. At the same time, in these areas, the impact of the ploughing phenomenon is intensified, especially in dry friction conditions when there are favourable conditions for revealing the scuffing mechanism (Figure 13a).

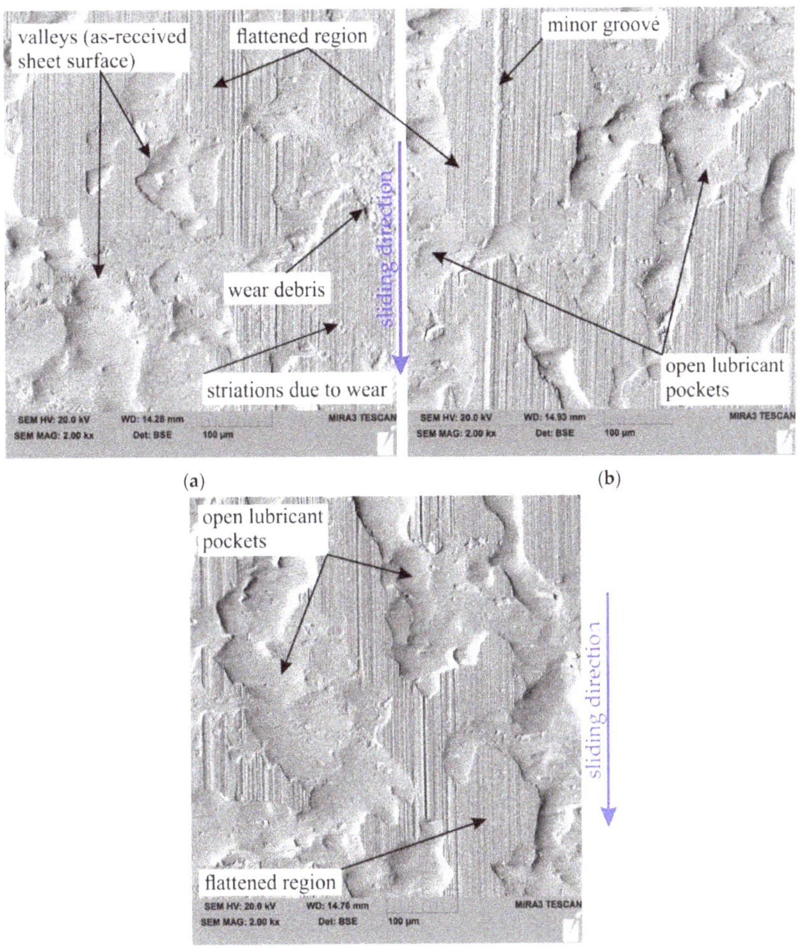

(a) (b) (c)

Figure 12. SEM micrographs of the sheet surface after friction tests carried out under the following conditions (nominal pressure P = 3 MPa): (**a**) dry friction, (**b**) 5W-30 oil and (**c**) 80W-90 oil.

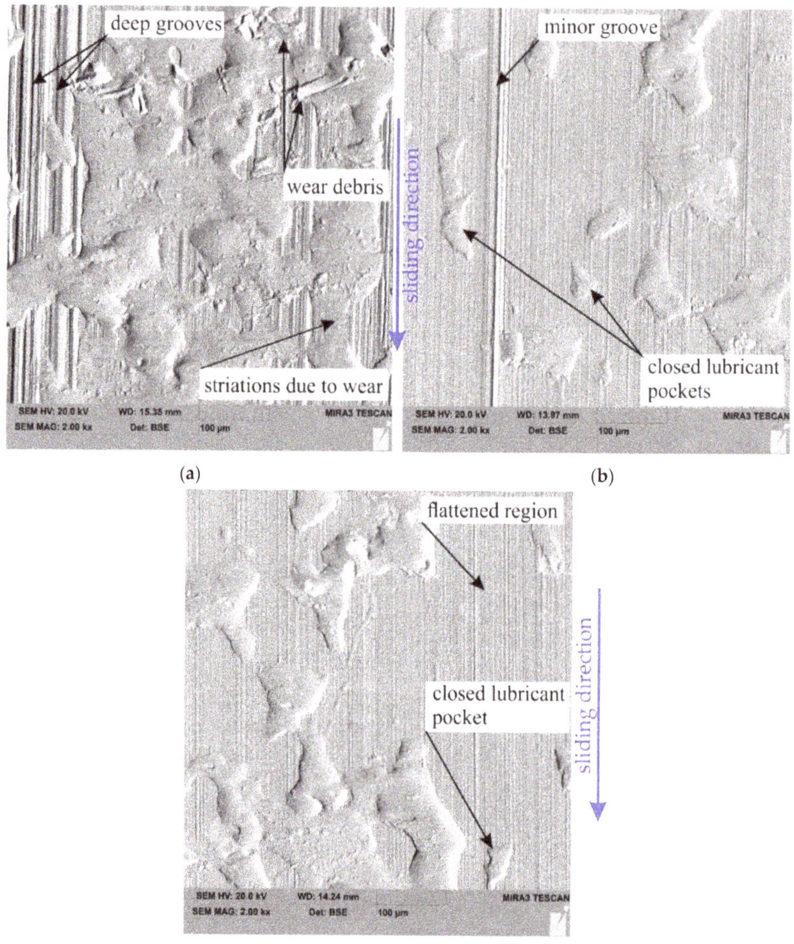

Figure 13. SEM micrographs of the sheet surface after friction tests carried out under the following conditions (nominal pressure P = 12 MPa): (**a**) dry friction, (**b**) 5W-30 oil and (**c**) 80W-90 oil.

4. Conclusions

This article presents the results of friction tests of low-carbon DC04 steel sheets destined for deep-drawing operations. Based on the strip drawing test that simulates the friction phenomena in the area of the blankholder of the stamping die, the following conclusions can be drawn:

- The value of the COF decreases with increasing nominal pressure for all analysed friction conditions.
- Use of 80W-90 gear oil ensured, depending on the value of the pressure, a reduction in the value of the COF in the range between 11.24% and 15.7% compared to dry friction conditions.
- In the entire range of nominal pressures examined, the engine oil decreased the value of the COF by only 3.84–8.87%.
- In dry friction conditions, the average roughness Sa and the root mean square roughness Sq decrease in the pressure range of 3–6 MPa, and then an increase in these parameters is observed as a result of the intensification of the ploughing mechanism.

On the other hand, when lubricating the sample surfaces there was a clear trend to a reduction of the Sa and Sq parameters with increasing nominal pressure in the whole range of pressures investigated.

- The occurrence of the flattening mechanism during cooperation of the relatively soft surface of the sheet metal with the hard surface of the countersamples causes a reduction in the highest peak of the surface Sp by over 50% in relation to the as-received surface.

Author Contributions: Conceptualization, M.S., K.S. and T.T.; methodology, M.S., K.S. and T.T.; software, M.S. and K.S.; validation, M.S., K.S. and T.T.; investigation, M.S., K.S. and T.T.; data curation, M.S., K.S. and T.T.; writing—original draft preparation, T.T.; writing—review and editing, M.S., K.S. and T.T. All authors have read and agreed to the published version of the manuscript.

Funding: This research received no external funding.

Institutional Review Board Statement: Not applicable.

Informed Consent Statement: Not applicable.

Data Availability Statement: The data presented in this study are available on request from the corresponding author.

Conflicts of Interest: The authors declare no conflict of interest.

References

1. Holmberg, K.; Erdemir, A. Influence of tribology on global energy consumption, costs and emissions. *Friction* **2017**, *5*, 263–284. [CrossRef]
2. Bazan, A.; Trzepieciński, T. Effect of friction conditions on the change of sheet surface roughness during deep drawing. *Zesz. Nauk. Pol. Rzesz.-Mech.* **2013**, *30*, 375–386. [CrossRef]
3. Trzepieciński, T.; Fejkiel, R.; Kubit, A. Experimental evaluation of value of friction coefficient in the drawbead region. *Zesz. Nauk. Pol. Rzesz.-Mech.* **2018**, *35*, 77–85. [CrossRef]
4. Nielsen, C.V.; Bay, N. Overview of friction modelling in metal forming processes. *Procedia Eng.* **2017**, *207*, 2257–2262. [CrossRef]
5. Trzepieciński, T.; Gelgele, H.L. Investigation of anisotropy problems in sheet metal forming using finite element method. *Int. J. Mater. Form.* **2011**, *4*, 357–369. [CrossRef]
6. Trzepieciński, T. 3D elasto-plastic FEM analysis of the sheet drawing of anisotropic steel sheet. *Arch. Civ. Mech. Eng.* **2010**, *10*, 95–106. [CrossRef]
7. Nie, N.; Su, L.; Deng, G.; Li, H.; Yu, H.; Tieu, A.K. A review on plastic deformation induced surface/interface roughening of sheet metallic materials. *J. Mater. Res. Technol.* **2021**, *15*, 6574–6607. [CrossRef]
8. Costa, H.L.; Hutchings, I.M. Effects of die surface patterning on lubrication in strip drawing. *J. Mater. Proc. Technol.* **2009**, *209*, 1175–1180. [CrossRef]
9. Wang, C.; Ma, R.; Zhao, J.; Zhao, J. Calculation method and experimental study of coulomb friction coefficient in sheet metal forming. *J. Manuf. Process.* **2017**, *27*, 126–137. [CrossRef]
10. Bay, N.; Olsson, D.D.; Andreasen, J.L. Lubricant test methods for sheet metal forming. *Tribol. Int.* **2008**, *41*, 844–853. [CrossRef]
11. Seshacharyulu, K.; Bandhavi, C.; Naik, B.B.; Rao, S.S.; Singh, S.K. Understanding Friction in sheet metal forming-A review. *Mater. Today: Proc.* **2018**, *5*, 18238–18244. [CrossRef]
12. Sanchez-Caballero, S.; Selles, M.A.; Pla-Ferrando, R.; Seguí, J.; Peydro, M.A. Research on the Friction Properties of DP600 Stainless Steel as a Function of Bending Angle and Pin Diameter. *Mater. Proc.* **2021**, *3*, 17.
13. Littlewood, M.; Wallace, J.F. The effect of surface finish and lubrication on the fictional variation involved in the sheet-metal-forming process. *Sheet Met. Ind.* **1964**, *41*, 925–1930.
14. Sigvant, M.; Pilthammar, J.; Hol, J.; Wiebenga, J.H.; Chezan, T.; Carleer, B.; van den Boogaard, T. Friction in sheet metal forming: Influence of surface roughness and strain rate on sheet metal forming simulation results. *Procedia Manuf.* **2019**, *29*, 512–519. [CrossRef]
15. Gierzyńska, M. *Tarcie, Zużycie i Smarowanie w Obróbce Plastycznej Metali*; Wydawnictwa Naukowo-Techniczne: Warszawa, Poland, 1983.
16. Kim, H.; Han, S.; Yan, Q.; Altan, T. Evaluation of tool materials, coatings and lubricants in forming galvanized advanced high strength steels (AHSS). *CIRP Annals-Manuf. Technol.* **2008**, *57*, 299–304.
17. Ludema, K.C. *Friction, Wear, Lubrication. A Textbook in Tribology*; CRC Press: Boca Raton, FL, USA, 2000.
18. Wihlborg, A.H. The Influence of Steel Sheet Surface Topography on Friction in Stamping. Ph.D. Thesis, Göteborg University, Göteborg, Sweden, 1997.

19. Trzepieciński, T.; FejkielR. On the influence of deformation of deep drawing quality steel sheet on surface topography and friction. *Tribol. Int.* **2017**, *115*, 78–88. [CrossRef]
20. Masters, L.G.; Williams, D.K.; Roy, R. Friction behaviour in strip draw test of pre-stretched high strength automotive aluminium alloys. *Int. J. Mach. Tools Manuf.* **2013**, *73*, 17–24. [CrossRef]
21. Evin, E.; Tomáš, M.; Výrostek, M. Verification the numerical simulation of the strip drawing test by its physical model. *Acta Mechanica Slovaca* **2016**, *20*, 14–21. [CrossRef]
22. Trzepieciński, T. A study of the coefficient of friction in steel sheets forming. *Metals* **2019**, *9*, 988. [CrossRef]
23. Trzepieciński, T. Tribological Performance of Environmentally Friendly Bio-Degradable Lubricants Based on a Combination of Boric Acid and Bio-Based Oils. *Materials* **2020**, *13*, 3892. [CrossRef]
24. Severo, V.; Vilhena, L.; Silva, P.N.; Dias, J.P.; Becker, D.; Wagner, S.; Cavaleiro, A. Tribological behaviour of W–Ti–N coatings in semi-industrial strip-drawing tests. *J. Mater. Process. Technol.* **2009**, *209*, 4662–4667. [CrossRef]
25. Tavares, A.F.; Lopes, A.P.O.; Mesquita, E.A.; Almeida, D.T.; Souza, J.H.C.; Costa, H.L. Effect of transfer layers on friction and wear mechanisms in strip drawing tests of commercially coated forming tools. *Wear* **2021**, *476*, 203733. [CrossRef]
26. Schell, L.; Emele, M.; Holzbeck, A.; Groche, P. Investigation of different lubricant classes for aluminium warm and hot forming based on a strip drawing test. *Tribol. Int.* **2022**, *168*, 107449. [CrossRef]
27. Sutcliffe, M.P.F.; Le, H.R.; Farrugia, D. Simulation of transfer layer formation in strip drawing of stainless steel. *Wear* **2003**, *254*, 523–531. [CrossRef]
28. *EN 10130:2006*; Cold Rolled Low Carbon Steel Flat Products for Cold Forming. Technical Delivery Conditions; British Standard: London, UK, 2006.
29. *EN ISO 6892-1:2020*; International Organization for Standardization: Geneva, Switzerland, 2020.
30. Domitner, J.; Silvayeh, Z.; Sabet, A.S.; Öksüz, K.I.; Pelcastre, L.; Hardell, J. Characterization of wear and friction between tool steel and aluminum alloys in sheet forming at room temperature. *J. Manuf. Proc.* **2021**, *64*, 774–784. [CrossRef]
31. Makhamov, A. Determination of the Friction Coefficient in the Flat Strip Drawing Test. *Engineering* **2021**, *13*, 595–604. [CrossRef]
32. Prakash, V.; Kumar, D.R. Performance evaluation of bio-lubricants in strip drawing and deep drawing of an aluminium alloy. *Adv. Mater. Process. Technol.* **2020**, 1–15. [CrossRef]
33. Roizard, X.; Pothier, J.M.; Hihn, J.Y.; Monteil, G. Experimental device for tribological measurement aspects in deep drawing process. *J. Mater. Processing Technol.* **2009**, *209*, 1220–1230. [CrossRef]
34. Le, H.R.; Sutcliffe, M.P.F. Measurements of friction in strip drawing under thin film lubrication. *Tribol. Int.* **2002**, *35*, 123–128. [CrossRef]
35. Cillaurren, J.; Galdos, L.; Sanchez, M.; Zabala, A.; de Argandoña, S.; Mendiguren, J. Contact pressure and sliding velocity ranges in sheet metal forming simulations. In Proceedings of the 24th International Conference on Material Forming ESAFORM, Liege, Belgium, 14–16 April 2021.
36. Erbel, S.; Kuczyński, K.; Marciniak, Z. *Obróbka Plastyczna na Zimno*; PWN: Warszawa, Poland, 1975.
37. Trzepieciński, T. *Wybrane Aspekty Oceny Tribologicznej Blach Odkształcanych Plastycznie*; Oficyna Wydawnicza Politechniki Rzeszowskiej: Rzeszów, Poland, 2013.
38. Trzepieciński, T.; Szpunar, M. Assessment of the effectiveness of lubrication of Ti-6Al-4V titanium alloy sheets using radial basis function neural networks. *Acta Polytech.* **2021**, *61*, 489–496. [CrossRef]
39. Payen, G.R.; Felder, E.; Repoux, M.; Mataigne, J.M. Infuence of contact pressure and boundary films on the frictional behaviour and on the roughness changes of galvanized steel sheets. *Wear* **2012**, *276–277*, 48–52. [CrossRef]
40. Kirkhorn, L.; Frogner, K.; Andersson, M.; Stahl, J.E. Improved tribotesting for sheet metal forming. *Procedia CIRP* **2012**, *3*, 507–512. [CrossRef]
41. Trzepieciński, T.; Bazan, A.; Lemu, H.G. Frictional characteristics of steel sheets used in automotive industry. *Int. J. Automot. Technol.* **2015**, *16*, 849–863. [CrossRef]
42. Vollertsen, F.; Hu, Z. Tribological Size Effects in Sheet Metal Forming Measured by a Strip Drawing Test. *CIRP Annals* **2006**, *55*, 291–294. [CrossRef]
43. ten Thije, R.H.W.; Akkerman, R.; van der Meer, L.; Ubbink, M.P. Tool-ply friction in thermoplastic composite forming. *Int. J. Mater. Form.* **2008**, *1*, 953–956. [CrossRef]
44. Jurkovic, M.; Jurkovic, Z.; Buljan, S. The tribological state test in metal forming processes using experiment and modelling. *J. Achiev. Mater. Manuf. Eng.* **2006**, *18*, 383–386.
45. Wang, D.; Yang, H.; Li, H. Advance and trend of friction study in plastic forming. *Trans. Nonferr. Met. Soc. China* **2014**, *24*, 1263–1272. [CrossRef]
46. Gåård, A. Wear in Sheet Metal Forming. Ph.D. Thesis, Kalstad University Studies, Karlstad, Sweden, 2018.

materials MDPI

Article

Optimizing the Thermomechanical Process of Nickel-Based ODS Superalloys by an Efficient Method

Wuqiang He [1,2], Feng Liu [1,2], Liming Tan [1,2], Zhihui Tian [3], Zijun Qin [1,2], Lan Huang [1,2], Xiangyou Xiao [1,2], Guowei Wang [1,2], Pan Chen [4,*] and Baogang Liu [5,*]

1 State Key Laboratory of Powder Metallurgy, Central South University, Changsha 410083, China; 183301019@csu.edu.cn (W.H.); liufeng@csu.edu.cn (F.L.); limingtan@csu.edu.cn (L.T.); zijun.qin@csu.edu.cn (Z.Q.); lhuang@csu.edu.cn (L.H.); xiangyou.xiao@csu.edu.cn (X.X.); wangguowei23@csu.edu.cn (G.W.)
2 Research Institute of Powder Metallurgy, Central South University, Changsha 410083, China
3 College of Science, Huazhong Agriculture University, Wuhan 430070, China; 2017301210207@webmail.hzau.edu.cn
4 School of Minerals Processing and Bioengineering, Central South University, Changsha 410083, China
5 School of Energy and Electromechanical Engineering, Hunan University of Humanities, Science and Technology, Loudi 417000, China
* Correspondence: panchen@csu.edu.cn (P.C.); liudd2016@126.com (B.L.); Tel.: +86-0731-8883-0938 (P.C.); +86-0731-8883-0938 (B.L.)

Abstract: Thermo-mechanical process of nickel-based oxide dispersion strengthened (ODS) superalloys is critical to produce desired components. In this study, an efficient method of consolidating powder is introduced to optimize the preparation process, microstructure and properties of nickel-based ODS superalloys. The influences of consolidation temperature, strain rate and ball milling time on the hardness of nickel-based superalloys were studied. The relationship among process, microstructure and hardness was established, the nanoparticles strengthening and grain boundary strengthening in nickel-based ODS superalloys were discussed. The results indicate that long ball milling time, moderately low consolidation temperature and high strain rates are beneficial to improving properties of nickel-based superalloys. Moreover, dispersion strengthening of nanoparticles and grain boundary strengthening play important roles in enhancing nickel-based ODS superalloys.

Keywords: nickel-based ODS superalloys; consolidation; strengthening; hardness

Citation: He, W.; Liu, F.; Tan, L.; Tian, Z.; Qin, Z.; Huang, L.; Xiao, X.; Wang, G.; Chen, P.; Liu, B. Optimizing the Thermomechanical Process of Nickel-Based ODS Superalloys by an Efficient Method. *Materials* **2022**, *15*, 4087. https://doi.org/10.3390/ma15124087

Academic Editors: Tomasz Trzepieciński and Valentin Ştefan Oleksik

Received: 21 April 2022
Accepted: 26 May 2022
Published: 9 June 2022

1. Introduction

Nickel-based oxide dispersion strengthened (ODS) superalloys have been considered as promising candidate materials such as gas turbines, heat exchanger tubes and nuclear reactors due to their excellent radiation resistance, comprehensive mechanical properties and high temperature creep strength [1–5]. These enhanced properties are mainly attributed to high number density nanoscale particles, which can pin grain boundaries and dislocations [6–8]. The addition such as YH_2 and Y_2O_3 is dissociated to Y during the mechanical alloying (MA) process, then Y reacts with and Al or Ti in the matrix to form nanoscale Y-Al-rich or Y-Ti-rich oxides (e.g., $YAlO_3$, $Y_3Al_5O_{12}$,$Y_4Al_2O_9$, $Y_2Ti_2O_7$, Y_2TiO_5) in consolidation process [9]. The hot extrusion (HEX) and hot isostatic pressing (HIP) are commonly used consolidation processes to prepare nickel-based ODS superalloys. However, the preparation of ball milling and consolidation of powder usually takes several months and thus it is costly, which limits the development of ODS superalloys.

It is well known that the properties of nickel-based superalloys including nickel-based ODS superalloys are strongly dependent on the processing parameters. For example, Kim et al. [10] investigated the influence of several processing parameters including milling temperature, rotation speed, and consolidation temperature on microstructure and properties of 14Cr ODS steels alloy. As milling temperature decreases, milling speed increases, and

HIP temperature decreases, the microstructures of samples become finer and more uniform, and considerable improvement of tensile strength was noticed. Kishimoto et al. [11] and Odette et al. [12,13] reported that with the increase of HIP temperature the grain size of the alloy grew significantly, and the dislocation density decreased, the average particle size of oxides increased and the density of oxides decreased. Tan et al. [14] studied the effects of HIP temperature and pressure on the grain in PM superalloy FGH96, and it was found that temperature and pressure played different roles in controlling prior particle boundary (PPB) precipitation and grain structure during HIP, the tendency of grain coarsening under high temperature could be inhibited by increasing HIP pressure which facilitated the re-crystallization. It is essential to optimize the processing parameters to obtain components with desired microstructure and properties. Therefore, to accelerate the development of nickel-based superalloys, some efficient experimental methods need to be developed [15].

In this study, a new efficient thermal consolidation method was used to prepare nickel-based superalloys with and without ODS. It takes only a few minutes and a few grams of powder to prepare a sample by this method, which dramatically shortens the preparation cycle and reduces the cost. In addition, the influences of consolidation temperature, strain rate and ball milling time on the hardness of nickel-based superalloys were studied, and the relationship of process, microstructure and properties was established, the strengthening mechanism of nickel-based superalloys with and without ODS were discussed.

2. Materials and Experiment Methods

The chemical composition of two nickel-based powders (PA1 and PA2) prepared by argon atomization in this work was given in Table 1. MA powder was prepared by ball milling PA powders and 0–0.6 wt.% YH_2 powders for 0–36 h with speed 350 rpm in high-energy planetary mill filled with high purity argon gas (99.999%). The ball-to-powder mass ratio was 8:1 and 0.5 wt.% ethanol was added as process control agent. The MIX powder was prepared by mixing MA and PA powder at ratio 1:2 (wt.%) for 12 h with speed 60 rpm. The laser particle size analyzer (Mastersizer 3000, Malvern Panalytical, Malvern, UK) was adopted to detected the particle size distribution of powder.

Table 1. Chemical compositions of PA1 and PA2 powder in wt.%.

	Al	Cr	Fe	Ti	Y	C	Ni
PA1	0.25	20.5	0.67	0.56	0	0.054	Bal.
PA2	0.25	21	0.85	0.57	0.68	0.059	Bal.

The powder was sealed in stainless steel container with 10 mm diameter and 15 mm height, vacuum electron beam welding is adopted to welded container under a vacuum of 10^{-2} torr. Thereafter, the container was compressed by Gleeble 3180D at temperatures ranging from 850 °C to 1150 °C with strain rates 0.1 s^{-1}–5 s^{-1}. Carbon sheets and lubricant were placed between specimen and dies to make the deformation more uniform during compression. The specimens were originally heated to certain temperatures with a rate of 5 °C/s and held for 3 min to homogenize temperature of specimen. To monitor the temperature during the compression, the thermal-couples were welded at the longitudinal center of the specimen surface. Finally, the specimen was compressed according to the pre-set procedure and immediately water quenched to freeze the deformation microstructure. All specimens were compressed to a total engineering strain of 80% with constant strain rates. The process flow diagram of specimens prepared is indicated in Figure 1, and the processing methods of specimens have been compared in Table 2.

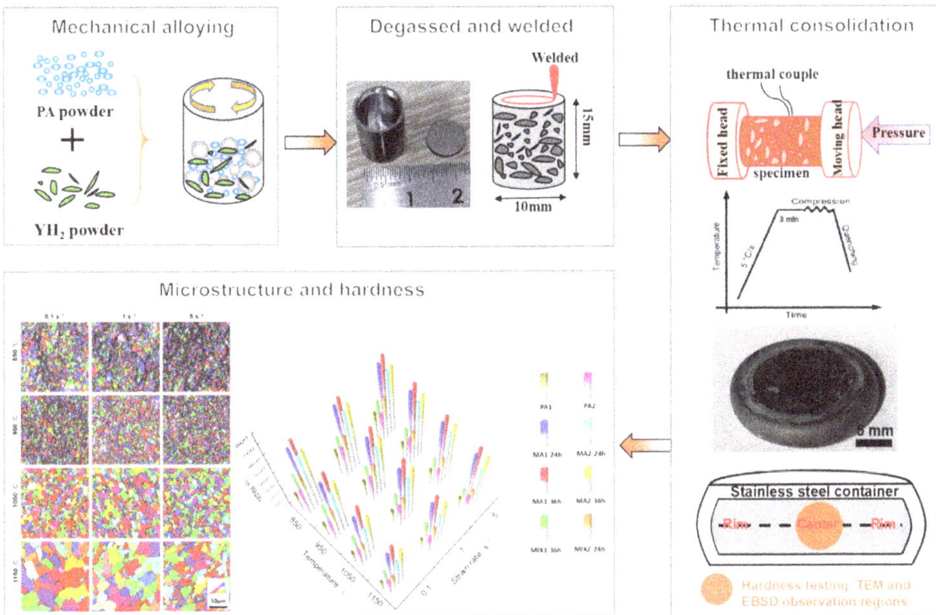

Figure 1. Flow-process diagram of samples preparation.

Table 2. Comparison of processing methods of specimens.

Specimen	Powder	Mill Time/h	Temperature/°C	Strain Rate/s^{-1}
PA1	PA1 powder	0		
MA1-24h	PA1 powder and 0.6%YH_2 powder	24		
MA1-36h	PA1 powder and 0.6%YH_2 powder	36		
MIX1-36h	MA1-36h powder and PA2 powder (ratio 1:2)	12 (mixed)	850/950/1050/1150	0.1/1/5
PA2	PA2 powder	0		
MA2-24h	PA2 powder	24		
MA2-36h	PA2 powder	36		
MIX2-24h	MA2-24h powder and PA2 (ratio 1:2)	12 (mixed)		

The phases of powder and alloys were investigated using X-ray diffractometer (XRD, D/MAX 2500, RIGAKU, Tokyo, Japan) with Cu Kα radiation of 0.15405 nm. The diffraction Angle (θ) ranges from 20° to 80° with scanning rate of 1 °/min.

The Vickers microhardness measurements were performed by using vickers hardness tester (THV-10, TEST-TECH, Shanghai, China) at a load of 3000 g and a dwell time of 10 s. The standard deviation was calculated based on 10 measurements.

The characterization of microstructure alloys was observed by optical microscopy (OM, DM4000M, LEICA, Germany) and field-emission SEM (Quanta 650, FEI, Hillsboro, OR, USA) equipped with an electron backscatter diffraction (EBSD), the central of samples was used to analyzed, as shown in Figure 1. The specimens were mechanically polished by abrasive papers and 50 nm aluminium oxide, the specimens for OM observation were etched by a solution of 100 mL ethanol + 100 mL HCl + 5 g $CuCl_2$, and the specimens for EBSD observation were vibration polished for 3–8 h. The vibration polishing was conducted on Buehler vibratory polisher VibroMet 2 (Buehler, USA) with 30% amplitude. The EBSD was conducted at 20 kV accelerating voltage, 0.05 or 1 μm step size, the data was analyzed via HKL Channel 5 software and the equivalent grain size was determined based on grain area A, $2\sqrt{A/\pi}$.

Transmission electron microscope (TEM, Themis Z 3.2, FEI, USA) observation with energy dispersive spectrometer (EDS) was used to characterize the structure of specimens, under 200 kV accelerating voltage. The TEM samples were polished to a thickness of 50 μm by abrasive papers, and sectioned to diameter 3mm. Then the slices were twin-jet electropolished at −25 °C and 40 V in the corrosive solution of 90% ethanol and 10% perchloric acid.

3. Results and Discussion

3.1. Morphologies of Powder Analysis

Figure 2a,c show the SEM images of precursor powders, it is evident that PA powder is spherical or subspherical, the EBSD inverse pole figure (IPF), grain distribution and grain misorientation are shown in Figure 2b–f. The grain size ranges from 0.5 μm to 18 μm, the average grain size of PA1 and PA2 powder is 4.8 μm and 5.8 μm, and the average grain misorientation of PA1 and PA2 powder is 33.8 and 31.4, respectively.

The microstructure and particle size distribution of powder with different ball milling time are shown in Figure 3. The particle size of PA powder ranges in 1–170 μm, the powder particle is irregular morphology after ball milling, because of the repeated and high-speed collision between powder and powder or powder and ball milling medium during MA. Large plastic deformation of powder after crushing during the ball milling process. Within ranges of 20–600 μm, the particle size increases with the increasing ball milling time, specifically, the average particle sizes of PA1 and PA2 powder are 182 μm and 224 μm after ball milling 24 h, and these of PA1 and PA2 powder increase to 205 μm and 245 μm after 36 h ball milling. The average particles size of powder increases due to welding of powders during ball milling.

3.2. Macroscopic Cracking Analysis

96 specimens were prepared at temperatures ranging from 850 °C to 1150 °C with strain rates of 0.1–5 s^{-1}. The shapes of the specimens after hot compression are shown in Figure 4. From macroscopic perspective, most of the specimens are regular pie with obvious bulging in the middle, which is due to the non-uniform deformation caused by the friction between the specimen and workpiece during hot compression [16]. Some specimens with obviously macroscopic failure features (marked by the red box) are observed, these specimens are compressed at the low temperature, the cracking percentage of the samples is 14.6% and 5.2% compressed at 850 °C and 950 °C, respectively. No macroscopic cracking was observed in samples consolidated at 1050 °C and 1150 °C.

Figure 5 depicts some typical features of PA1 and MA1-36h alloys consolidated at different temperatures with strain rate of 0.1 s^{-1}. The microscopic cracks along the PPB are observed in PA1 alloys and MA1-36h alloys at 850 °C corresponding to Figure 5a,c, respectively. Generally, PPB are difficult to be broken and eliminated at low temperature, which facilitates the crack nucleation and propagation during thermal consolidation [17]. The microcrack along the PPB are mainly caused by local stress concentration [18,19]. With the increase of consolidation temperature, the thermal activation of the material increases, the atomic diffusion rate accelerates, and the kinetic energy of the atoms increases, weakening the binding force between atoms, the softening effect of the material becomes obvious. The PPB is broken and eliminated quickly at elevated temperature and high stress, replaced by dynamic recrystallized grains, and good metallurgical bonding between the powder particles is achieved [20]. No PPB and macroscopic cracks are observed in PA1 and MA1-36h alloy consolidated at 1050 °C.

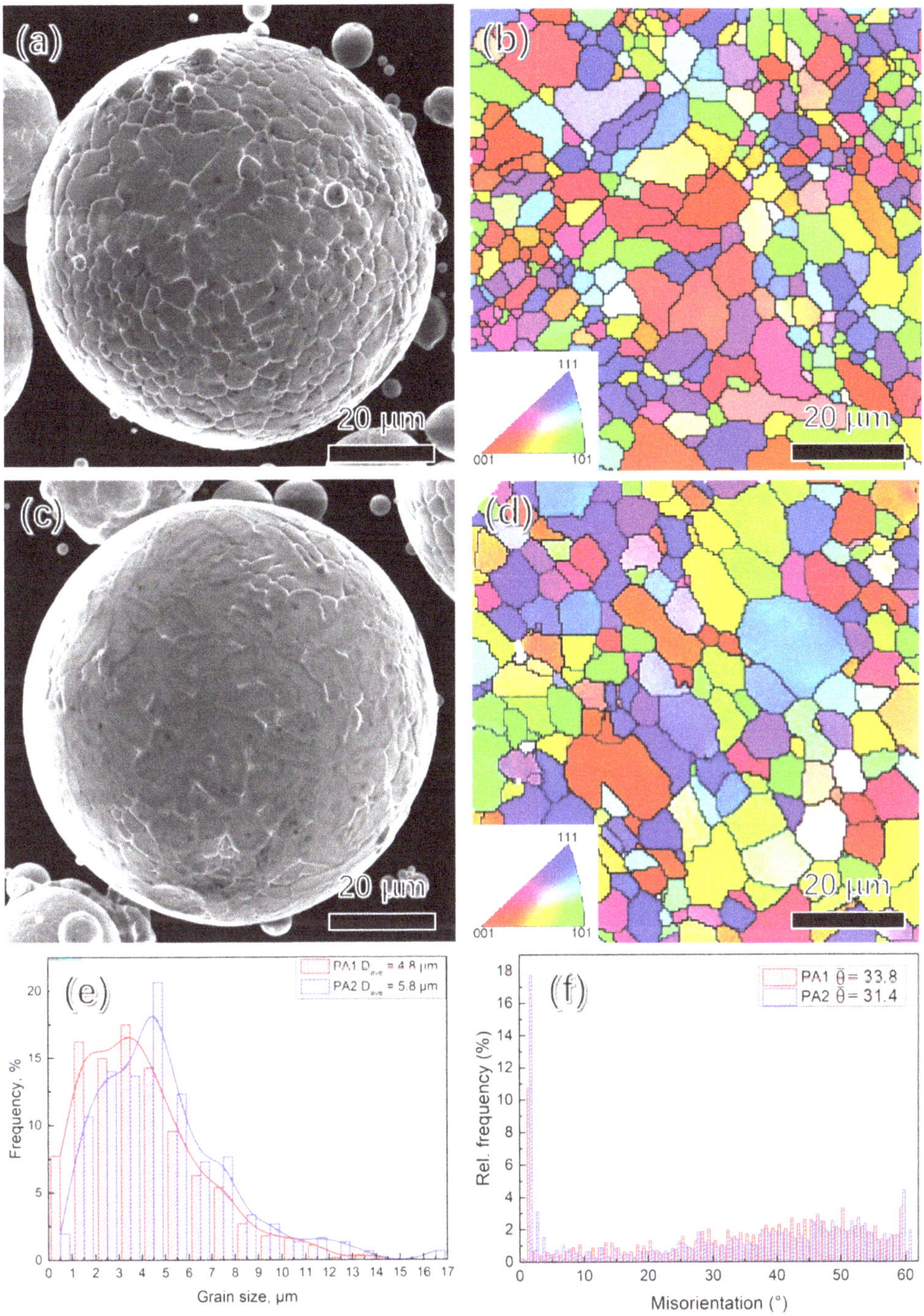

Figure 2. The SEM and EBSD IPF images of powders: (**a**,**b**) PA1; (**c**,**d**) PA2; (**e**) grain size distribution; (**f**) grain misorientation.

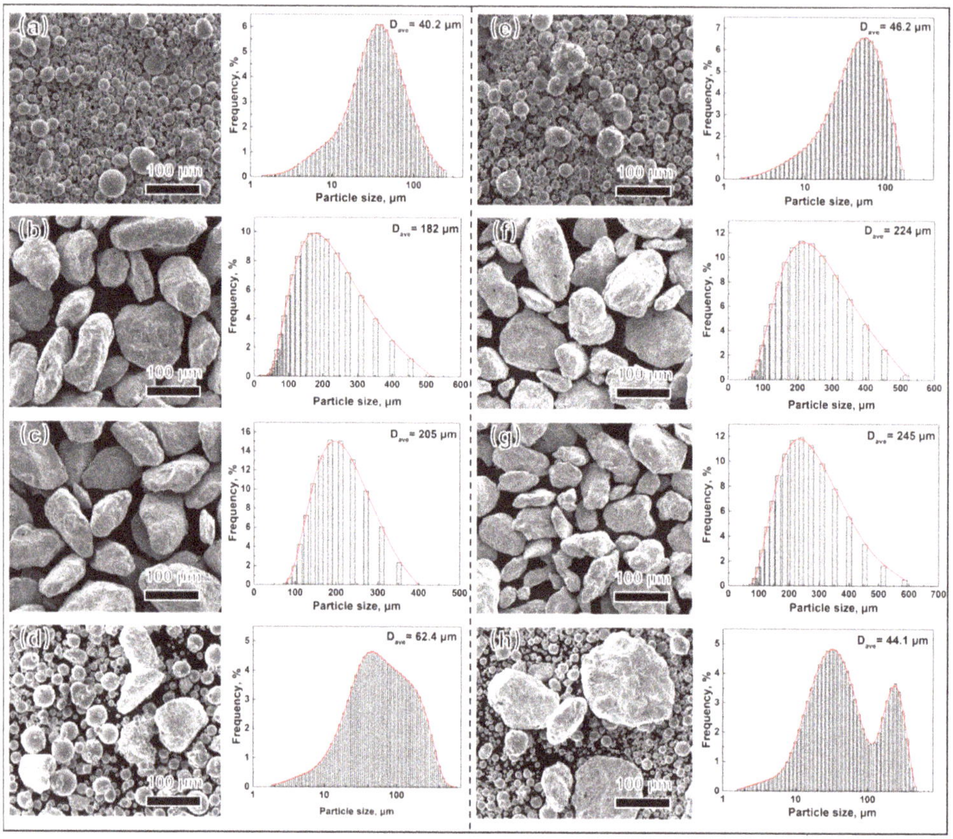

Figure 3. The microstructural feature and particle size distribution for (**a**) PA1, (**b**) MA1-24h, (**c**) MA1-36h, (**d**) MIX1-36h, (**e**) PA2, (**f**) MA2-24h, (**g**) MA2-36h, (**h**) MIX2-24h, respectively.

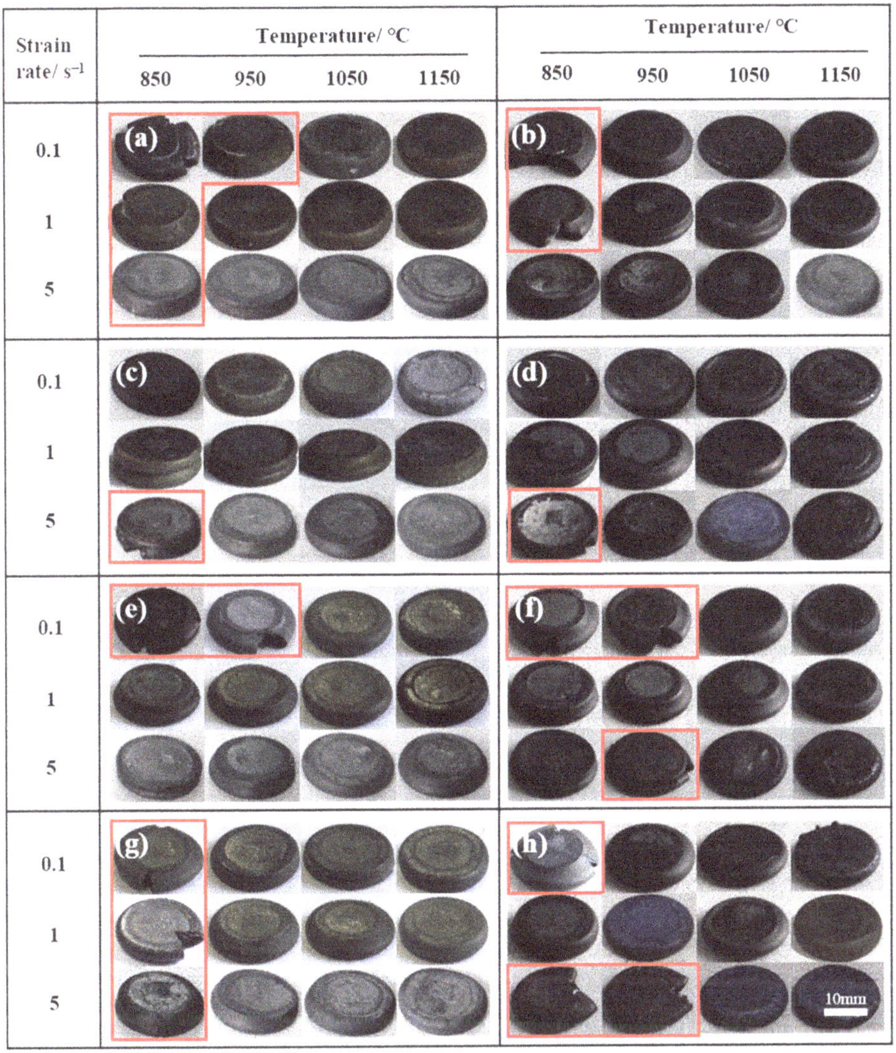

Figure 4. The shapes of specimens compressed at different conditions: (**a**) PA1; (**b**) PA2; (**c**) MA1-24 h; (**d**) MA2-24 h; (**e**) MA1-36 h; (**f**) MA2-36 h; (**g**) MIX1-36 h; (**h**) MIX2-24 h.

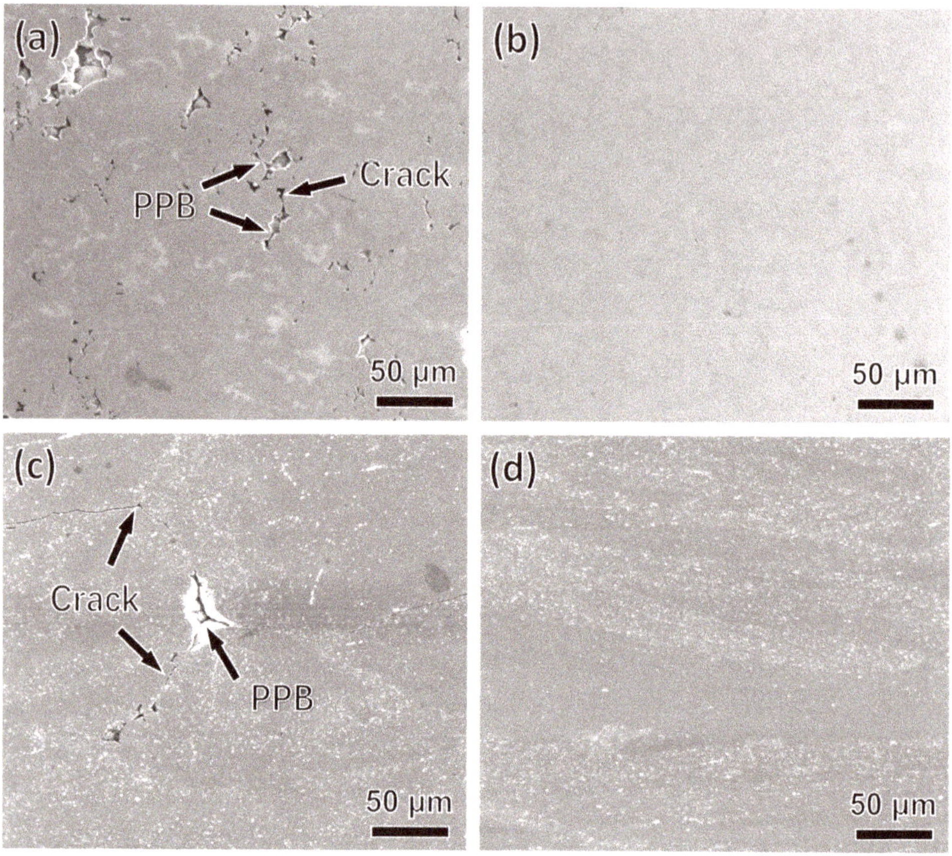

Figure 5. The SEM images of failure specimens consolidated at different temperatures with strain rate 0.1 s^{-1}: (**a**) PA1: 850 °C; (**b**) PA1: 1050 °C; (**c**) MA1-36h: 850 °C; (**d**) MA1-36h: 1050 °C.

The formation of crack in hot compression are described by the three steps in the schematic, as shown in Figure 6. At initial stages, loose powder began to slide and rearrangement under low pressure. Thereafter, as the stress increase, the plastic deformation of powder gradually happens, when the stress reaches the yield point of the powder. As hot compress continue, strain increases gradually, dynamic recrystallization (DRX) occurs. However, when hot compression temperature is low, PPB can hardly be broken, and gradually is stretched into ellipsoid as strain increases [20], leading to the concentration of stress at PPB, microcracks generates and expands along the PPB, leading to the formation of macroscopic cracks eventually.

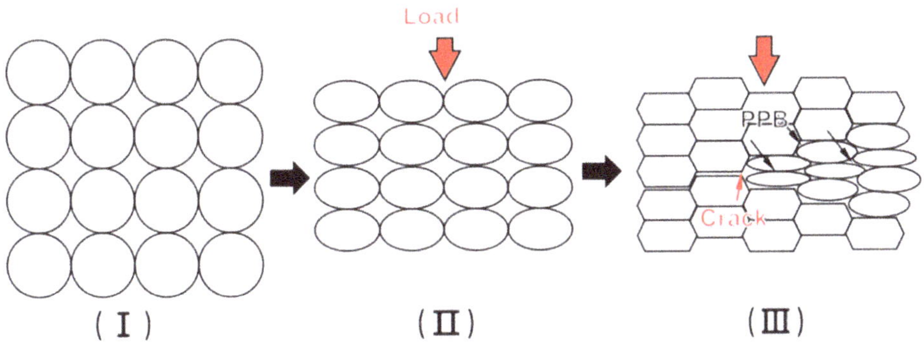

Figure 6. Schematic illustration of formation of compression cracks during hot compression. (**I**) prior to deformation, (**II**) deformation, (**III**) the formation of cracks.

3.3. XRD of the Powder and Alloys Analysis

Figure 7a shows the XRD patterns of precursor and ball milling powder. γ-Ni and YH_2 diffraction peaks are visible in the mixture powder of PA and YH_2 powder. After ball milling 36 h, only γ-Ni diffraction peak are observed in MA powder, YH_2 is unstable and dissociated to Y and H during the MA, which is consistent with other studies [21]. Meanwhile, the impact energy generated by the ball milling results in a reduction of grain size and an accumulation of the lattice strain, which leads to the decrease of diffraction peak intensity, increase of width and angular deviation [22]. The XRD patterns of PA1 and MA1-36h alloy consolidated at 1050 °C/0.1 s^{-1} is illustrated in Figure 7b, the diffraction peaks of $Y_4Al_2O_9$ and TiO_2 are detected in MA1-36h alloy, only the diffraction of TiO_2 can be found in PA1 alloy.

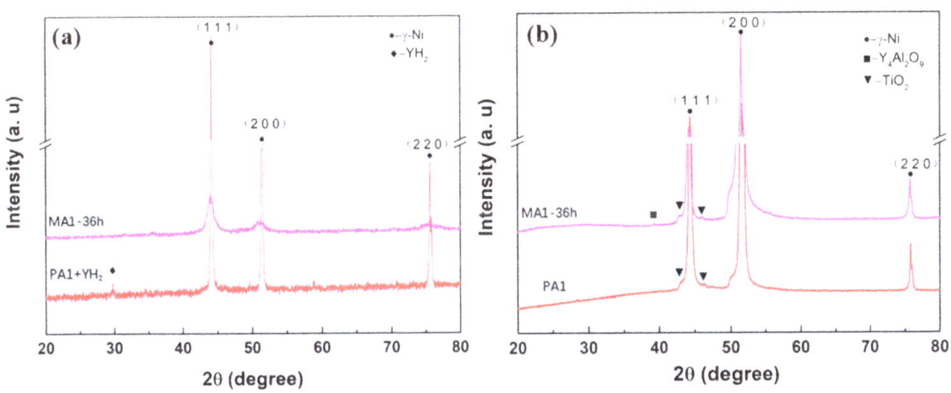

Figure 7. XRD patterns of (**a**) powders milled for different times and (**b**) samples consolidated at 1050 °C/0.1 s^{-1}.

3.4. Microstructure Characterization

The grains at center regions of PA1 and MA1-36h alloy consolidated at different conditions are presented by EBSD IPF, as shown in Figures 8 and 9. Generally, the samples consolidated at higher temperatures and lower strain rates have large average grain size. Significant grain growth happens when the temperature excesses 950 °C, especially at low strain rate. As the consolidation temperature increases from 850 °C to 1150 °C at strain rate of 0.1 s^{-1}, the average grain size of PA1 alloy increases from 2.76 μm to 16.5 μm, and that of the MA1-36h alloy increases from 0.18 μm to 0.31 μm. The grain evolution is generally controlled by dynamic recovery (DRV), DRX, and grain growth processes [23–26]. For the PA1 alloy, comparing with the initial microstructure of precursor powder of PA1 (Figure 2b), grains are not uniform and some large grains only deform under stress rates of 0.1 s^{-1} and 1 s^{-1} at 850 °C. Grains of alloy consolidated at 950 °C are more uniform than that of the samples consolidated at 850 °C, with the increase of thermal activation. Grain growth plays a dominating role during the microstructure evolution when the temperature excesses 950 °C. Comparing with microstructure of the PA1 alloys, the grains of MA1-36h alloys consolidated at 850–950 °C keep fine during consolidation, and the average grain size fluctuates slightly. The grains grow gradually in samples as the consolidation temperature increasing to 1050 °C. Moreover, it is clear that the grains decrease from micron to submicron by comparing PA1 alloy with MA1-36 alloy, the grains are refined by ball milling. The grain growth rate of PA1 alloy is more rapidly than that of MA1-36h alloy with the temperature above 1050 °C, mechanism behind that will be discussed thereafter. In terms of strain rate, the strain rate has ambiguous effects on the grain evolution at 850–950 °C, as the consolidation temperature excesses 1050 °C, the average grain size of samples consolidated at 5 s^{-1} is smaller than that of other strain rates, and at a certain temperature, grains in alloys deformed at 5 s^{-1} are more uniform than these of the counterparts consolidated at other strain rates. Accordingly, the consolidation temperature of near 1050 °C with strain rate of 5 s^{-1} is suggested, to obtain homogeneous distributed fine-grains.

Figure 10 shows the grain misorientations of PA1 and MA1-36h alloys consolidated at different conditions. The influence of consolidation conditions on the grain evolution can be reflected by grain boundary misorientation [27]. The dislocation occurs in the process of consolidation, which leads to the misorientation in grains, and forming low angle grain boundaries (LAGBs, <15°). The higher the dislocation density, the larger the proportion of LAGBs. DRX is triggered, when the dislocation density reaches critical dislocation density. The dislocation is absorbed during the grain growth, resulting in the decrease of dislocation density, the fraction of LAGBs decreases. With sufficient driving force and time, DRX continues to occur and the dislocation density decreases further. For the PA1 alloy, comparing with PA1 powder (Figure 2), the average grain misorientation of samples decreases, as shown in Figure 10a. Generally, As the consolidation temperature increases from 850 °C to 1050 °C, the thermal activation increases, the fraction of high angle grain boundary (HAGBs, >15°) and the average grain misorientation of samples increases. However, the average grain misorientation decreases at 1150 °C, it is speculated that the grain growth leads to the decrease of the grain quantity in the same area. For the MA1-36h alloy, the average grain misorientation of samples increases with the increase of temperature except 1150 °C/1 s^{-1} (Figure 10b).

In addition, Figure 11 shows the TEM bright field images of PA1 and MA1-36h alloy consolidated at 1150 °C/5 s^{-1}. Figure 11a shows a large-grain region with a high dislocation density in the PA1 alloy. The dislocations have long, relatively straight-line segments, with sharp corners and a serrated appearance in the other locations. Figure 11b illustrates the microstructure of MA1-36h alloy, significant nanoparticles and dislocations are noticed in sample, the dislocations interact with the nanoparticles, indicating that they are strongly pinned by nanoparticles. The nanoparticles are distributed in grain and grain boundary, ranging in size from a few nanometers to tens of nanometers.

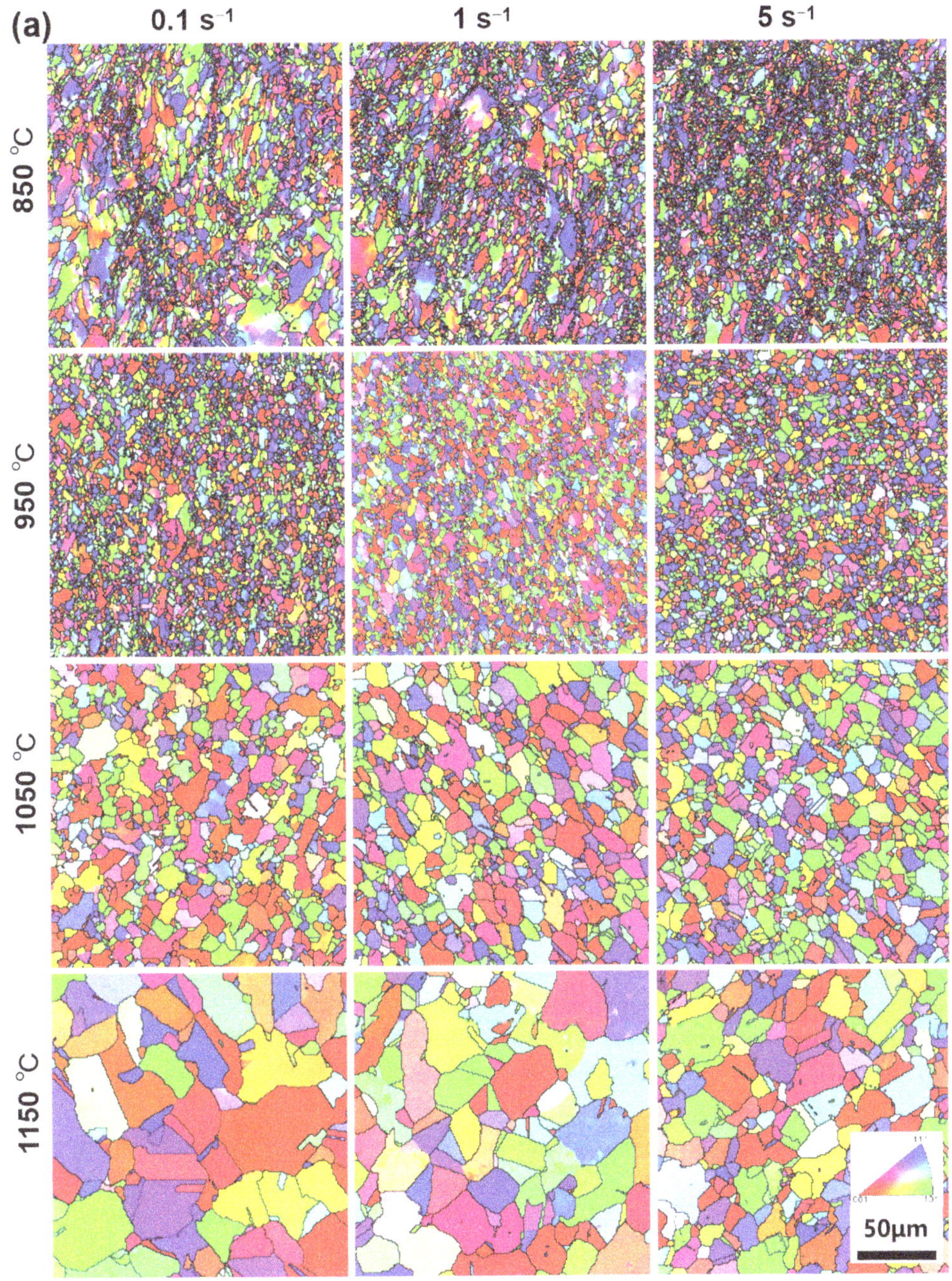

Figure 8. *Cont.*

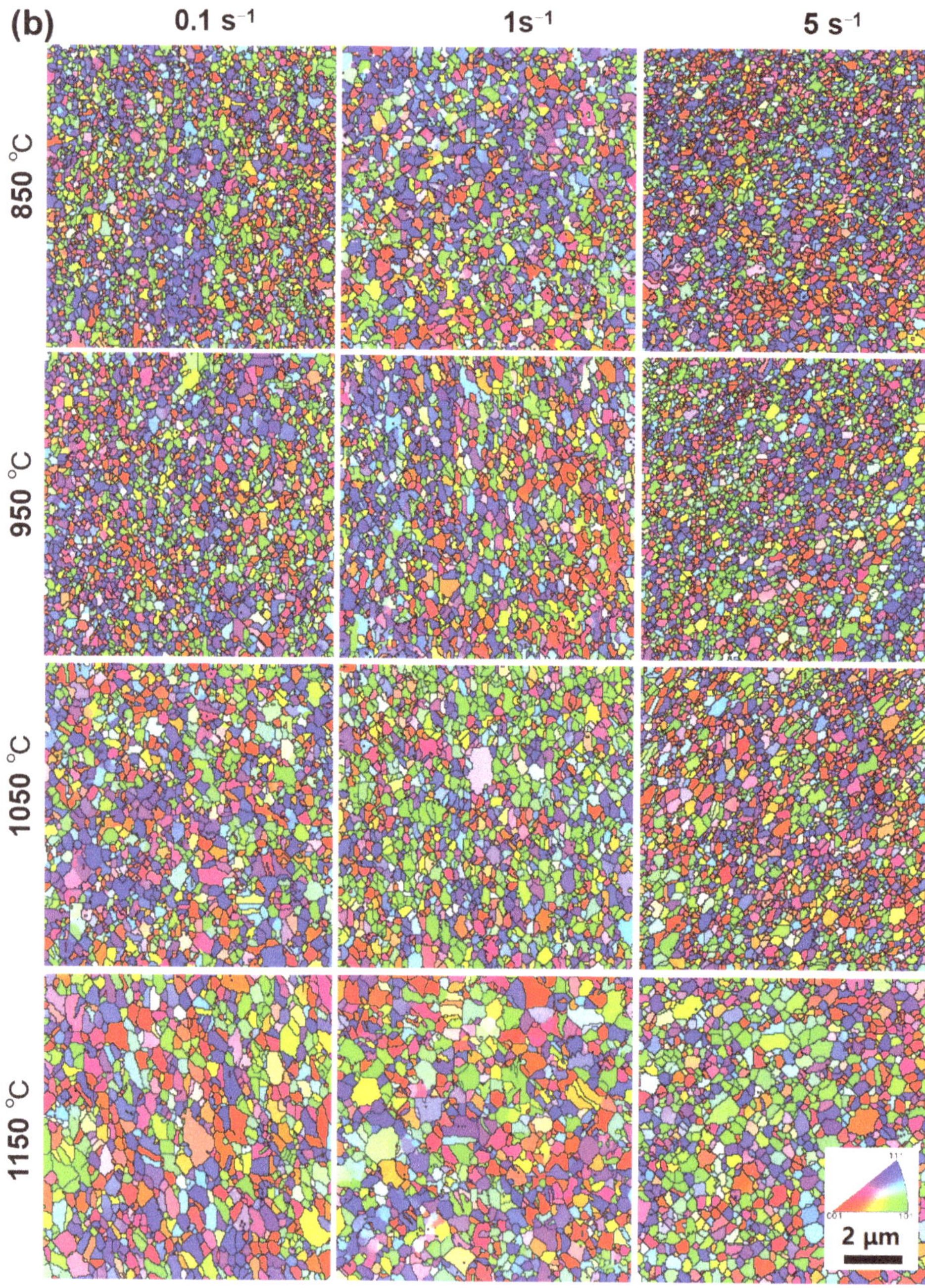

Figure 8. EBSD maps at center region of (**a**) PA1 and (**b**) MA1-36h alloys consolidated at different conditions.

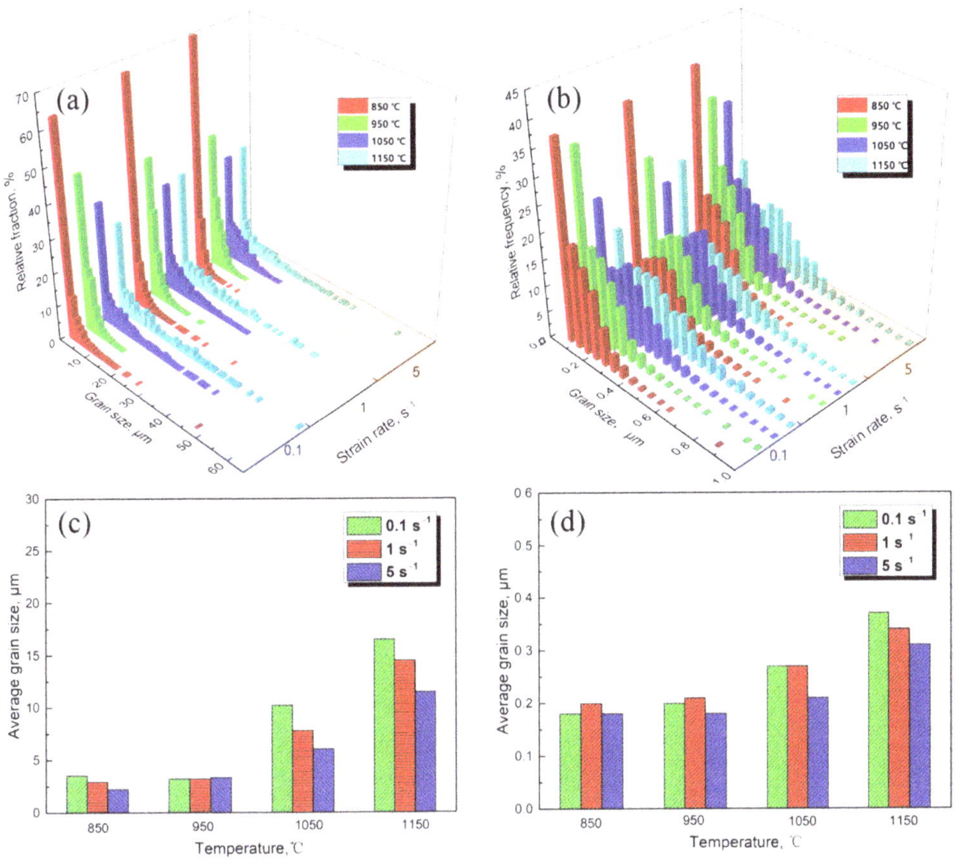

Figure 9. The equivalent grain size distributions and average grain sizes of (**a**,**c**) PA1 and (**b**,**d**) MA1-36h alloys consolidated at different conditions.

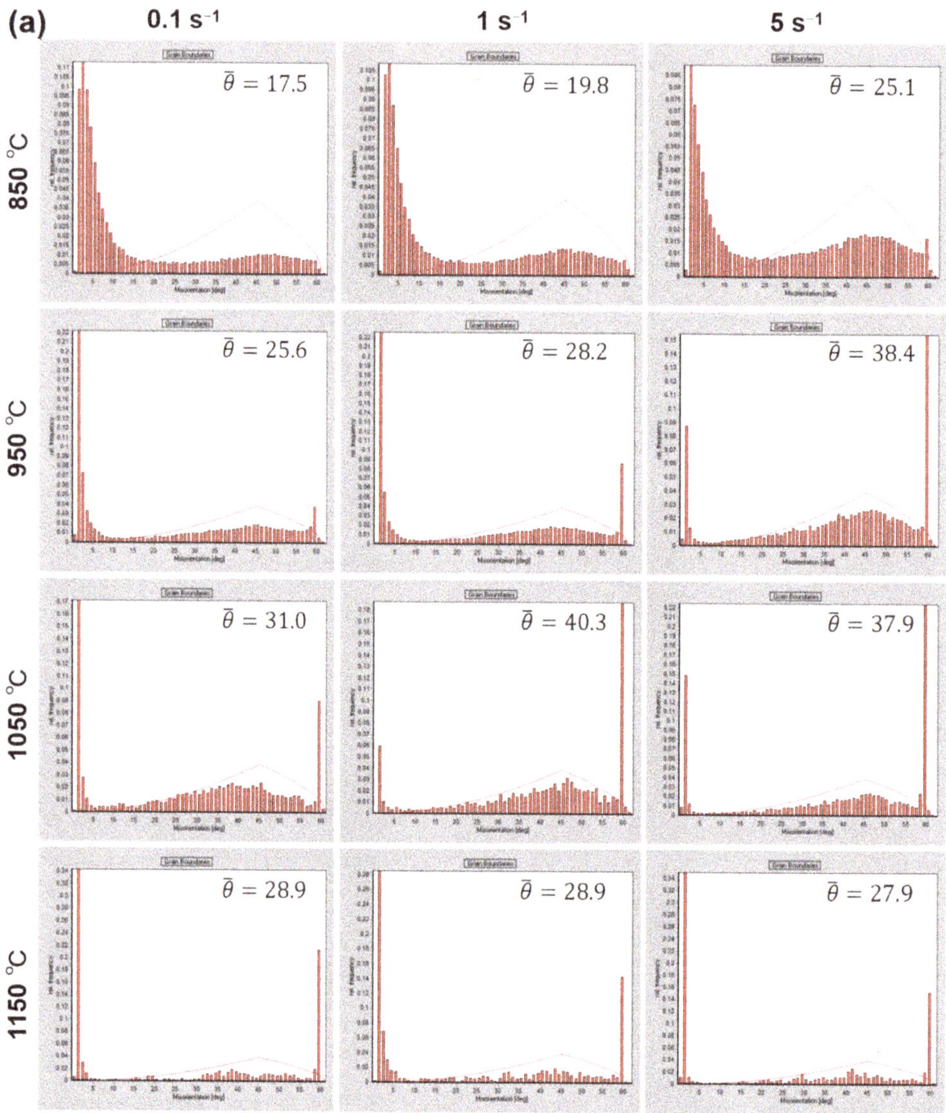

Figure 10. *Cont.*

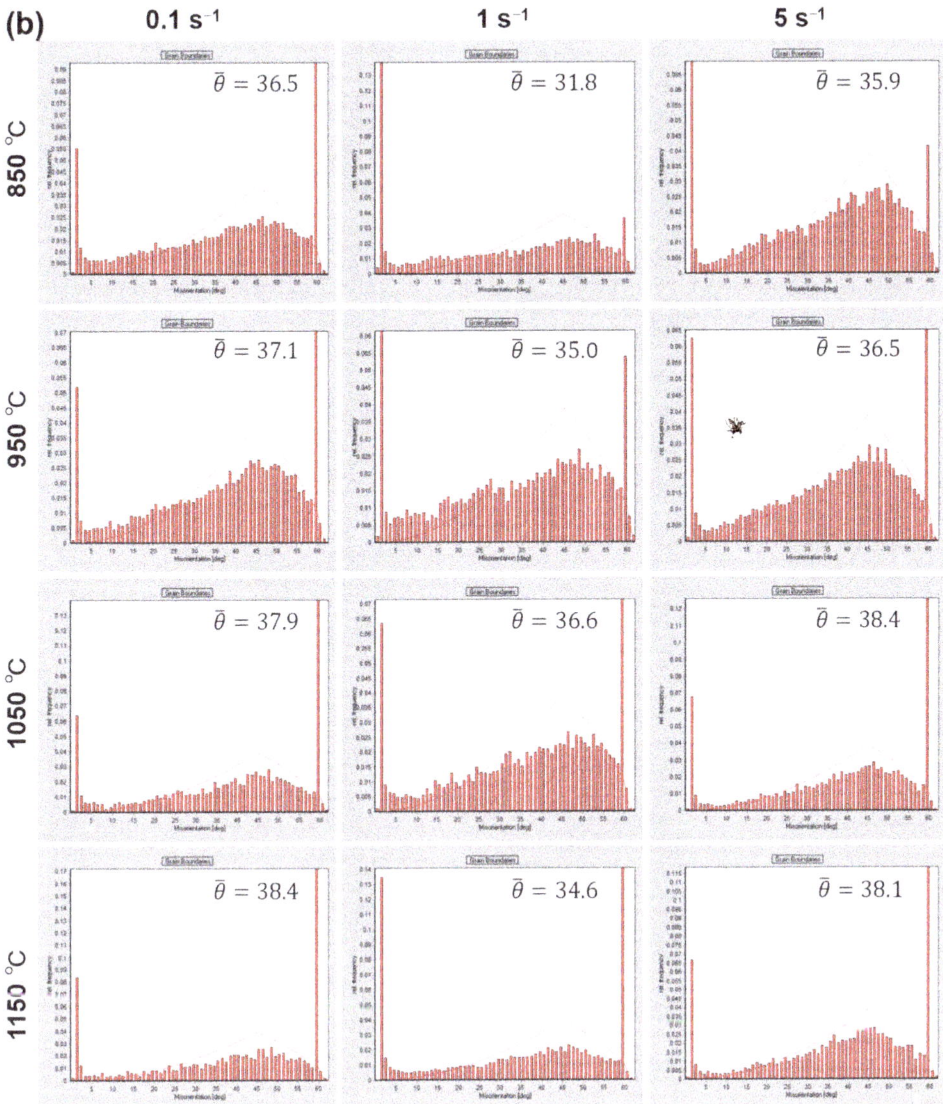

Figure 10. The grain misorientations of (**a**) PA1 and (**b**) MA1-36h alloys consolidated at different conditions.

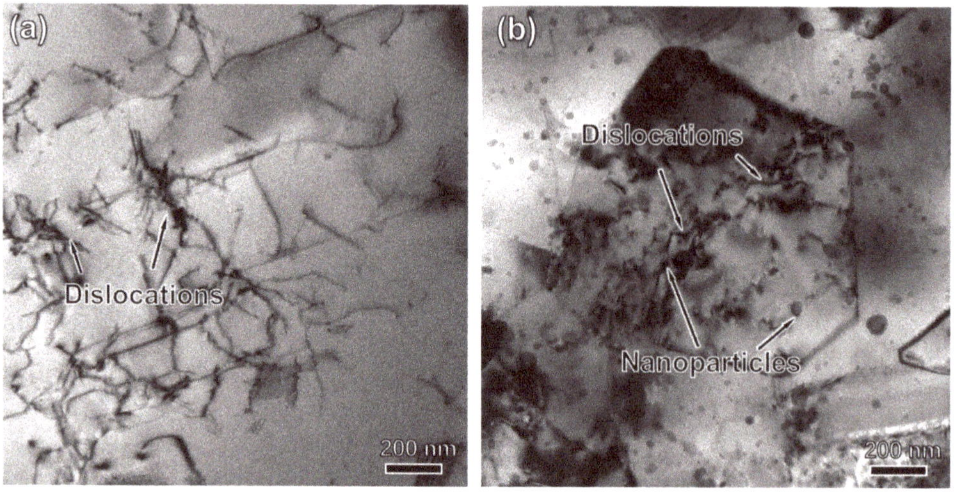

Figure 11. TEM photographs of (**a**) PA1 and (**b**) MA1-36h alloy compressed at 1150 °C/5 s^{-1}.

The grain growth rate of MA1-36h alloy is retarded comparing with PA1 alloy. Since nanoparticles detected as $Y_4Al_2O_9$ by XRD are dispersed in MA1-36h alloy, which prevent DRV and grain growth via pinning dislocations and grain boundaries [28], the grain boundaries of MA1-36 h alloy are more stable than that of PA1 alloy, even at very high temperature [29].

The local misorientation images of PA1alloy and MA1-36h alloy consolidated at different conditions illustrate the hardening status of grains, as presented in Figure 12. In general, the specimens consolidated at higher temperature shows lower local misorientation. The strain hardening of PA1 consolidated at 850 °C is significant, as shown in Figure 12a, which is related to dislocation accumulation in materials. With the increase of temperature, the strain hardening gets relieved by DRV at higher temperature, which also facilitates the DRX via subgrain formation [30]. In addition, as the strain rate increases from 0.1 s^{-1} to 5 s^{-1}, the PA1 alloy presents lower local misorientation except at 1050 °C. However, the strain rate has ambiguous effects on the local misorientation evolution in MA1-36h alloy, as shown in Figure 12b.

EBSD maps in Figures 13 and 14 depict the deformed, substructured and recrystallized grains in the specimens consolidated at different conditions. In general, with the increase of consolidation temperature, the frequency of deformed grains deduces gradually, while fractions of the recrystallized grains get larger at the same strain rates. In general, the deformed grains of PA1 alloy decreases gradually with consolidation strain rate increasing from 0.1 s^{-1} to 5 s^{-1}. Astonishingly, the substructured grains in MA1-36h alloy consolidated at strain rates of 1 s^{-1} are more than the counterparts consolidated at strain rates of 0.1 s^{-1} and 5 s^{-1}.

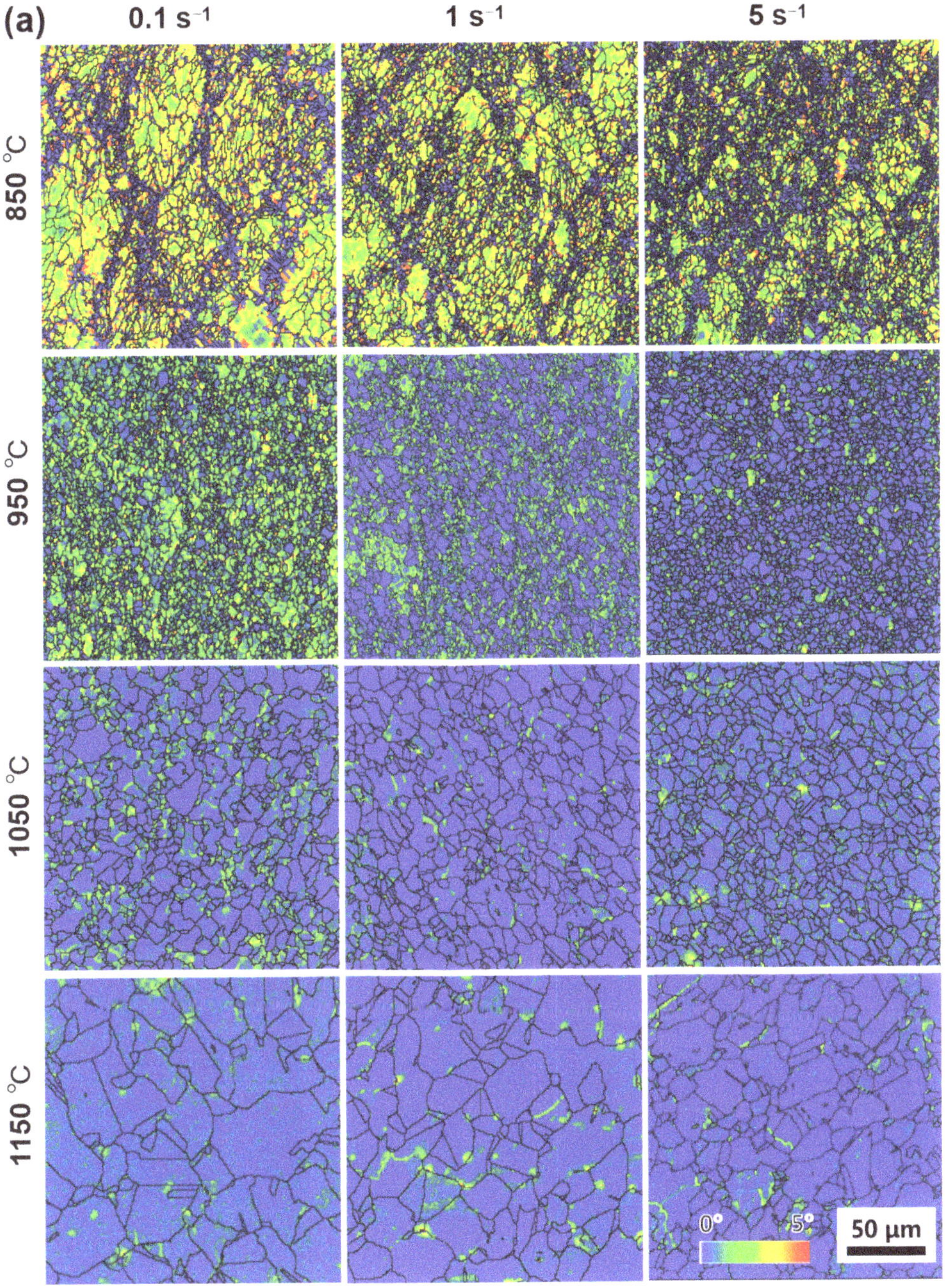

Figure 12. *Cont.*

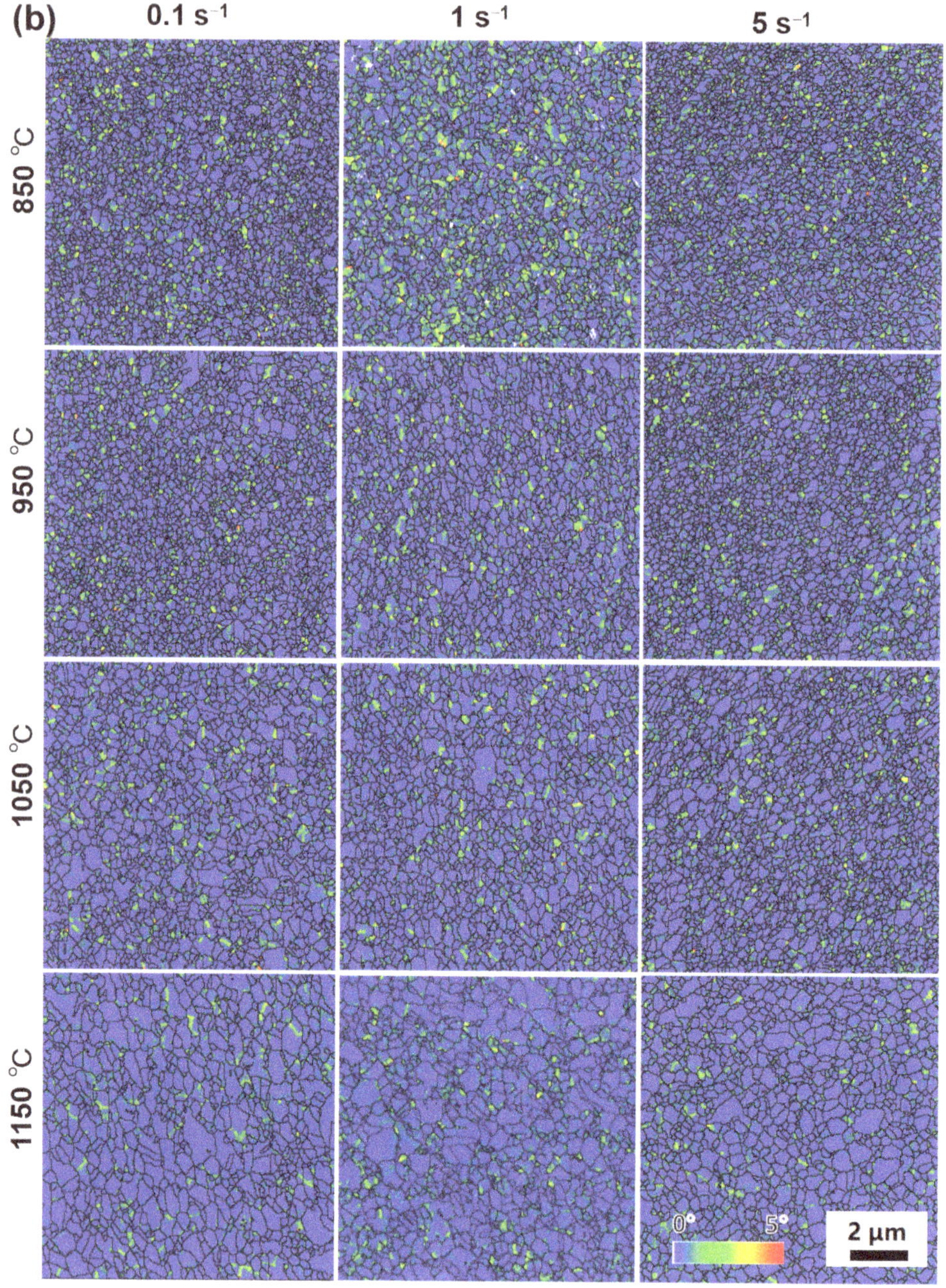

Figure 12. EBSD maps showing local misorientation in grains of (**a**) PA1 and (**b**) MA1-36h alloys consolidated at different conditions, wherein different colors ranging from 0° to 5° correspond to different misorientations as indicated by the color bar.

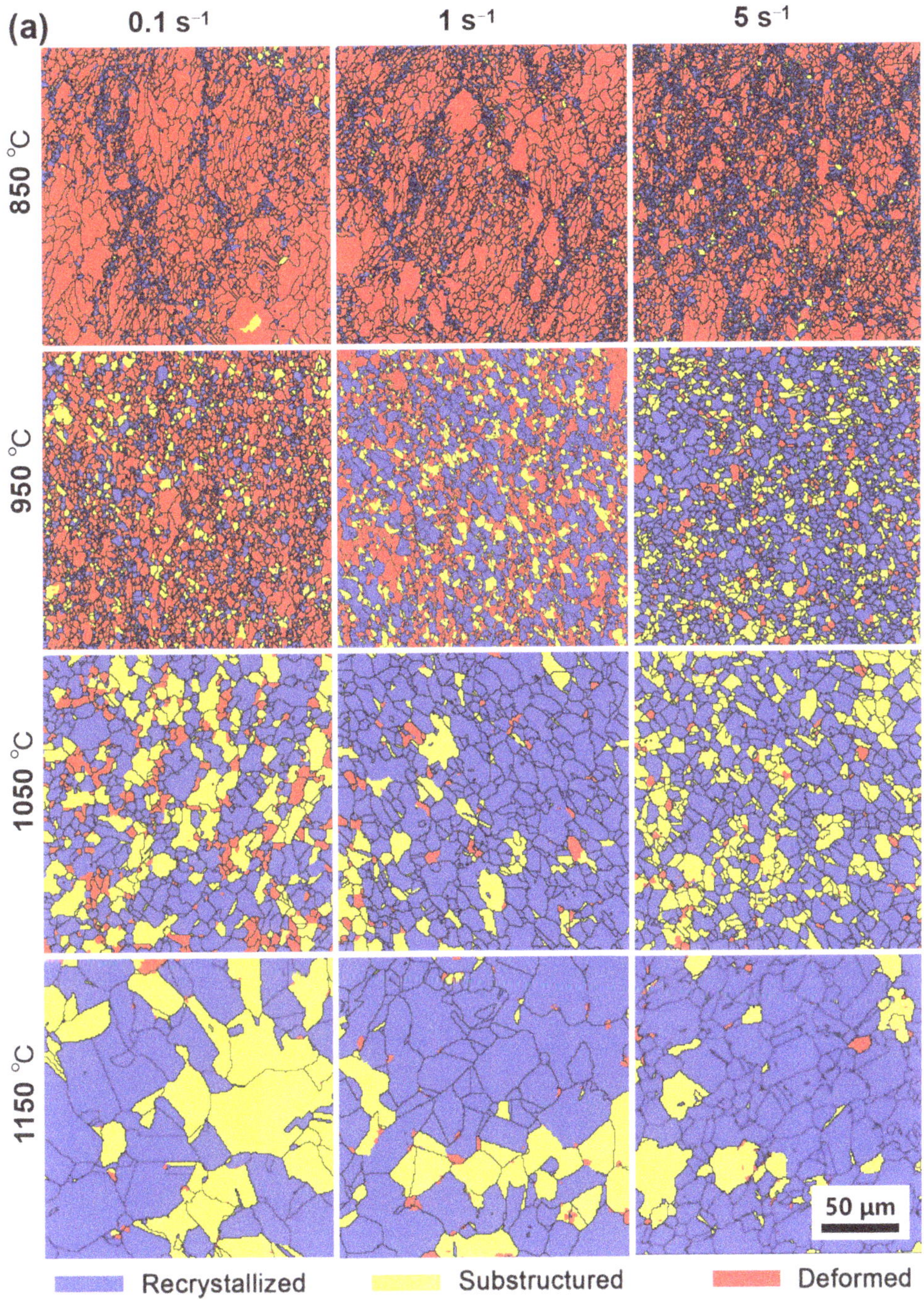

Figure 13. *Cont.*

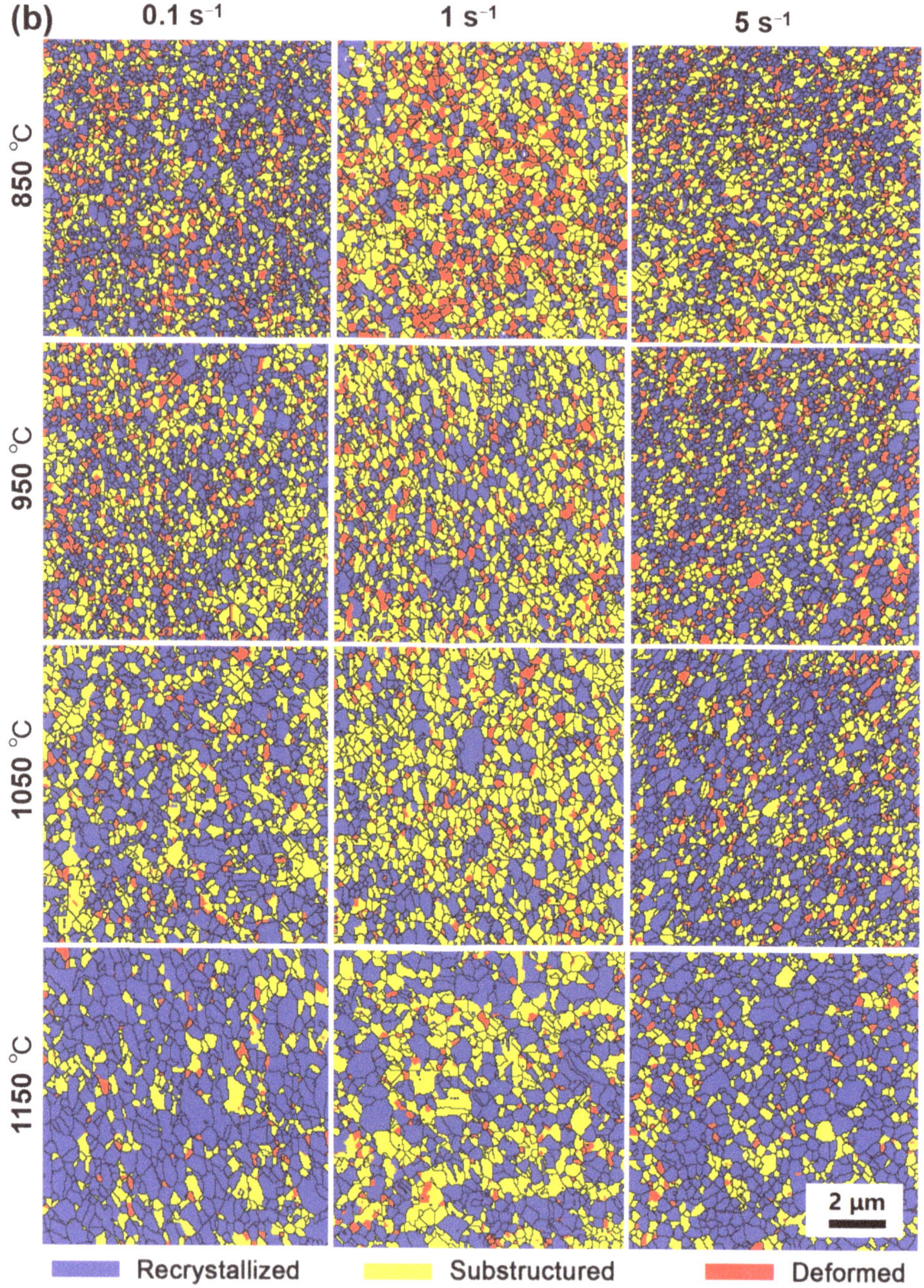

Figure 13. EBSD images showing the distributions of recrystallized, substructured, and deformed grains in (**a**) PA1 and (**b**) MA1-36h alloys consolidated at different conditions.

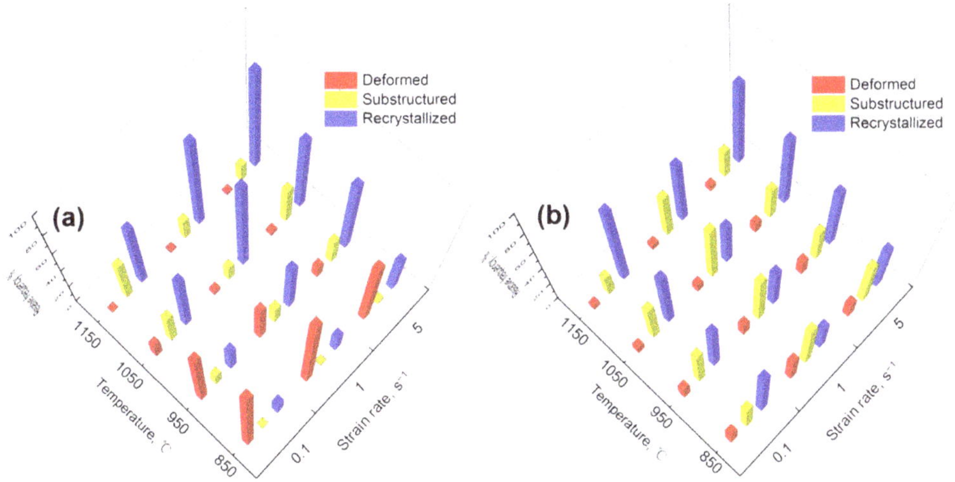

Figure 14. Relative frequency of deformed, substructured and recrystallized grains in (**a**) PA1 and (**b**) MA1-36h alloys consolidated at different conditions.

3.5. *The Variation of Hardness*

The hardness of specimens manufactured at different conditions is shown in Figure 15. The hardness of PA alloy and MA alloy ranges in 154.8–344.9 HV and 421.1–697.3 HV, respectively. As expected, the hardness of MA alloy is significantly higher than that of PA alloy, and the hardness increases with the ball milling time prolonging. In addition, the hardness of samples is also strongly dependent upon temperature. The hardness of samples decreases with the increase of the consolidation temperature. For example, the hardness of PA1 alloy decreases from 288.6 HV to 166.3 HV, with the consolidation temperature increases from 850 °C to 1150 °C at strain rate of 5 s^{-1}. The effect of strain rate on the alloy hardness is also presented. In general, the hardness of samples increases with increasing strain rate, especially at 1050–1150 °C. High deformation rate is conducive to the formation of high-density dislocation, higher energy storage and refine grain size, improving the properties of the alloy, which is consistent with the other literature reports [31–38].

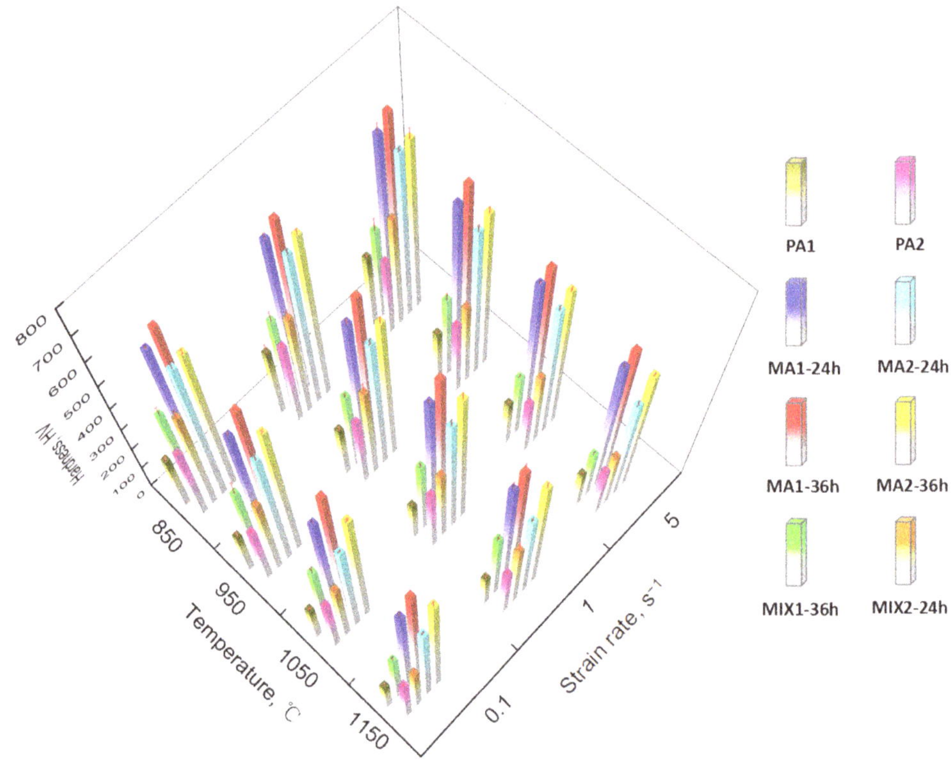

Figure 15. The hardness of alloys prepared at different ball milling time and consolidation process.

4. Discussion

4.1. Microstructure Evolution during Consolidation

The grain evolution during thermal consolidation is generally controlled by DRV, DRX and grain growth [23–26]. The strain hardening is significant in specimens compressed at 850 °C, adequate energy is stored in materials by dislocation accumulation to trigger DRX, as illustrated by the local misorientation maps in Figure 12. As temperature increasing, the strain hardening gets relieved, the improved temperature enhances the formation of DRX nucleation via subgrain formation, the initiation of DRX is faster at higher temperature [30,39]. In general, the improved temperature increases the average grain size of samples based on Figures 8 and 9. The improved temperature can also enhance the DRV and grain growth which relief the dislocation accumulation and consume the stored energy, making it less sufficient to trigger sufficient recrystallization.

In addition, in terms of strain rate, the grain growth is relatively slackened and average grain size drops during consolidation as the strain rate increases from 0.1 s^{-1} to 5 s^{-1}, specially at 1050–1150 °C. The higher strain rate inhabits DRV and limits the time of boundary migration, DRX nucleation via coalescence of subgrain and the strain induced boundary migration is retarded [39,40].

4.2. Strengthening Mechanism of Nickel-Based ODS Superalloys

The hardness H_V of samples can be divided into the matrix hardness H_0, grain boundary strengthening H_g, solid solution strengthening H_{ss}, oxides strengthening H_p and dislocation strengthening H_d, it can be expressed by [41]:

$$H_V = H_0 + H_g + H_{ss} + H_P + H_d \quad (1)$$

In this work, the difference of PA alloy and MA alloy can be estimated by equation:

$$\Delta H_V = \Delta H_g + \Delta H_{Ps} + \Delta H_{Dis} \tag{2}$$

The contribution of dislocation strengthening is given by Equations (3) and (4) [42,43]:

$$\Delta H_d = \alpha \mathrm{MGb} \sqrt{\rho_d} \tag{3}$$

$$\rho_d = 2\theta / \mu b \tag{4}$$

where M is Taylor factor equal to, G is shear modulus, b is Burgers vector, α is the dislocation strengthening coefficient, ρ_d is the dislocation density, μ is the unit length, θ is the misorientation angle estimated by the local misorientation maps (Figure 12).

The grain boundary strengthening is related to the average grain size D, which can be determined by the Hall-Petch equation [44]:

$$\Delta H_g = \frac{k}{\sqrt{D}} \tag{5}$$

where k is the Halle-Petch strength constant. It is clear that the hardness of MA1-36 alloy is higher than that of PA1 alloy by comparing with the average grain size based on Figures 8 and 9.

In addition, nano-oxides plays an important role in ODS superalloy, as shown in Figure 11. Based on Orowan strengthening mechanism [45,46], homogeneously dispersed oxide particles can suppress effectively dislocation movement, as well as grain boundaries migration [47,48]. Therefore, high hardness of MA1-36h alloy comes mainly from grain boundary strengthening and nano-oxides strengthening.

5. Conclusions

In summary, nickel-based superalloys with and without ODS have been fabricated by ball milling and efficient consolidation method. The relationship among the preparation process, microstructure and properties of alloy are studied. Basically, the following conclusions can be reached:

(1) The PPB of nickel-based superalloys is difficult to be broken and eliminated at low temperature, which facilitates the crack nucleation and propagation during thermal consolidation. With increasing of temperature, the thermal activation of the material increases, the PPB is broken and eliminated quickly at elevated temperature and high stress.

(2) The grain size is sensitive to the consolidation temperature, the average grain size increases with the increase of consolidation temperature. The average grain size of samples consolidation at strain rate 5 s^{-1} decreases and more uniform than low strain rate, since higher strain rate inhabits DRV and limits the time of boundary migration.

(3) The hardness of nickel-based superalloys decreases with the increase of the consolidation temperature, strain rate, and the hardness increases after ball-milled for longer time. In addition, the hardness of nickel-based ODS superalloys is significantly higher than that of nickel-based superalloys without ODS due to the grain boundary strengthening and nano-oxides strengthening.

(4) Basically, in order to obtain fine-grains, excellent properties and less cracking risk, consolidation temperature of near 1050 °C and strain rate of 5 s^{-1} are suggested.

Author Contributions: Experimental design and project administration, W.H., F.L., L.T., L.H., B.L., P.C., Z.T., Z.Q., X.X. and G.W.; investigation and Formal analysis, Z.T., Z.Q., X.X., G.W., B.L. and W.H.; writing—review & editing, W.H. and P.C. All authors have read and agreed to the published version of the manuscript.

Funding: This work was supported by The National Science and Technology Major Project (Grant No. 2017-VI-0008-0078), The Natural Science Foundation of China (Grant No. 91860105, and 52074366), and the Science and Technology Innovation Program of Hunan Province (2021RC3131). Lan Huang acknowledges the Changsha Municipal Natural Science Foundation (kq2014126).

Institutional Review Board Statement: Not applicable.

Informed Consent Statement: Not applicable.

Data Availability Statement: The data used to support the findings of this study are available from the corresponding author upon request.

Acknowledgments: This work was supported by the Project supported State Key Laboratory of Powder Metallurgy, Central South University, Changsha, China.

Conflicts of Interest: The authors declare no conflict of interest.

References

1. Verhiest, K.; Almazouzi, A.; Wispelaere, N.D.; Petrov, R.; Claessens, S. Development of oxides dispersion strengthened steels for high temperature nuclear reactor applications. *J. Nucl. Mater.* **2009**, *385*, 308–311. [CrossRef]
2. De Carlan, Y.; Bechade, J.L.; Dubuisson, P.; Seran, J.L.; Billot, P.; Bougault, A.; Cozzikaa, T.; Doriota, S.; Hamona, D.; Henry, J.; et al. CEA developments of new ferritic ODS alloys for nuclear applications. *J. Nucl. Mater.* **2009**, *386–388*, 430–432. [CrossRef]
3. Ukai, S.; Harada, M.; Okada, H.; Inoue, M.; Nomura, S.; Shikakura, S.; Nishidab, T.; Fujiwarab, M.; Asabe, K. Tube manufacturing and mechanical properties of oxide dispersion strengthened ferritic steel. *J. Nucl. Mater.* **1993**, *204*, 74–80. [CrossRef]
4. Benjamin, J.S.; Volin, T.E. The mechanism of mechanical alloying. *Metall. Trans.* **1974**, *5*, 1929–1934. [CrossRef]
5. Dong, H.; Yu, L.; Liu, Y.; Liu, C.; Li, H.; Wu, J. Effect of hafnium addition on the microstructure and tensile properties of aluminum added high-Cr ODS steels. *J. Alloy. Compd.* **2017**, *702*, 538–545. [CrossRef]
6. Park, J.; Jang, J.; Kim, T.K.; Kim, S.-J.; Ahn, J.-H. Formation of Nanostructures in Ni-22Cr-11Fe-1X (X = Y_2O_3, TiO_2) Alloys by High-Energy Ball-Milling. *J. Nanosci. Nanotechnol.* **2011**, *11*, 6213–6218. [CrossRef]
7. Srolovitz, D.J.; Luton, M.J.; Petkovic-Luton, R.; Barnett, D.M.; Nix, W.D. Diffusionally modified dislocation-particle elastic interactions. *Acta Metall.* **1984**, *32*, 1079–1088. [CrossRef]
8. Arzt, E.; Rösler, J. The kinetics of dislocation climb over hard particles—II. Effects of an attractive particle-dislocation interaction. *Acta Metall.* **1988**, *36*, 1053–1060. [CrossRef]
9. Chinnappan, R. Thermodynamic Stability of Oxide Phases of Fe-Cr Based ODS Steels via Quantum Mechanical Calculations. *Procedia Eng.* **2014**, *86*, 788–798. [CrossRef]
10. Kim, J.H.; Byun, T.S.; Lee, J.H.; Min, J.Y.; Kim, S.W.; Park, C.H.; Lee, B.H. Effects of processing condition on the microstructural and tensile properties of 14Cr-based oxide dispersion strengthened alloys. *J. Nucl. Mater.* **2014**, *449*, 300–307. [CrossRef]
11. Kishimoto, H.; Alinger, M.J.; Odette, G.R.; Yamamoto, T. TEM examination of microstructural evolution during processing of 14CrYWTi nanostructured ferritic alloys. *J. Nucl. Mater.* **2004**, *329–333*, 369–371. [CrossRef]
12. Odette, G.R. Recent Progress in Developing and Qualifying Nanostructured Ferritic Alloys for Advanced Fission and Fusion Applications. *Jom* **2014**, *66*, 2427–2441. [CrossRef]
13. Odette, G.R.; Alinger, M.J.; Wirth, B.D. Recent developments in irradiation-resistant steels. *Annu. Rev. Mater. Res.* **2008**, *38*, 471–503. [CrossRef]
14. Tan, L.; He, G.; Liu, F.; Li, Y.; Jiang, L. Effects of Temperature and Pressure of Hot Isostatic Pressing on the Grain Structure of Powder Metallurgy Superalloy. *Materials* **2018**, *11*, 328. [CrossRef] [PubMed]
15. Liu, F.; Wang, Z.; Wang, Z.; Zhong, J.; Zhao, L.; Jiang, L.; Zhou, R.; Liu, Y.; Huang, L.; Tan, L.; et al. High-Throughput Method–Accelerated Design of Ni-Based Superalloys. *Adv. Funct. Mater.* **2022**. [CrossRef]
16. Tan, L.; Li, Y.; He, G.; Liu, F.; Nie, Y.; Jiang, L. Optimized hot workability of a powder metallurgy nickel-base superalloy. *Mater. Charact.* **2019**, *147*, 340–352. [CrossRef]
17. Somani, M.C.; Muraleedharan, K.; Prasad, Y.V.R.K.; Singh, V. Mechanical processing and microstructural control in hot working of hot isostatically pressed P/M IN-100 superalloy. *Mater. Sci. Eng. A* **1998**, *245*, 88–99. [CrossRef]
18. He, G.; Liu, F.; Huang, L.; Jiang, L. Hot deformation behaviors of a new hot isostatically pressed nickel based powder metallurgy superalloy. *J. Mater. Res.* **2016**, *31*, 3567–3579. [CrossRef]
19. He, G.; Liu, F.; Huang, L.; Jiang, L. Analysis of forging cracks during hot compression of powder metallurgy nickel-based superalloy on simulation and experiment. *Adv. Eng. Mater.* **2016**, *18*, 1823–1832. [CrossRef]
20. Tan, L.; Li, Y.; Liu, C.; Yang, C.; Ding, H.; Huang, L.; Liu, F.; Qin, Z.; Jiang, L. The evolution history of superalloy powders during hot consolidation and plastic deformation. *Mater. Charact.* **2018**, *140*, 30–38. [CrossRef]
21. Tang, Q.X.; Ukai, S.; Oono, N.; Hayashi, S.; Leng, B.; Sugino, Y.; Han, W.; Okuda, T. Oxide Particle Refinement in 4.5 mass%Al Ni-Based ODS Superalloys. *Mater. Trans.* **2012**, *53*, 645–651. [CrossRef]
22. Lee, M.K.; Park, J.J.; Rhee, C.K. Synthesis and structural properties of Ni-20Cr-Y2O3 nanocomposite alloy prepared by a very high energy mechanical milling. *Mater. Chem. Phys.* **2012**, *137*, 129–134. [CrossRef]

23. McCarley, J.; Tin, S. Utilization of hot deformation to trigger strain induced boundary migration (SIBM) in Ni-base superalloys. *Mater. Sci. Eng. A* **2018**, *720*, 189–202. [CrossRef]
24. Kumar, S.S.S.; Raghu, T.; Bhattacharjee, P.P.; Appa Rao, G.; Borah, U. Evolution of microstructure and microtexture during hot deformation in an advanced P/M nickel base superalloy. *Mater. Charact.* **2018**, *146*, 217–236. [CrossRef]
25. Konkova, T.; Rahimi, S.; Mironov, S.; Baker, T.N. Effect of strain level on the evolution of microstructure in a recently developed AD730 nickel based superalloy during hot forging. *Mater. Charact.* **2018**, *139*, 437–445. [CrossRef]
26. Huang, K.; Marthinsen, K.; Zhao, Q.; Logé, R.E. The double-edge effect of second-phase particles on the recrystallization behaviour and associated mechanical properties of metallic materials. *Prog. Mater. Sci.* **2018**, *92*, 284–359. [CrossRef]
27. He, G.; Tan, L.; Liu, F.; Huang, L.; Huang, Z.; Jiang, L. Revealing the role of strain rate during multi-pass compression in an advanced polycrystalline nickel base superalloy. *Mater. Charact.* **2017**, *128*, 123–133. [CrossRef]
28. Zener, C. Theory of growth of spherical precipitates from solid solution. *J. Appl. Phys.* **1949**, *20*, 950–953. [CrossRef]
29. De Castro, V.; Marquis, E.A.; Lozano-Perez, S.; Pareja, R.; Jenkins, M.L. Stability of nanoscale secondary phases in an oxide dispersion strengthened Fe–12Cr alloy. *Acta Mater.* **2011**, *59*, 3927–3936. [CrossRef]
30. Huang, K.; Logé, R.E. A review of dynamic recrystallization phenomena in metallic materials. *Mater. Des.* **2016**, *111*, 548–574. [CrossRef]
31. Sakasegawa, H.; Ohtsuka, S.; Ukai, S.; Tanigawa, H.; Fujiwara, M.; Ogiwara, H.; Kohyama, A. Particle size effects in mechanically alloyed 9Cr ODS steel powder. *J. Nucl. Mater.* **2007**, *367–370*, 185–190. [CrossRef]
32. Ohtsuka, S.; Ukai, S.; Sakasegawa, H.; Fujiwara, M.; Kaito, T.; Narita, T. Nano-mesoscopic structural characterization of 9Cr-ODS martensitic steel for improving creep strength. *J. Nucl. Mater.* **2007**, *367–370*, 160–165. [CrossRef]
33. Read, H.G.; Murakami, H. Microstructural influences on the decomposition of an Al-containing ferritic stainless steel. *Appl. Surf. Sci.* **1996**, *94–95*, 334–342. [CrossRef]
34. Ukai, S.; Ohtsuka, S. Low cycle fatigue properties of ODS ferritic–martensitic steels at high temperature. *J. Nucl. Mater.* **2007**, *367–370*, 234–238. [CrossRef]
35. Lee, J.S.; Kimura, A.; Ukai, S.; Fujiwara, M. Effects of hydrogen on the mechanical properties of oxide dispersion strengthening steels. *J. Nucl. Mater.* **2004**, *329–333*, 1122–1126. [CrossRef]
36. Yoshitake, T.; Ohmori, T.; Miyakawa, S. Burst properties of irradiated oxide dispersion strengthened ferritic steel claddings. *J. Nucl. Mater.* **2002**, *307–311*, 788–792. [CrossRef]
37. Ohtsuka, S.; Ukai, S.; Fujiwara, M.; Kaito, T.; Narita, T. Improvement of 9Cr-ODS martensitic steel properties by controlling excess oxygen and titanium contents. *J. Nucl. Mater.* **2004**, *329–333*, 372–376. [CrossRef]
38. Miller, M.K.; Kenik, E.A.; Russell, K.F.; Heatherly, L.; Hoelzer, D.T.; Maziasz, P.J. Atom probe tomography of nanoscale particles in ODS ferritic alloys. *Mater. Sci. Eng. A* **2003**, *353*, 140–145. [CrossRef]
39. Tan, L.; Huang, Z.; Liu, F.; He, G.; Wang, X.; Huang, L.; Zhang, Y.; Jiang, L. Effects of strain amount and strain rate on grain structure of a novel high Co nickel-based polycrystalline superalloy. *Mater. Des.* **2017**, *131*, 60–68. [CrossRef]
40. Wusatowska-Sarnek, A.M.; Miura, H.; Sakai, T. Nucleation and microtexture development under dynamic recrystallization of copper. *Mater. Sci. Eng. A* **2002**, *323*, 177–186. [CrossRef]
41. Xu, H.; Lu, Z.; Ukai, S.; Oono, N.; Liu, C. Effects of annealing temperature on nanoscale particles in oxide dispersion strengthened Fe-15Cr alloy powders with Ti and Zr additions. *J. Alloy. Compd.* **2017**, *693*, 177–187. [CrossRef]
42. Yan, P.; Yu, L.; Liu, Y.; Liu, C.; Li, H.; Wu, J. Effects of Hf addition on the thermal stability of 16Cr-ODS steels at elevated aging temperatures. *J. Alloy. Compd.* **2018**, *739*, 368–379. [CrossRef]
43. Jorge-Badiola, D.; Iza-Mendia, A.; Gutiérrez, I. Study by EBSD of the development of the substructure in a hot deformed 304 stainless steel. *Mater. Sci. Eng. A* **2005**, *394*, 445–454. [CrossRef]
44. Bui, Q.H.; Dirras, G.; Ramtani, S.; Gubicza, J. On the strengthening behavior of ultrafine-grained nickel processed from nanopowders. *Mater. Sci. Eng. A* **2010**, *527*, 3227–3235. [CrossRef]
45. Rosenberg, J.; Piehler, H. Calculation of the Taylor Factor and Lattice Rotations for Bcc Metals Deforming by Pencil Glide. *Metall. Trans.* **1971**, *2*, 257–259. [CrossRef]
46. Wasilkowska, A.; Bartsch, M.; Messerschmidt, U.; Herzog, R.; Czyrska-Filemonowicz, A. Creep mechanisms of ferritic oxide dispersion strengthened alloys. *J. Mater. Process. Technol.* **2003**, *133*, 218–224. [CrossRef]
47. Xu, H.; Li, W.; Sha, X.; Meng, J.; Kang, C.; Wang, W.; Zang, X.; Wang, Z. Effects of Zr addition on the microstructural stability of 15Cr-ODS steels under elevated-temperature annealing. *Fusion Eng. Des.* **2019**, *138*, 231–238. [CrossRef]
48. Zhang, Y.; Ouyang, T.; Liu, D.; Du, J.; Wang, Y.; Zhang, C.; Suo, J. 9Cr-ODS steel composite material reinforced by Ta layers. *J. Alloy. Compd.* **2016**, *682*, 294–301. [CrossRef]

materials

Article

Numerical and Experimental Analysis of Titanium Sheet Forming for Medical Instrument Parts

Wojciech Więckowski [1], Maciej Motyka [2,*], Janina Adamus [3], Piotr Lacki [3] and Marcin Dyner [4]

1. Faculty of Mechanical Engineering and Computer Science, Czestochowa University of Technology, 69 Dabrowskiego St., 42-201 Czestochowa, Poland; wojciech.wieckowski@pcz.pl
2. Faculty of Mechanical Engineering and Aeronautics, Rzeszow University of Technology, 12 Powstancow Warszawy Ave., 35-959 Rzeszow, Poland
3. Faculty of Civil Engineering, Czestochowa University of Technology, 69 Dabrowskiego St., 42-201 Czestochowa, Poland; janina.adamus@gmail.com (J.A.); piotr@lacki.com.pl (P.L.)
4. Faculty of Science and Technology, Jan Dlugosz University in Czestochowa, 13/15 Armii Krajowej Ave., 42-200 Czestochowa, Poland; m.dyner@ujd.edu.pl

* Correspondence: motyka@prz.edu.pl; Tel.: +48-17-7432-416

Abstract: The paper analyses the forming of the surgical instrument handles made of Grade 2 titanium sheets. Sheet metal forming is a technology ensuring high strength and light weight of products. Replacing stainless steels with titanium further reduces instrument weight and additionally provides the required resistance to corrosive environments typical for surgeries. The low instrument weight is important to prevent fatigue of surgeons and allow them to maintain high operational accuracy during long term surgeries. The numerical analysis of the technological process was performed in order to adapt it to forming tool handles using titanium sheets instead of steel sheets. The numerical calculations were experimentally verified. It was found that, in the case of titanium handles, it is necessary to use a blank holder in the first forming operation to eliminate sheet wrinkling in the flange area. The shape and dimensional accuracy of the drawn part after trimming were high enough and the 4th forming operation became unnecessary. Moreover, the process modification included lubrication using rapeseed oil with the addition of boric acid, which effectively prevents the galling of titanium on the working surfaces of the steel tools and ensures a more uniform distribution of plastic strains in the drawn part.

Keywords: sheet titanium forming; Grade 2 titanium; environmentally benign lubricant; numerical analysis; medical instruments

Citation: Więckowski, W.; Motyka, M.; Adamus, J.; Lacki, P.; Dyner, M. Numerical and Experimental Analysis of Titanium Sheet Forming for Medical Instrument Parts. *Materials* **2022**, *15*, 1735. https://doi.org/10.3390/ma15051735

Academic Editor: Daolun Chen

Received: 24 January 2022
Accepted: 22 February 2022
Published: 25 February 2022

1. Introduction

Medical instruments are mostly made of highly-alloyed stainless steels, including martensitic, ferritic and austenitic steels. Only in justified cases, materials such as platinum, titanium, tantalum and their alloys are used [1]. In the case of gripping parts, used for holding instruments in the operator's hand, aluminium alloys and, recently, plastics, as well as composite materials are also used. In order to increase their corrosion resistance, as well as antibacterial and antiviral properties, instruments are subjected to surface treatment [2]. In [3], Xue et al. summarised the latest advances in surface modification techniques, especially titanium and its alloys, for biomedical applications. The surface techniques include plasma spray, physical vapor deposition, sol-gel, micro-arc oxidation, etc. The conditions in which medical instruments are used, namely long-term contact with organic tissue, body fluids and disinfectants are the reasons why the materials used for medical instruments should have good corrosion resistance. As emphasised by Gherlone et al. [4], in the light of the COVID19 pandemic, when CoV-2019 can be transmitted not only directly from person to person by respiratory drops, but also by direct contact with contaminated material, it is important that medical instruments were made of materials resistant to disinfectants. Frequent sterilisation accelerates corrosive processes and induces corrosive wear of the tools.

The second very important factor taken into account when choosing materials for medical instruments is their weight. The light weight of surgical instruments is of particular importance during long-term surgical operations, often requiring high precision. In addition, parts of medical instruments, such as the working and gripping parts and connectors, must exhibit appropriate mechanical properties ensuring the transfer of loads occurring during the work of instruments, without changing the functional characteristics of individual parts of the instruments. The normative recommendations relating to materials used for medical instruments are described in [5,6].

The use of gripping parts made of thin-walled elements is a solution that allows the weight of medical instruments to be reduced. Among the various manufacturing techniques such as machining, casting or 3D printing, metal forming provides the greatest strength to thin-walled elements [7]. Products manufactured by sheet metal forming, despite the small thickness of the walls of the drawn part, are characterised by high strength due to strain hardening taking place during cold working. In work [8,9], discussing the latest trends in metal working, sheet metal forming is indicated as a process enabling the production of light and durable products, so desired in many areas of life, including the medical industry. To advance the rehabilitation efficacy and shorten the recovery period of patients, Cheng et al. [10] proposes incremental sheet titanium forming for the production of biomedical implants. However, this flexible forming technique still requires a great deal of effort to increase the geometric accuracy to make the drawn part fit properly. Scratches and marks, caused by the forming tool, pose problems which are difficult to overcome, so the process is still in the experimental phase.

An additional reduction in the mass of drawn parts, while maintaining appropriate strength, can be obtained by using sheets of light and high-strength materials such as titanium and its alloys. Up to 600 °C, titanium alloys have the highest strength-to-weight ratio among all construction metals [11]. Biocompatibility is an additional advantage of titanium regarding medical applications, especially due to postoperative complications in the case of using implants and surgical instruments made of materials sensitising patients, which the authors of article [12] pay attention to. According to [13,14], titanium does not cause an undesirable biological interaction in contact with living tissue. The beginnings of using titanium and its alloys for structural elements date back to the end of the 1940s. With the development of implants in the 1960s, the first surgical titanium instruments also appeared [15]. Somewhat later, in the 1980s, titanium started to be used to reduce the weight of car parts [16]. However, due to the high cost of producing and processing titanium materials, their use in civil applications is still limited, mainly to produce implants and high-end sports equipment, as well as in the automotive industry for high-end cars and racing ones. Furthermore, although at the beginning of the 1980s titanium sheets began to enter the construction market mainly as roofing materials, curtain walls and ventilation shafts [17,18], the aviation industry is still the largest customer of titanium products. Unfortunately, at ambient an temperature, sheets made of materials with high specific strength, such as titanium alloys, are characterised by low formability. Among titanium materials, only Grade 1 and Grade 2 commercially pure titanium sheets can be relatively easily formed in ambient temperatures with typical dies used for sheet steel forming [11,19]. Other commercially pure titanium (Grade 3 and 4) as well as titanium alloys belong to hard-to deform materials in ambient temperatures. To solve this problem, hot forming is applied. Wang et al. [20] propose Fast light Alloys Stamping Technology (FAST), where fast heating is employed to maintain the post-form strength but not impairing the formability. Moreover, Tian and Li [21] studied the hot stamping process for forming complex-shaped titanium alloy components, combining heat treatment and fast stamping. Although hot forming improves the drawability, it requires more expenditure on heating and the use of protective atmospheres to prevent increased gas (oxygen, hydrogen and nitrogen) absorption by titanium.

The literature review shows that works in the field of cold forming titanium and its alloys, especially sheet metal forming, are almost unique materials, what is also emphasised in the work [22] from 2021. Scientific works mainly concern the study of titanium properties and microstructure, heat and surface treatment, as well as corrosion problems. These publications mainly emphasise the advantages of titanium and its alloys compared to steel, especially the favourable mechanical strength in relation to titanium density, high resistance to most common corrosive environments and good biocompatibility. Usually, the research concerns the drawn part with simple geometry, such as a spherical cup. There is no detailed information on the process parameters of cold forming titanium sheets. In [22], plastic behaviour and formability of the commercially pure titanium CP Grade 2 hcp-titanium T40 was examined. The Authors paid attention to the significant influence of sheet anisotropy on the occurrence of instability and distribution of plastic strains. Chinapareddygari et al. [23] assessed the stretchability of Grade 4 commercially pure titanium sheets on the basis of tensile tests (yield strength, ultimate tensile strength, elongation), the plastic strain ratio (*r*-value) and forming limit diagram (FLD). Lin et al. [24], when analysing the deep drawing behaviour of commercially pure titanium at room temperature, pointed out that earing is a result of planar anisotropy and tension-compression asymmetry (TCA). TCA reduced the thickness of the deep drawn sections, increased the earing ratio, and influenced the drawing force. The earing profiles depended strongly on the texture evolution. Pham [25,26] examined both experimentally and numerically the anisotropic yielding behaviour and the distortional hardening behaviour of commercially pure titanium sheets during the hydraulic bulge test. Chen and Chiu [27] stated that the low ductility of commercially pure titanium at room temperature results from its hexagonal close-packed crystal structures, and thermal activation is required to increase its ductility and formability. Mohanraj and Elangovan [28] also showed that the formability of Grade 2 titanium sheets can be improved at elevated temperatures. However, the manufacturing process at room temperature is always preferred for the reason of cost-effectiveness. Therefore, scientists are constantly looking for new methods of the cold forming of sheets of hard-to-deform materials. Adamus and Lacki [29,30] indicate the possibility of improving the drawability of titanium sheets thanks to the use of semi-flexible tools. Aydogan et al. [31] used hydroforming with a membrane diaphragm to shape Grade 2 commercially pure titanium. They pointed out that hydroforming reduces the springback value of the sheets and gives a more uniform strain distribution comparing to other forming methods due to the smooth action of fluid pressure. However, hydroforming is an inefficient and very demanding process. It requires high-capacity presses, a qualified workforce, etc. Kumar [32] underlined that material formability depends on intrinsic material properties and prevailing process parameters, especially lubricating conditions. Kakulite [33] added that galling is one of the major disadvantages of friction, which is commonly observed in sheet metal forming, and that the use of suitable lubricants and coatings can be two of the easiest, most economical and efficient methods to mitigate galling. Adamus and Dyja [34] achieved good results in decreasing the coefficient of friction during sheet-titanium forming thanks to the application of lubricants based on vegetable oils with an additive of boric acid. In [35], the authors concluded that anti-adhesive coatings put on the working parts of the dies are additional protection against galling when the lubricant film breaks under the high unit pressure occurring in the sheet-titanium forming process. The authors' experience to date has shown that tool geometry, especially the die and punch fillet radii and the blank holder force are also key factors in sheet metal forming.

In this paper, an analysis of the forming process of the gripping part of a surgical instrument previously made of steel 1.4021 [36] was carried out. Grade 2, the most commonly used commercially pure titanium in technology and medicine, was selected as the material for handle production. This choice was made due to its biocompatibility, quite good formability resulting from the ratio of yield point to tensile strength of about 0.7 and elongation of about 20%, low density of 4500 kg/m^3, almost half that of steel, high corrosion resistance, and good weldability. As Grade 2 titanium belongs to α alloys which undergo strain hardening sheet thinning occurring during plastic forming, and will be compensated by material hardening. Due to the fact that the drawn part has an elongated rather complex shape, it is important to take into account the sheet plastic anisotropy and the appropriate location of the blank in relation to the rolling direction.

Theoretical analyses of the sheet metal forming processes were carried out with the commercial PAMStamp 2G [37] programme, dedicated to sheet metal forming processes. The results of the numerical analyses were experimentally verified.

The novelty of the work refers to the use of grade 2 titanium on parts usually made of steel, which allows for a weight reduction of about 47%, and the use of technological lubricant based on vegetable oil with an additive of boric acid, which effectively lowered the frictional resistance during forming from 0.4 in dry conditions to 0.1 with lubrication. This lubricant completely eliminates the galling phenomenon, which allows for forming titanium handles with the use of rigid steel tools. What is very important also, is that the lubricant does not pose a threat to the environment. An additional advantage of titanium products is the possibility of colouring them by performing anode oxidisation to the titanium, thanks to which it is easy to distinguish tools during a surgical procedure.

2. Materials and Methods

2.1. Aim and Scope of Work

In the work, the cold forming of the components of a surgical instrument handle, which are presented in Figure 1, is analysed. The main goal of the study is to modify the technological process, which is currently used for forming steel handles, in order to make it possible to produce them from titanium sheets.

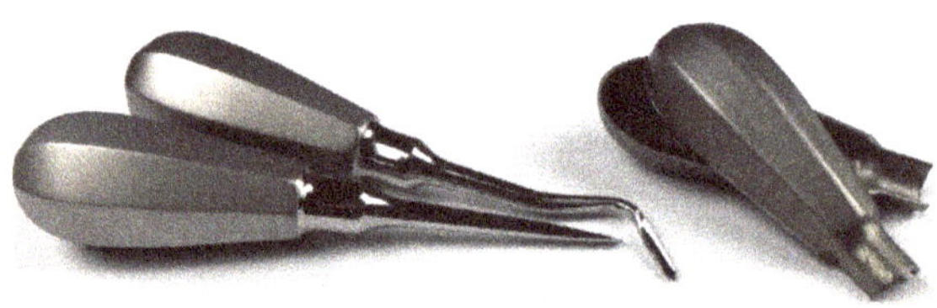

Figure 1. View of handle parts of surgical instruments.

The handle consists of two parts formed from sheet metal, which are then joined by welding. In order to reduce the instrument weight, it was proposed to make these parts from Grade 2 commercially pure (CP) titanium sheet (Table 1), produced by Kobe Steel, LTD, Kakogawa, Japan according to EN 10204 -Material Certificate 3.1 [38]. Its chemical composition was confirmed using energy-dispersive X-ray spectroscopy (EDS)—Ti and Fe—99.8 and <0.1 wt.%, respectively (Figure 2). Until now, these parts were made of 1.4021 steel sheet (Stahl Krebs, Solingen, Germany).

The analyses include:

– experimental forming of the drawn parts in industrial conditions;
– numerical simulations of the sheet metal forming process, using the PAMStamp 2G programme, based on the finite element method (FEM).

Mechanical properties (Young's modulus E, yield stress $R_{p0.2}$, tensile strength R_m) and technological ones (Lankford coefficient r, strength coefficient K and strain hardening exponent n) of the analysed sheet were determined in the static uniaxial tensile test in accordance with [39]. The tests were carried out on Zwick Z50 (Zwick Roell Polska Sp.zoo. Sp.k., Wroclaw, Poland) tensile testing machine. To determine forming limit diagram (FLD) of the sheet, ARAMIS 5M System was used. It is a non-contact measuring system based on digital image correlation and photogrammetry. To determine coefficient of friction, a strip drawing test was applied. Strip drawing test is used to model friction condition in sheet metal forming, i.e., between the punch/die and deformed sheet.

Table 1. Nominal chemical composition of Grade 2 CP titanium.

Element	Ti	Fe	O	C	N	H
Content, wt.%	≥98.9	≤0.30	≤0.25	≤0.09	≤0.03	≤0.015

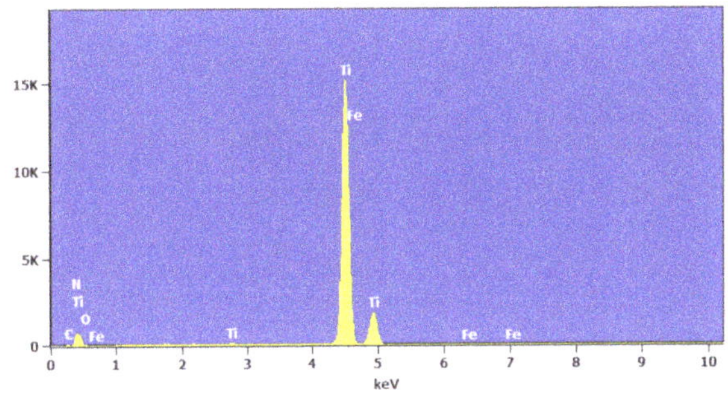

Figure 2. EDS spectrum of Grade 2 CP titanium.

2.2. Experimental Tests

Tests of sheet metal forming were performed on typical steel rigid tools used for the production of steel sheet handles. The tool was mounted in UPV80 hydraulic press (Figure 3) with pressure of 80 T, which is used in the Factory of Medical Instruments CHIRMED, Rudniki, Poland for mass production of different drawn parts. The manufacturing process of the handle part to date used in the factory consists of four operations: initial and supplementary forming, trimming the flange and final forming. All the forming operations are performed in a tool consisting of a die and a punch. Rectangular Grade 2 titanium sheet blanks with dimensions of 85 mm × 57 mm × 1 mm were used to form the drawn parts. The forming speed was 5 mm/s. To reduce the forming resistance and to prevent titanium sticking to the working surface of the steel tools in the subsequent forming operations, a vegetable oil-based lubricant with the addition of boric acid was used. According to [34], using this lubricant when forming titanium sheets significantly reduces the coefficient of friction between the deformed material and the working surface of the tool from 0.4 in the absence of lubrication to 0.1 in the presence of the lubricant.

Figure 3. View of forming tool installed in UPV80 hydraulic press.

2.3. Numerical Model

Numerical models of the individual forming operations were developed on the basis of the actual geometry of the forming tools. In the calculations, the process parameters used in the experimental tests were assumed. The Solidworks System was used to create the tool geometry. Then the geometry was exported to the PamStamp 2G System. The working surfaces of the tools and the sheet metal surface were meshed with four-node shell elements (see Figure 4).

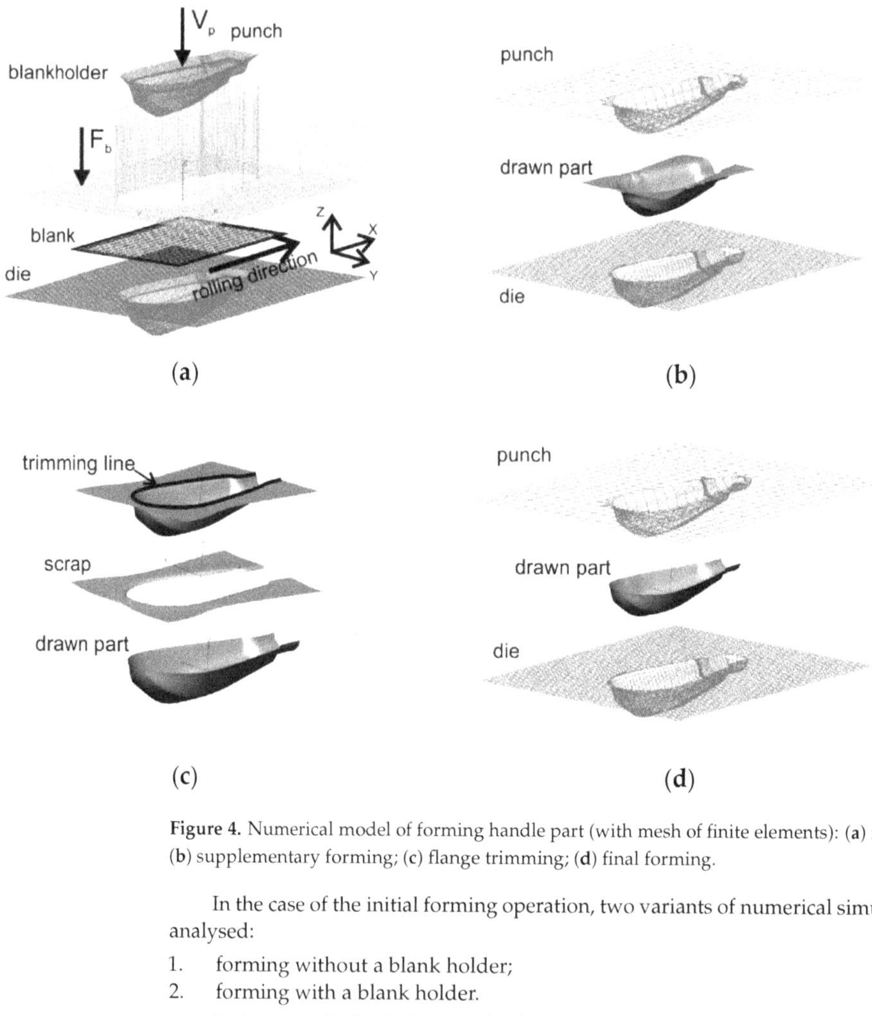

Figure 4. Numerical model of forming handle part (with mesh of finite elements): (**a**) initial forming; (**b**) supplementary forming; (**c**) flange trimming; (**d**) final forming.

In the case of the initial forming operation, two variants of numerical simulations were analysed:

1. forming without a blank holder;
2. forming with a blank holder.

In the numerical calculations, the friction and lubrication conditions were described by the coefficient of friction. In the case of forming without the blank holder, the coefficient of friction $\mu = 0.1$ was assumed on the contact surfaces between the die and the deformed sheet to reflect the lubrication conditions prevailing during tests in industrial conditions. In the case of forming with the blank holder, two different coefficients of friction were assumed:

- $\mu = 0.1$ for lubricated contact surfaces, i.e., on the contact "die-deformed material-blank holder";
- $\mu = 0.4$ for unlubricated contact surfaces.

In the analysed numerical models, the following boundary conditions were assumed: the dies were deprived of all degrees of freedom, while the punch and the blank holder could move in the Z direction. Velocity vector *Vp* was assigned to the punch, while force vector *Fb* was applied to the blank holder. The blank sheet had all degrees of freedom. The rolling direction of the sheet was assumed in accordance with the *X* axis parallel to the longer edge of the blank.

In the numerical calculations, an elastic-plastic Hill model for the deformed material with nonlinear strain hardening described by Hollomon's law was adopted:

$$\sigma_p = K\varepsilon^n, \quad (1)$$

where: *K*—strength coefficient, *n*—strain hardening exponent.

The basic strength parameters, including the strength coefficient and strain hardening exponent of the Grade 2 titanium sheet, were determined in the static uniaxial tensile test in accordance with the PN-EN ISO 6892-1 standard. Normal anisotropy ratios r_α were also determined experimentally in the tensile test of sheet metal strips cut along (0°), transversely (90°) and at an angle of 45° in relation to the rolling direction of the sheet, in accordance with [40]. The material data for the Grade 2 sheet, which were assumed in the calculations, are given in Table 2.

Table 2. Material data for Grade 2 commercially pure titanium sheet t = 1 mm.

Young's modulus E, GPa		105.00
Yield stress $R_{p0.2}$, MPa		354.30
Tensile strength R_m, MPa		472.40
Poisson's ratio ν, -		0.34
Density ρ, g/cm^3		4.50
Lankford coefficient	r_0, -	2.49
	r_{45}, -	4.50
	r_{90}, -	5.20
Strength coefficient K, MPa		724.40
Strain hardening exponent n, -		0.144

The forming limit diagram (FLD) of a Grade 2 titanium sheet with thickness t = 1 mm, which is presented in Figure 5, was determined experimentally according to the methodology described in [41].

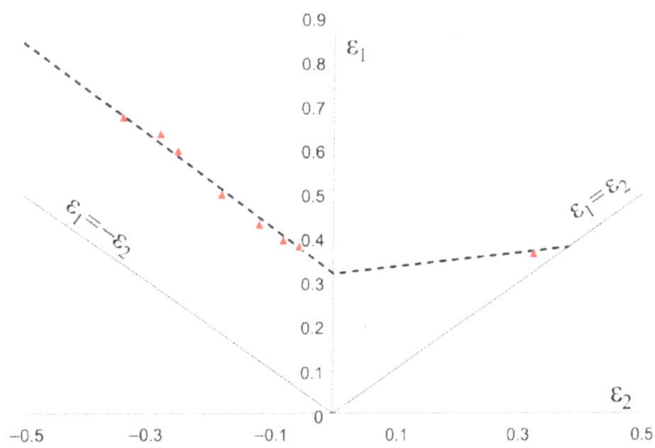

Figure 5. Forming limit diagram for Grade 2 titanium sheet, t = 1.0 mm.

3. Results

3.1. Experimental Test Results

The results of the forming process are presented in Figure 6.

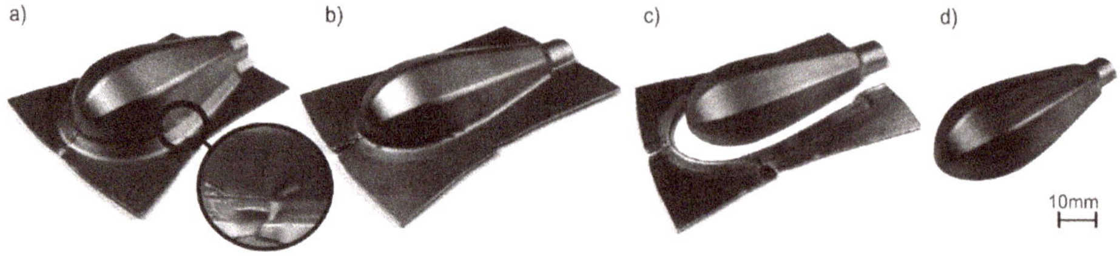

Figure 6. Subsequent stages of forming handle part from Grade 2 titanium sheet: (**a**) initial forming with magnification of wrinkling area; (**b**) supplementary forming; (**c**) trimming; (**d**) final forming.

The experimental tests showed that it is possible to produce the handle parts of Grade 2 titanium sheet with standard steel tools. However, the lack of a blank holder in the initial forming process results in the wrinkling of the sheet, both in the flange of the drawn part and on its surface (Figure 7a). On the other hand, forming with the blank holder in dry conditions very often leads to fracture (Figure 7b). Moreover, the experiments showed a high tendency to create titanium protrusions on the working surfaces of the steel die, which in turn leads to galling (Figure 7b) and deterioration of the quality of the drawn part surface. Titanium protrusions cause scratches and dents that cannot be removed in subsequent forming operations. Therefore, lubrication was necessary to separate the rubbing surfaces (i.e., the deformed material—titanium—and the die). Lubricant based on rapeseed oil with an additive of boric acid [42] was used. The use of the blank holder and technological lubricant, made it possible to obtain a good-quality titanium drawn part (Figure 7c).

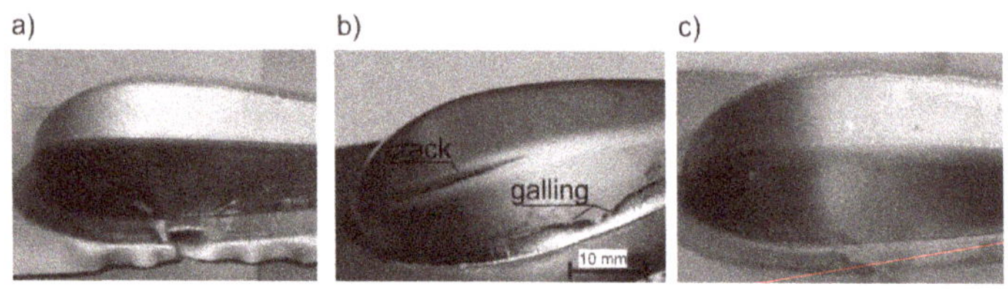

Figure 7. Results of forming: (**a**) with no blank holder—visible wrinkling, (**b**) with blank holder in dry conditions—visible crack and marks of galling, (**c**) with blank holder and lubrication—there are no defects.

To eliminate this defect, it was decided to modify the forming process using numerical simulations.

3.2. Numerical Calculations Results-Discussion

The results of the numerical calculations in the form of plastic strain distribution, as well as thickness distribution in the drawn parts for all the stages of forming the titanium handle are presented. Additionally, the distribution of quality zones in comparison with the FLD of the deformed sheet are given. The simulation results of forming the handle parts in conditions corresponding to the experimental tests, i.e., without a blank holder during initial forming and with lubrication of the die surface, which are presented in Figure 8, are in good agreement with the experimental tests (also see Figure 6).

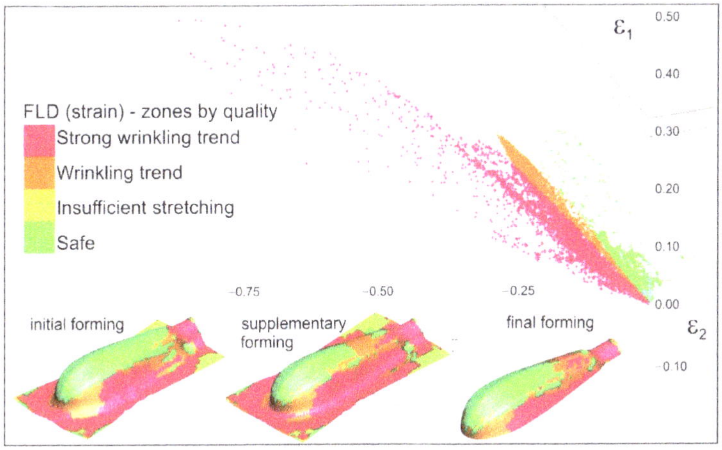

Figure 8. Strain distribution (zones by quality) in comparison with forming limit diagram.

During the initial forming operation, the areas with a strong tendency to wrinkle are visible in the flange as well as on the drawn part surface. This tendency also continues during subsequent forming operations. The use of the blank holder, as demonstrated by the numerical simulations, eliminated the wrinkling phenomenon. Simulations were carried out under various friction conditions between the sheet material and the working surfaces of the tool, i.e., with and without lubrication. The results are presented in Figures 9–12. The thickness distributions in the drawn parts after the initial, supplementary and final forming operations are presented in Figures 9 and 10, respectively, for forming in dry conditions and with lubrication.

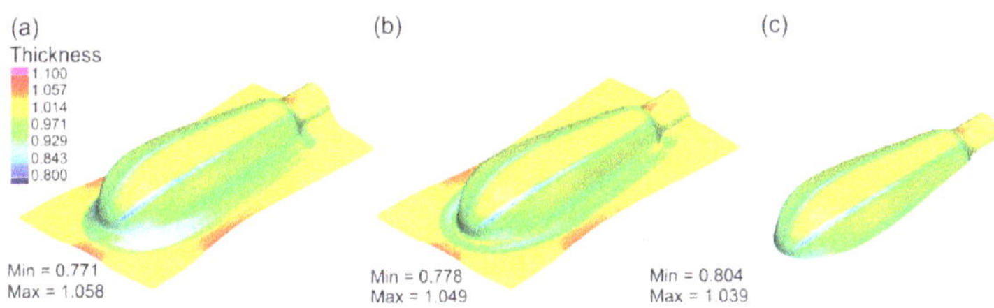

Figure 9. Thickness distribution in drawn part wall after: (**a**) initial forming; (**b**) supplementary forming; (**c**) final forming in dry conditions.

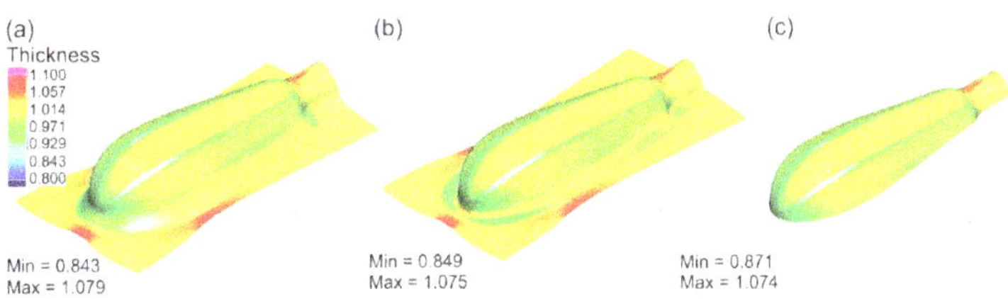

Figure 10. Thickness distribution in drawn part wall after: (**a**) initial forming; (**b**) supplementary forming; (**c**) final forming performed with lubricant.

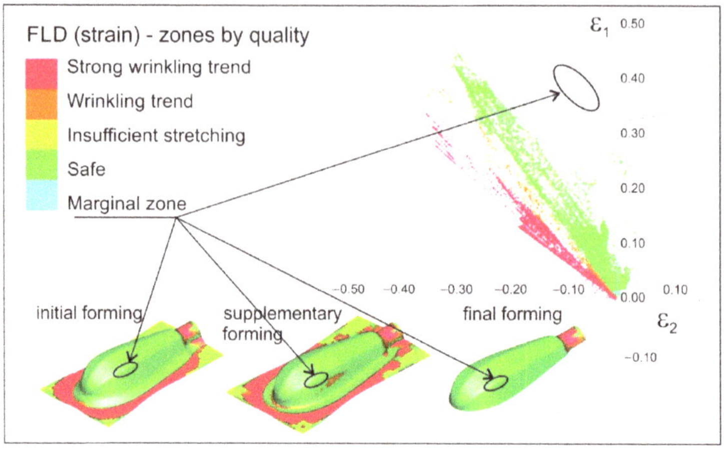

Figure 11. Plastic strain distribution in comparison with forming limit diagram (variant: forming with blank holder and without lubrication).

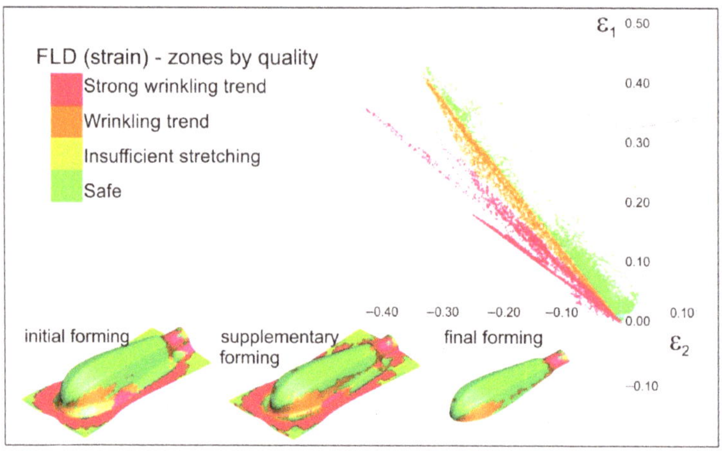

Figure 12. Plastic strain distribution in comparison with forming limit diagram (variant: forming with blank holder and with lubrication).

Greater thinning of the drawn part walls is observed in the case of forming in dry conditions. The greatest thinning occurs on the edges of the handle, in the area with the small fillet radius. The application of lubricant ($\mu = 0.1$) causes a more uniform thickness distribution—the wall thinning is smaller.

During forming without lubrication, significant material deformation along the edge of the drawn part occurs, which results in excessive thinning of the sheet and can lead to its fracture. This is reflected in the points in the marginal zone just below the forming limit curve (Figure 11). In this area, cracking occurred during the actual forming process (see Figure 7b).

Introducing lubricant between the rubbing surfaces decreases the frictional resistance. All the points representing strain values are below the FLD in the safe zone (see Figure 12). This means that the risk of cracking is eliminated.

On the other hand, the tendency to wrinkling grows because lubrication decreases frictional resistance in the contact zone: "die-deformed material–blank holder" makes material flow to be easier. In the actual forming process, when both lubrication and the blank holder were used to tend to, wrinkling was not observed. Moreover, the lubricant prevented the sticking of the deformed material (titanium) to the forming tools, as evidenced by the smooth surface of the drawn part shown in Figure 13.

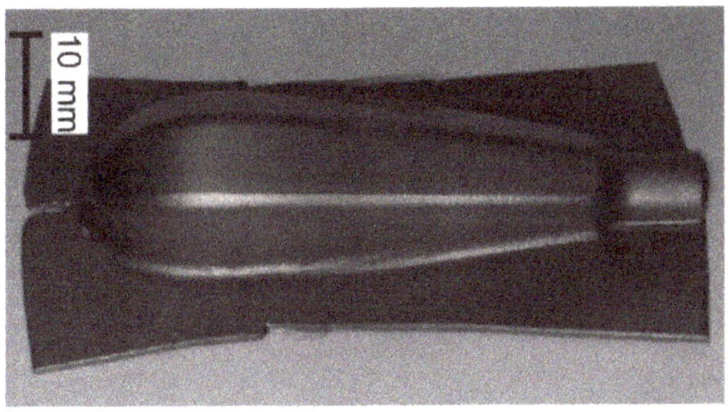

Figure 13. View of drawn part formed in presence of lubricant and blank holder.

Moreover, the numerical analyses show that the forming parameters, such as the blank holder force and lubrication, have an impact on the dimensional accuracy of the obtained parts. The values of springback occurring after flange trimming are shown in Figure 14.

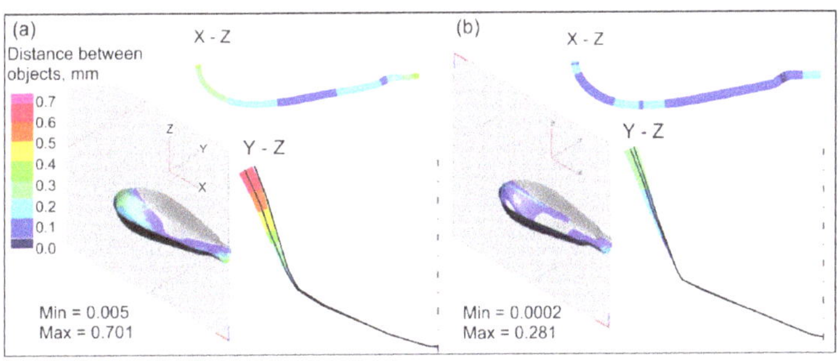

Figure 14. Springback after flange trimming: (**a**) initial forming without blank holder, (**b**) initial forming with blank holder and lubrication.

The highest springback is observed after trimming the flange in the case of forming without the blank holder (Figure 14a). The maximal distance between the nominal and formed drawn part is about 0.7 mm. Applying the blank holder force of 20 kN reduced this distance to about 0.3 mm. No effect of lubrication on springback was observed. If the forming process is performed with the blank holder, the forth operation, i.e., final forming, increases the shape and dimensional accuracy of the drawn part only slightly. Thus, if it were decided to modernise the process and introduce the blank holder in the initial forming operation, then final forming after trimming can be omitted. The use of the blank holder is advisable in view of eliminating sheet wrinkling, which makes it difficult to further form the handle parts of the instrument.

3.3. Experimental Verification of Numerical Results

The results of the numerical calculations, namely the wall thicknesses of the drawn part, were compared with thicknesses of the actual drawn part (Figure 15). For this purpose, the final drawn part was cut with a wire EDM machine, and then the wall thicknesses were measured at the selected points (along the wall of the drawn part every 1 mm—Figure 15c) using a light microscope.

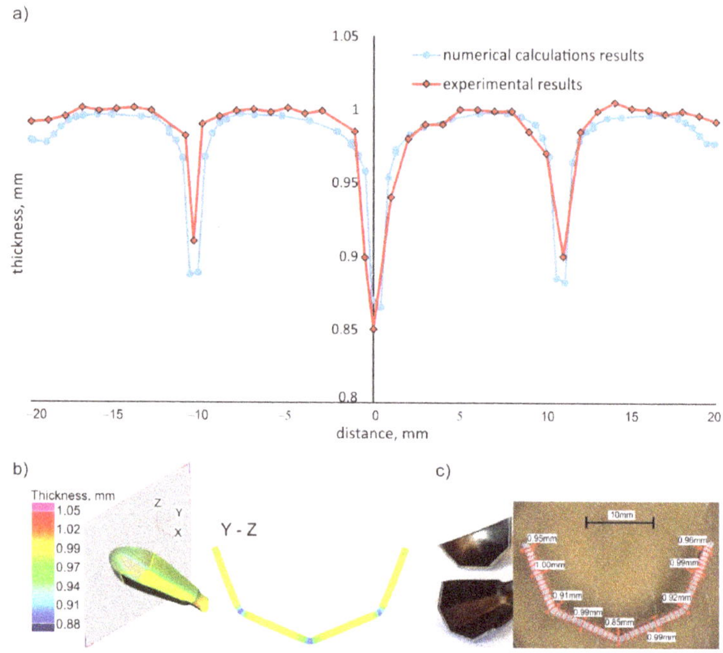

Figure 15. Thickness distribution: (**a**) comparison of experimental and numerical results; (**b**) numerical calculation results; (**c**) experimental verification.

Modification of the technological process, consisting in the use of blank holder force in the first operation (Fb = 20 kN) and lubrication with rapeseed oil with an additive of boric acid (coefficient of friction $\mu = 0.1$), resulted in obtaining a titanium drawn part with the required shape and dimensions. The maximal difference in thickness of the drawn part walls between the numerical and experimental results is about 2%. Moreover, there was no tendency of the sheet to wrinkle or scratches on the outer surface of the drawn part (see the forming results in Figure 13). Additionally, forming with the proposed lubrication protects the working surfaces of the tool against the galling of the titanium.

4. Conclusions

On the basis of the experimental and numerical studies the following conclusions can be drawn:

1. There is a possibility of forming shallow drawn parts, such as the analysed handle of surgical instruments, of Grade 2 CP titanium sheets using standard steel tools, usually used for forming steel drawn parts.
2. The use of Grade 2 CP titanium sheet reduces the handle weight of by about 47%, which significantly improves the comfort of using the tools during surgery.
3. The blank holder force has a dominant influence on the dimensional accuracy of the drawn part; according to the numerical calculations, the blank holder force equals 20 kN.
4. Although the numerical calculations show a small effect of the lubrication on the dimensional accuracy of the drawn parts, the experimental tests demonstrate that lubrication must be used to avoid galling.
5. The blank holder force and lubrication synergy result in a more uniform plastic strain distribution (compare Figures 11 and 12), thus decreasing the risk of fracture of the drawn part.

6. The use of lubricant facilitates flow of the deformed material and improves smoothness of the drawn part surface, which improves the quality of the product.
7. The lubricant, applied to the sheet surface according to patented method [42], is environmentally friendly and has no harmful effect on people

Author Contributions: Conceptualisation, W.W. and J.A.; methodology, W.W., J.A., P.L., M.M. and M.D.; formal analysis, W.W., J.A., P.L. and M.M.; investigation, W.W., J.A., P.L., M.M. and M.D.; resources, P.L. and M.D.; writing—original draft preparation, W.W. and J.A.; writing—review and editing, W.W., J.A. and M.M.; visualisation, W.W. and J.A.; supervision, J.A. All authors have read and agreed to the published version of the manuscript.

Funding: This research received no external funding.

Institutional Review Board Statement: Not applicable.

Informed Consent Statement: Not applicable.

Data Availability Statement: The data presented in this study are available on request from the corresponding author.

Conflicts of Interest: The authors declare no conflict of interest.

References

1. Paszenda, Z.; Tyrlik-Held, J. Surgical Instruments. In *Polish: Instrumentarium Chrurgiczne*; Printing House of the Silesian University of Technology: Gliwice, Poland, 2003.
2. Gierzyńska-Dolna, M.; Adamus, J.; Lacki, P. Surface treatment of the implants and tools made of the titanium alloys. *Adv. Mater. Sci.* **2007**, *7*, 19–24.
3. Xue, T.; Attarilar, S.; Liu, S.; Liu, J.; Song, X.; Li, L.; Zhao, B.; Tang, Y. Surface Modification Techniques of Titanium and its Alloys to Functionally Optimize Their Biomedical Properties: Thematic Review. *Front. Bioeng. Biotechnol.* **2020**, *8*, 603072. [CrossRef] [PubMed]
4. Gherlone, E.; Polizzi, E.; Tetè, G.; Capparè, P. Dentistry and COVID-19 pandemic: Operative indications post-lockdown. *New Microbiol.* **2021**, *44*, 1–11. [PubMed]
5. *ISO 7153-1:2016*; Surgical Instruments. Materials. Part 1: Metals; ISO: Geneva, Switzerland, 2016.
6. *PN-EN 10088-1:2014-12*; Stainless Steels. Part 1: List of Stainless Steels; British Standard Institution: London, UK, 2014.
7. Adamus, J. Forming of the titanium implants and medical tools by metal working. *Arch. Mater Sci. Eng.* **2007**, *28*, 313–316.
8. Gronostajski, Z.; Pater, Z.; Madej, L.; Gontarz, A.; Lisiecki, L.; Lukaszek-Solek, A.; Luksza, J.; Mróz, S.; Muskalski, Z.; Muzykiewicz, W.; et al. Recent development trends in metal forming. *Arch. Civ. Mech. Eng.* **2019**, *19*, 898–941. [CrossRef]
9. Trzepieciński, T. Recent developments and trends in sheet metal forming. *Metals* **2020**, *10*, 779. [CrossRef]
10. Cheng, Z.; Li, Y.; Xu, C.; Liu, Y.; Ghafoor, S.; Li, F. Incremental sheet forming towards biomedical implants: A review. *J. Mater. Res. Technol.* **2020**, *9*, 7225–7251. [CrossRef]
11. Adamus, J. *Selected Problems of Forming Parts from Steel, Aluminum and Titanium Sheets*; In Polish: Wybrane Problemy Kształtowania Wytłoczek z Blach Stalowych, Aluminiowych i Tytanowych; Polish Scientific Publishers: Warsaw, Poland, 2018.
12. Vinci, R.; Teté, G.; Lucchetti, F.R.; Capparé, P.; Gherlone, E.F. Implant survival rate in calvarial bone grafts: A retrospective clinical study with 10 year follow-up. *Clin. Implant Dent. Relat. Res.* **2019**, *21*, 662–668. [CrossRef]
13. Harloff, T.; Hönle, W.; Holzwarth, U.; Bader, R.; Thomas, P.; Schuh, A. Titanium allergy or not? "Impurity" of titanium implant materials. *Health* **2010**, *2*, 306–310. [CrossRef]
14. Ali, S.; Rani, A.M.A.; Baig, Z.; Ahmed, S.W.; Hussain, G.; Subramaniam, K.; Hastuty, S.; Rao, T.V.V.L.N. Biocompatibility and corrosion resistance of metallic biomaterials. *Corros. Rev.* **2020**, *38*, 381–402. [CrossRef]
15. Jackson, M.J.; Ahmed, W. (Eds.) Titanium and Titanium Alloy Applications in Medicine. In *Surface Engineered Surgical Tools and Medical Devices*; Springer: Berlin/Heidelberg, Germany, 2007.
16. Takebe, H.; Takahashi, K. Applications of titanium and its alloys for weight reduction of automobiles. *Nippon Steel Tech. Rep.* **2019**, *122*, 181–185.
17. Adamus, J. Applications of titanium sheets in modern building construction. *Adv. Mat. Res.* **2014**, *1020*, 9–14. [CrossRef]
18. *Features of Titanium Building Materials*; Nippon Steel Corporation: Tokyo, Japan, 2019; Available online: https://www.nipponsteel.com/product/catalog_download/pdf/T003en.pdf (accessed on 22 February 2022).
19. Beal, J.D.; Boyer, R.; Sanders, D. Forming of titanium and titanium alloys. In *ASM Handbook*; Semiatin, S.L., Ed.; 14B: Metalworking: Sheet, Forming; ASM International: Cleveland, OH, USA, 2006; pp. 656–669. [CrossRef]
20. Wang, K.; Kopec, M.; Chang, S.; Qu, B.; Liu, J.; Politis, D.J.; Wang, L.; Liu, G. Enhanced formability and forming efficiency for two-phase titanium alloys by Fast light Alloys Stamping Technology (FAST). *Mater. Des.* **2020**, *194*, 108948. [CrossRef]
21. Tian, F.; Li, L. Investigation of the Feasibility of a Novel Heat Stamping Process for Producing Complex-shaped Ti-6Al-4V Panel Components. *Procedia Manuf.* **2020**, *47*, 1374–1380. [CrossRef]

22. Revil-Baudard, B.; Cazacu, O.; Massoni, E. Room-temperature plastic behavior and formability of a commercially pure titanium: Mechanical characterization, modeling, and validation. *Int. J. Solids Struct.* **2021**, *69*, 195–206. [CrossRef]
23. Chinapareddygari, T.M.; Ravishankar, C.; Thangaraj, K.; Albert, S.K.; Borah, U.; Vesangi, A.K.; Gupta, R.K. Stretchability of commercial purity titanium sheet. *Met. Mater. Trans. A.* **2019**, *50*, 5602–5613. [CrossRef]
24. Lin, P.; Hao, Y.G.; Zhang, B.Y.; Chi, C.Z.; Cui, X.L.; Shen, J.; Gao, D.S. Planar anisotropy, tension-compression asymmetry, and deep drawing behavior of commercially pure titanium at room temperature. *J. Mater. Eng. Perform.* **2019**, *28*, 1734–1744. [CrossRef]
25. Pham, Q.T.; Lee, M.G.; Kim, Y.S. Effect of distortional hardening behaviour on material responses of pure titanium sheets during hydraulic bulge test. *J. Phys. Conf. Ser.* **2018**, *1063*, 1–6. [CrossRef]
26. Pham, Q.T.; Lee, M.G.; Kim, Y.S. Characterization of the isotropic-distortional hardening model and its application to commercially pure titanium sheets. *Int. J. Mech. Sci.* **2019**, *160*, 90–102. [CrossRef]
27. Chen, F.-K.; Chiu, K.-H. Stamping formability of pure titanium sheets. *J. Mater. Process Tech.* **2005**, *170*, 181–186. [CrossRef]
28. Mohanraj, R.; Elangovan, S. Thermal modeling and experimental investigation on the influences of the process parameters on warm incremental sheet metal forming of titanium grade 2 using electric heating technique. *Int. J. Adv. Manuf. Tech.* **2020**, *110*, 255–274. [CrossRef]
29. Adamus, J.; Lacki, P. Investigation of sheet-titanium forming with flexible tool—Experiment and simulation. *Arch. Met. Mater.* **2012**, *57*, 1247–1252. [CrossRef]
30. Adamus, J.; Lacki, P. Possibility of the increase in titanium sheets' drawability. *Key Eng. Mat.* **2013**, *549*, 31–38. [CrossRef]
31. Aydogan, C.S.; Hatipoglu, H.A.; Keles, O. Springback prediction of CP2 titanium sheets in hydroforming with membrane diaphragm process with finite element method. *MATEC Web Int. Conf. New Form. Technol. (ICNFT 2018).* **2018**, *190*, 5. [CrossRef]
32. Kumar, A. Experimental analysis of formability of commercially pure titanium under PTFE lubricated conditions. *IOP Conf. Ser. Mater. Sci. Eng.* **2018**, *455*, 1–7. [CrossRef]
33. Kakulite, K.K.; Kandasubramanian, B. Rudiment of 'galling: Tribological phenomenon' for engineering components in aggregate with the advancement in functioning of the anti-galling coatings. *Surf. Interfaces* **2019**, *17*, 1–19. [CrossRef]
34. Adamus, J.; Dyja, K.; Więckowski, W. Lubricants based on vegetable oils as effective lubricating agents in sheet-titanium forming. *Key Eng. Mat.* **2016**, *687*, 163–170. [CrossRef]
35. Adamus, J.; Lackner, J.M.; Major, Ł. A study of the impact of anti-adhesive coatings on the sheet-titanium forming processes. *Arch. Civ. Mech. Eng.* **2013**, *13*, 64–71. [CrossRef]
36. Więckowski, W.; Adamus, J.; Dyner, M. Sheet metal forming using environmentally benign lubricant. *Arch. Civ. Mech. Eng.* **2020**, *20*, 1–14. [CrossRef]
37. PamStamp 2G, v. 2011. In *User's Guide*; ESI Group: Paris, France, 2011.
38. *EN 10204:2004*; Metallic Products—Types of Inspection Documents; Inspection Certificate 3.1; British Standard Institution: London, UK, 2004.
39. *PN-EN ISO 6892-1:2020-05*; Standard: Metallic Materials. Tensile Testing. Part 1: Method of Test at Room Temperature; Polish Committee for Standardization: Warsaw, Poland, 2020.
40. *PN-EN ISO 10113:2014-08*; Standard: Metallic Materials. Sheet and Strip; Determination of Plastic Strain Ratio. Polish Committee for Standardization: Warsaw, Poland, 2014.
41. Adamus, J.; Dyja, K.; Motyka, M. Experimental and theoretical determination of forming limit curve. *Arch. Met. Mater.* **2015**, *60*, 1881–1885. [CrossRef]
42. Method of Applying Lubricant on Surface of Sheet Made of Hard to Deform Material Before Cold Sheet Metal. Forming. Patent No PL229731, 21 October 2021.

materials

MDPI

Communication

Investigation of the Size Effects on the Elongation of Ti-2.5Al-1.5Mn Foils with Digital Image Correlation Method

Chunju Wang [1], Weiwei Zhang [2] and Zhenwu Ma [3,*]

[1] Robotics and Microsystems Center, School of Mechanical and Electrical Engineering, Soochow University, Suzhou 215131, China; cjwang@suda.edu.cn
[2] Institute of Electronic Engineering, China Academy of Engineering Physics, Mianyang 621999, China; zhangweiwei0509103@163.com
[3] College of Mechanical Engineering, Suzhou University of Science and Technology, Suzhou 215009, China
* Correspondence: mazw@usts.edu.cn

Abstract: The increasing demand for parts with a large specific surface area such as fuel panels has put forward higher requirements for the plasticity of foils. However, the deformation characteristics of foils is hard to be illustrated in-depth due to their very short deformation process. In this paper, the digital image correlation method was applied to investigate the influence of size effect on the elongation of Ti-2.5Al-1.5Mn foils. The results showed that the elongation of Ti-2.5Al-1.5Mn foils increased with the decrease in the ratio of thickness-to-grain diameter (t/d value). Then, the macro deformation distribution of foils was analyzed, combined with their microstructure characteristics, and it was found that the increasing influence of individual grain heterogeneity leads to the earlier formation of a concentrated deformation zone, which changes the deformation mode of foils. The concentrated deformation increases with the decrease in t/d value, thus dominating the trend of the foil elongation. Furthermore, the homogeneous deformation and concentrated deformation can be divided into two different zones by a certain critical t/d value. These results provide a basis for understanding and further exploration of the deformation behavior of titanium foils.

Keywords: foil forming; size effect; digital image correlation; concentrated deformation; critical t/d value

Citation: Wang, C.; Zhang, W.; Ma, Z. Investigation of the Size Effects on the Elongation of Ti-2.5Al-1.5Mn Foils with Digital Image Correlation Method. *Materials* **2021**, *14*, 7353. https://doi.org/10.3390/ma14237353

Academic Editors: Virgil-Florin Duma and Andrzej Dziedzic

Received: 29 October 2021
Accepted: 29 November 2021
Published: 30 November 2021

1. Introduction

With the miniaturization trend of mechanical equipment, the requirements for micro-parts are increasing significantly [1–5]. Among many manufacturing processes for micro-parts, the foil-forming process has attracted the most attention due to the high production efficiency and material utilization, endowing it with competitive advantage in manufacturing parts with a large specific surface area such as fuel panels and corrugated plates [6,7]. However, as the foil thickness decreases continuously, the deformation behavior of foils exhibits a strong dependence on dimension parameters, resulting in the size effects. The appearance of the size effects leads to the deterioration of plastic properties of foils, implying in-depth analyses on the deformation characteristics of foils.

The elongation is an important index to measure the plasticity of foils. In recent years, extensive studies have been devoted to the influence of size effects on the foil elongation. In general, they can be divided into two major subsets: the thickness effect and the grain size effect. The thickness effect on the elongation of CuZn20 foils has been studied via the tensile experiment, showing that the elongation decreases with the decrease in foil thickness [8]. This result can be explicated by the surface grain theory, in which the surface grain acts as a weakening factor during the deformation due to a lack of constraint of the grain boundaries [9,10]. Therefore, as the foil thickness reduces, the proportion of surface grains increases, so the fracture strength of the foil decreases, leading to less elongation. Similar phenomena have also been observed in the CuZn15 foils, CuZn36 foils, aluminum foils, etc. [11–13]. On the other hand, the effect of grain size on deformation behavior has been studied with pure copper foils and increased elongation, with the decrease in grain size having been demonstrated [14]. It can be explained by the fact that the grain boundary area is larger in materials with more small grains, so the distance of the crack propagation is longer before fracture, contributing to more plastic deformation of materials [15]. Consistent experimental results have been reported on copper alloy foils, stainless steel 304 foils, etc. [16,17].

To represent the interactive effects between the thickness effect and the grain size effect, the ratio of thickness-to-grain diameter (t/d value) has been proposed and applied widely, because it can interpret the experimental phenomenon of smaller sample size results in weaker deformation behavior [18–20]. However, as the foil thickness decreases to an order of magnitude of grain size, the heterogeneity of individual grains will dominate the deformation behavior of foils. Therefore, the individual grain heterogeneity should be taken into account when analyzing the deformation behavior of foils [21–23].

Taken together, most previous studies focus on the influence mechanism of size effects on the foil elongation from the perspective of grain numbers or surface grains, but there is a relative lack of research on the foil macroscopic deformation characteristics. In fact, the explicit deformation mode of a certain foil depends on comprehensively analyzing its macro and micro deformation characteristics, contributing to the illustration of the influence mechanism of size effects on the foil elongation. However, the deformation process of foils is very short, making it difficult to investigate its macroscopic deformation characteristics. The digital image correlation (DIC) technology is an optical measurement based on the calibration information of digital images. DIC has the advantages of non-contact, real-time, full-field, and local strain measurement. Therefore, it is an effective and accurate technology for studying lightweight bio-inspired composite materials [24] and measuring the full-field high-temperature thermal strain [25] and the strain in the foil deformation process. The Ti-2.5Al-1.5Mn foil is a kind of popular foil in the fuel panels and micro aircraft fields, but there is still a lack of in-depth study on its size effects during the micro-forming process. In this work, the macro and micro deformation characteristics of Ti-2.5Al-1.5Mn foil were first studied with the DIC technology. Then, the influence mechanism of size effects on the elongation of Ti-2.5Al-1.5Mn foils was explored. This study would provide a new perspective for understanding the influence of size effects on the deformation characteristics of foils.

2. Materials and Methods

Ti-2.5Al-1.5Mn foils with a thickness of 50 μm were recrystallized and annealed at a vacuum of 1.5×10^{-3} Pa at different temperatures and times, so various grain sizes were obtained. Then, Ti-2.5Al-1.5Mn foils with specific t/d values (Table 1) were selected for study. The microstructures along thickness direction of foils were observed with an optical microscope (OM, Novel MR5000), as shown in Figure 1. Tensile specimens were processed by the Wire Electrical Discharge Machining with annealed Ti-2.5Al-1.5Mn foils.

Table 1. Annealing parameters of Ti-2.5Al-1.5Mn foils.

Number	Annealing Parameters	Grain Size d (μm)	t/d
a	700 °C, 1.5 h	3.72 ± 1.23	13.43
b	720 °C, 2.5 h	5.34 ± 2.13	9.37
c	750 °C, 3.5 h	6.52 ± 2.46	7.68
d	800 °C, 3.5 h	9.73 ± 3.15	5.14
e	850 °C, 7 h	15.06 ± 4.67	3.32

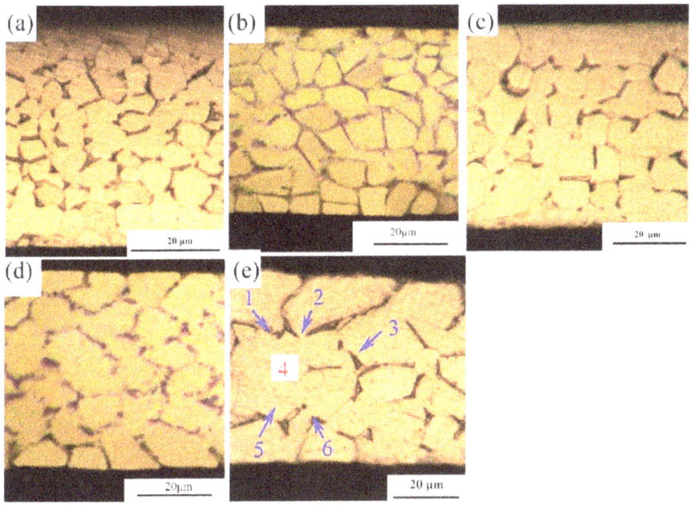

Figure 1. Microstructures of Ti-2.5Al-1.5Mn foils: (**a**) t/d = 13.43, (**b**) t/d = 9.37, (**c**) t/d = 7.68, (**d**) t/d = 5.14, (**e**) t/d = 3.32.

The tensile test for various Ti-2.5Al-1.5Mn foils was conducted at room temperature by using the material testing machine (Shimadzu, AGS-X 5KN, Kyoto, Japan). The cross-head velocity was 1 mm/min, and the strain rate was less than $1 \times 10^{-3}\ s^{-1}$. The initial measuring length and width for all specimens were kept constant as 50 and 12.5 mm, respectively. Six specimens were tested for each group.

As shown in Figure 2, the tensile deformation of Ti-2.5Al-1.5Mn foils throughout the process was captured with the DIC technology. The DIC system contains a material testing machine, a charge coupled device (CCD) camera with a 1/2-inch CMOS sensor and a 2.2 × 2.2 μm pixel size (Saga, SJ-UH500), and the GOM Correlate software. The measurement error in the DIC was ±0.01 pixels, so the influence of measurement error on deformation analysis can be neglected. The data acquisition scheme is as follows:

(1) First, a layer of white paint was sprayed on the specimen surface as the background color, and black speckles were evenly sprayed so achieve each 50% of black and white. Then, specimens were installed in the material testing machine.
(2) At the same time, when the material testing machine started running, the CCD camera, which was controlled by the GOM Correlate software, collected specimen images at the frequency of 1 image per second.
(3) The changes in marked speckle information were calculated via the GOM Correlate software, and then, the displacement and strain field information were obtained.

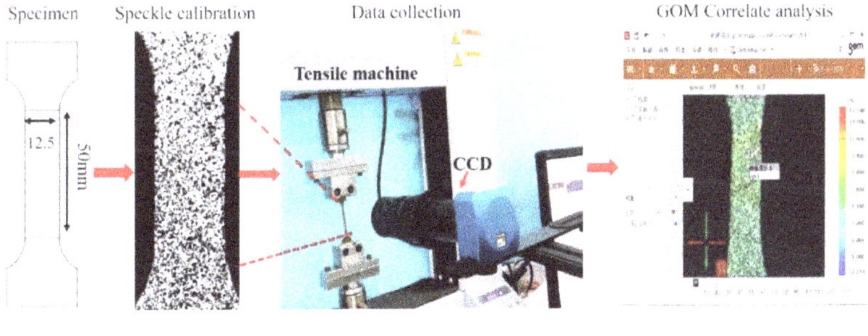

Figure 2. Analysis of the foil deformation process with the DIC technology.

3. Results and Discussion

First, the elongation at the break of Ti-2.5Al-1.5Mn foils was evaluated with respect to the t/d value referring to the current size effect analysis method. As shown in Figure 3a, the foil elongation increased with the decrease in the t/d value. When the t/d value decreased from 13.43 to 3.32, the elongation of foils increased by 4.3%. This result cannot be explicated by the surface grain theory [9–13].

Due to the thickness of foil specimens being constant at 50 μm, the elongation of Ti-2.5Al-1.5Mn foils was evaluated with respect to the grain size. According to existing research results, for most of materials, the larger the grain size, the smaller the material elongation is [14–17]. However, as shown in Figure 3b, the elongation of Ti-2.5Al-1.5Mn foils increased as the grain size increased.

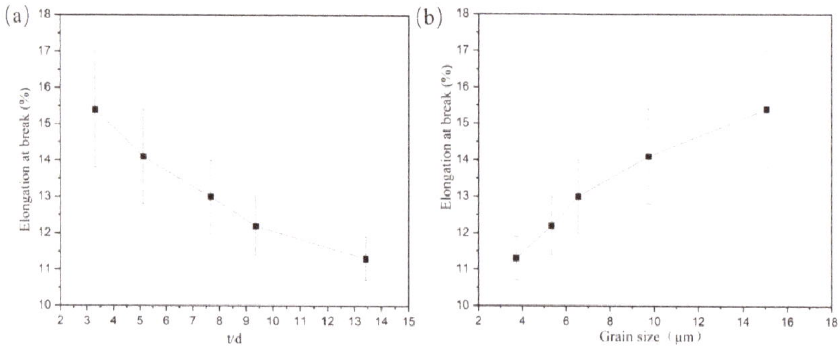

Figure 3. The variation of the elongation of Ti-2.5Al-1.5Mn: (**a**) with t/d value, (**b**) with grain size.

Furthermore, the scatter of the elongation between different Ti-2.5Al-1.5Mn foil specimens in the same experiment was investigated. In order to quantitatively study the scatter level of the foil elongation, the coefficient of variation (C.V) value of elongation was introduced. As shown in Figure 4, the C.V value of elongation increased with the decrease in t/d value, and the degree of this increase intensified markedly, indicating that the uniformity of the elongation of Ti-2.5Al-1.5Mn foils was becoming worse as the t/d value reduced. From the microstructures of various foils (Figure 1), the number of grains in the deformation area of the foil decreased as the annealing temperature increased. Since during the deformation process, surface grains were very easy to deform, the properties of internal grains played a significant role in the foil deformation behavior. The microhardness test for individual grains in the deformation area of foils indicated that the hardness distribution of the foils is disorderly, and the hardness distribution of foils exhibits block characteristics, which means that the hard- and soft-oriented grains are randomly distributed in the deformation area of the foil [8]. Because of the heterogeneity of individual grain properties, that is, different orientation and shape properties, the influence of individual grain heterogeneity was stronger for foils with fewer internal grains [20,21]. As shown in Figure 1e, the foil specimen with the smallest t/d value only owned three grains in the thickness direction, so its entire deformation behavior heavily depended on the deformation characteristics of individual internal grains, ultimately leading to the poor consistency of foil elongation. The drastically increased scatter behavior indicated that individual grain heterogeneity could be a critical starting point to explore the mechanism of size effects on the elongation of Ti-2.5Al-1.5Mn foils.

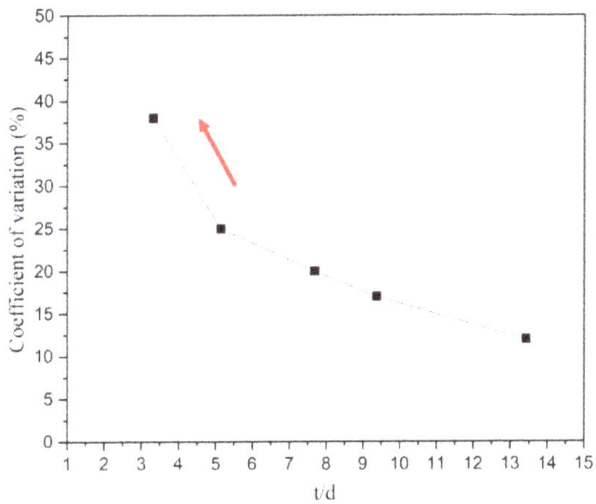

Figure 4. The change in C.V (δ) with the t/d value.

Compared with hard-oriented grains, soft-oriented grains tend to deform easily due to their crystal structure characteristics. It is known that the deformation performs in the manner of minimizing energy consumption. The specimen with fewer grains is prone to having a section with a large proportion of soft grains [26,27]. Therefore, a concentrated deformation zone is easy to form if the number of grains in the thickness direction is less than three. As shown in Figure 1e, if the internal grain 4 is a soft-oriented grain, the grains 1, 2, 3, 5, and 6 will deform in coordination with grain 4 in all directions to form a concentrated deformation zone. Once this concentrated deformation zone is developed, it will lead to the occurrence and concentration of remarkable deformation. Next, the DIC method is applied to study the influence of the concentrated deformation zone on the deformation behavior of foils.

A typical equivalent strain profiles during the deformation process of foils were collected with the DIC method, as shown in Figure 5. The original state of speckle calibration is shown in Figure 5a. Next, in Figure 5b,c, the overall strain distribution is relatively uniform, indicating that the foil is dominated by the homogeneous deformation throughout the region at this stage. In Figure 5d, the strain distribution of the foil appears partitioned, suggesting the occurrence of deformation stage transition. In Figure 5e,f, the strain distribution exhibits the characteristics of blockiness and concentration, while the rest of the regions are basically not deformed, which proves that local deformation is prominent at this time. The graphics of each specimen in each deformation stage corresponded to its engineering strain.

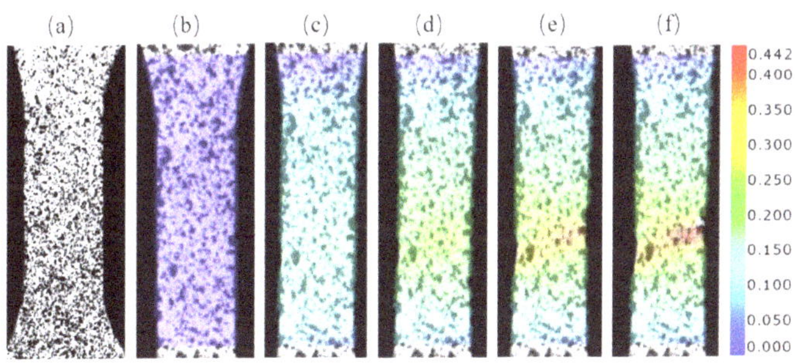

Figure 5. A typical equivalent strain profile of foils at different stages. (**a**) Original state, (**b**,**c**) homogeneous deformation stage, (**d**) transition of deformation stage, (**e**,**f**) concentrated deformation stage.

Therefore, combining the formation of concentrated deformation zone and the collected strain distribution characteristics with the DIC method, the deformation process of foils is further subdivided into the homogeneous deformation stage and the concentrated deformation stage. The critical point is set to the beginning of strain partition in this research. Accordingly, the foil deformation amount can be divided into the homogeneous deformation amount and the concentrated deformation amount. As shown in Figure 6, in the homogeneous deformation stage, the homogeneous deformation amount of foil specimens decreases with the decrease in t/d value, while in the concentrated deformation stage, the concentrated deformation amount of foil specimens increases rapidly with the decrease in t/d value. After calculating the difference values of homogeneous deformation amount and concentrated deformation amount between foils with adjacent t/d values, it can be seen that when the t/d value decreases from 5.14 to 3.32, the homogeneous deformation amount and concentrated deformation of foils amount both change markedly, suggesting the enhanced influence of concentrated deformation zone at this time. By comparing the variation degree of homogeneous deformation amount and concentrated deformation amount, it can be found that as the t/d value decreases, the increase in concentrated deformation amount is greater than the decrease in homogeneous deformation amount. Therefore, the sum of homogeneous deformation amount and concentrated deformation amount is greater for the foil specimen with a small t/d value. This is the reason why the elongation of Ti-2.5Al-1.5Mn foils increases with the decrease in the t/d value in this experiment.

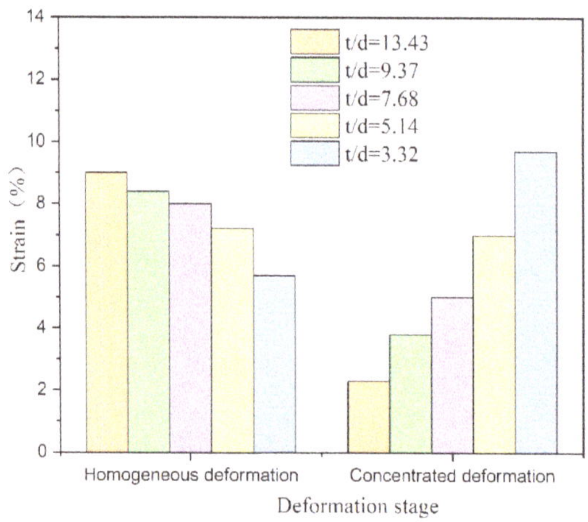

Figure 6. The homogeneous deformation and the concentrated deformation of various Ti-2.5Al-1.5Mn foils.

Ti-2.5Al-1.5Mn foil is a close-packed hexagonal lattice metal, and there are only three slip systems that can be activated at room temperature, so its deformation behavior is very sensitive to the change in deformation state. For the foil specimen with a large t/d value, the concentrated deformation zone is not easy to form because there are numerous grains inside the deformation zone [8,17]. After the deformation begins, the movable dislocations in the soft-oriented grains first produce slip stacking, and then, the dislocation slip is transferred to other grains easily due to the coordination effect between grains. In other words, both hard-oriented and soft-oriented grains in the deformation area are involved in the deformation, gradually producing the dislocation pile-up and deformation hardening [17,18], until forming the concentrated deformation zone. Subsequently, the foil enters the concentrated deformation stage. Since these grains have already produced severe dislocation stacking and deformation hardening in the homogeneous deformation stage, the amount of further deformation that can be generated in the concentrated deformation stage is very small. Finally, the foil quickly reaches the limit and fractures in the concentrated deformation zone. Therefore, the foil with a large t/d value has a large amount of homogeneous deformation but a small amount of concentrated deformation.

For the specimen with a small t/d value, the concentrated deformation zone is easy to form because there are a few grains in the deformation area [17]. After the deformation begins, the movable dislocations in soft-oriented grains produce slip stacking. Due to the poor coordination between grains, it is difficult to transfer the dislocation slip in soft-oriented grains to adjacent grains [27,28], resulting in the premature formation of concentrated deformation zone in this process. Subsequently, the foil quickly enters the concentrated deformation stage. Because most grains only deform slightly or do not deform in the homogeneous deformation stage, they still have a strong ability to further deform. In addition, the larger the individual grain size, the larger the area of concentrated deformation zone, which is composed of core soft-oriented grains and adjacent multiple grains, so greater deformation can be generated. This is the reason why Ti-2.5Al-1.5Mn foils that own smaller t/d value and larger grain size can produce a greater concentrated deformation amount.

The percentages of homogeneous deformation and concentrated deformation of Ti-2.5Al-1.5Mn foils with various t/d values were calculated and summarized in Figure 7. The results showed that the proportion of homogeneous deformation and concentrated deformation changes in significantly different trends, resulting in the formation of two zones, which are divided by a critical t/d value. For Ti-2.5Al-1.5Mn foils, this critical value is close to 5. In zone 2, the homogeneous deformation accounts for a large proportion and dominates the variation trend of the foil elongation. Subsequently, the proportion of concentrated deformation increases as the t/d value decreases. In zone 1, the concentrated deformation accounts for a large proportion and dominates the variation trend of the foil elongation. Due to the randomness of the position and deformation ability of the concentrated deformation zone, the uniformity of parts manufactured by using foils with the t/d value in zone 1 is poor. Taken together, it can be concluded that the plasticity of the Ti-2.5Al-1.5Mn foil with a thickness of 50 µm can be elevated by increasing its grain size, so as to produce parts with large specific surface area. However, foils with the t/d value in zone 1 should be avoided for manufacturing parts that require high forming uniformity.

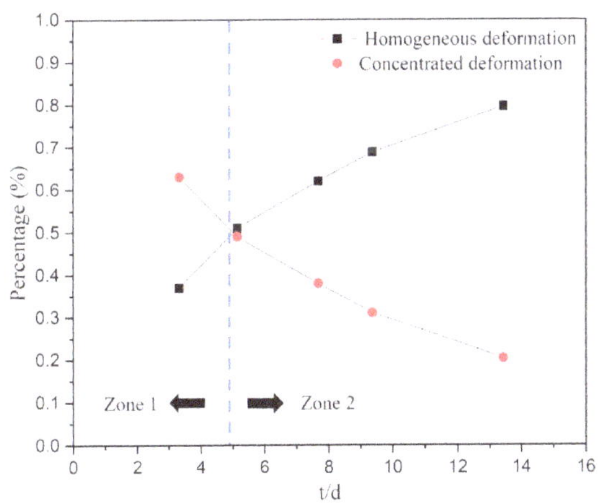

Figure 7. The percentage of homogeneous deformation amount and concentrated deformation amount of various Ti-2.5Al-1.5Mn foils.

4. Conclusions

In this work, the influence mechanism of size effect on the elongation of Ti-2.5Al-1.5Mn foil with a close-packed hexagonal crystal structure was studied with the DIC method, and the following conclusions can be drawn:

(1) The influence of individual grain heterogeneity increases during the deformation process, which results in the premature formation of concentrated deformation zones, contributing to the increased role of concentrated deformation in the foil elongation.
(2) The proportion of homogeneous deformation amount and concentrated deformation amount can be divided into two different zones by a certain critical t/d value, which is near 5 for Ti-2.5Al-1.5Mn foils.
(3) It is feasible to increase the plasticity of Ti-2.5Al-1.5Mn foils with a thickness of 50 μm by increasing the grain size, but foils with the t/d value in zone 1 should be avoided for forming parts that require high uniformity.

Author Contributions: Writing—original draft preparation, methodology, C.W.; conceptualization, investigation, W.Z.; supervision, project administration, Z.M. All authors have read and agreed to the published version of the manuscript.

Funding: This research was funded by the Natural Science Foundation of Jiangsu Province (BK20192007), the National Natural Science Foundation of China (No. 51905362, 51875128, 51805497), The Natural Science Foundation of Jiangsu Higher Education Institutions of China (No. 19KJB460022), Six Talent Peaks in Jiangsu Province (GDZB-069), and the Fundamental Research Funds for the Central Universities (Grant No. HIT. NSRIF. 2020043).

Institutional Review Board Statement: Not applicable.

Informed Consent Statement: Not applicable.

Data Availability Statement: The data cannot be shared because it also belongs to an ongoing research.

Conflicts of Interest: The authors declare no conflict of interest.

References

1. Feng, F.; Li, J.; Chen, R.; Peng, Y.; Su, H.; Zhang, Q.; Huang, P.; Zheng, Z. Effect of Die Geometry on the Formability of 5052 Aluminum Alloy in Electromagnetic Impaction Deformation. *Materials* **2018**, *11*, 1379. [CrossRef]
2. Wang, C.J.; Shan, D.B.; Zhou, J.; Guo, B.; Sun, L.N. Size effects of the cavity dimension on the microforming ability during coining process. *J. Mater. Process. Technol.* **2007**, *187*, 256–259. [CrossRef]
3. Wang, C.J.; Wang, H.Y.; Chen, G.; Zhu, Q.; Zhang, G.W.; Cui, L.J.; Zhang, P. Size effects affected uniaxial tensile properties and formability in rubber pad micro-forming process of pure nickel thin sheets. *Int. J. Mech. Sci.* **2020**, *182*, 105757–105763. [CrossRef]
4. Gu, X.; Wang, X.; Ma, Y.J.; Zhang, H.K.; Lu, J.X.; Wang, K.Y.; Liu, H.X. Investigation on grain size effect and forming mechanism of laser shock hydraulic micro-forming of copper foil. *Int. J. Adv. Manuf. Technol.* **2021**, *114*, 1049–1064. [CrossRef]
5. Huang, K.L.; Yang, Y.; Yin, D.Q.; Yang, X.Z.; Qin, Y. Effect of Electric Current Magnitude on Particles Deformation during the Densification Process of Micro-FAST. *Rare. Metal. Mat. Eng.* **2015**, *44*, 365–369.
6. Ma, Z.W.; Tong, G.Q.; Chen, F.; Wang, Q.; Wang, S.C. Grain size effect on springback behavior in bending of Ti-2.5Al-1.5Mn foils. *J. Mater. Process. Technol.* **2015**, *224*, 11–17. [CrossRef]
7. Peng, L.F.; Hu, P.; Lai, X.M.; Mei, D.Q.; Ni, J. Investigation of micro/meso sheet soft punch stamping process—Simulation and experiments. *Mater. Des.* **2009**, *30*, 783–790. [CrossRef]
8. Ma, Z.W.; Peng, X.; Wang, C.J.; Cao, Z.Y. Modeling of material deformation behavior in micro-forming under consideration of individual grain heterogeneity. *Trans. Nonferrous Met. Soc. China* **2020**, *30*, 2994–3005. [CrossRef]
9. Wang, C.J.; Wang, C.J.; Guo, B.; Shan, D.B.; Guan, H. Size effect on flow stress in uniaxial compression of pure nickel cylinders with a few grains across thickness. *Mater. Lett.* **2013**, *106*, 294–296. [CrossRef]
10. Vollertsen, F.; Hu, Z.; Schulzeniehoff, H.; Theiler, C. State of the art in micro forming and investigations into micro deep drawing. *J. Mater. Process. Technol.* **2004**, *151*, 70–79. [CrossRef]
11. Michel, J.F.; Picart, P. Size effects on the constitutive behavior for brass in sheet metal forming. *J. Mater. Process. Technol.* **2003**, *141*, 439–446. [CrossRef]
12. Raulea, L.V.; Goijaerts, A.M.; Govaert, L.E.; Baaijens, F.P.T. Size effects in the processing of thin metal sheets. *J. Mater. Process. Technol.* **2001**, *115*, 44–48. [CrossRef]
13. Wang, C.J.; Liu, Y.; Guo, B.; Shan, D.B.; Zhang, B. Acoustic softening and stress superposition in ultrasonic vibration assisted uniaxial tension of copper foil: Experiments and modeling. *Mater. Des.* **2016**, *112*, 246–253. [CrossRef]

14. Liu, J.G.; Fu, M.W.; Chan, W.L. A constitutive model for modeling of the deformation behavior in microforming with a consideration of grain boundary strengthening. *Comp. Mater. Sci.* **2012**, *55*, 85–94. [CrossRef]
15. Takeuchi, S. The mechanism of the inverse Hall-Petch relation of nanocrystals. *Scr. Mater.* **2001**, *44*, 1483–1487. [CrossRef]
16. Mahabunphachai, S.; Ko, M. Investigation of size effects on material behavior of thin sheet metals using hydraulic bulge testing at micro/meso-scales. *Int. J. Mach. Tools Manuf.* **2008**, *48*, 1014–1029. [CrossRef]
17. Chan, W.L.; Fu, M.W.; Lu, J.; Liu, J.G. Modeling of grain size effect on micro deformation behavior in micro-forming of pure copper. *Mater. Sci. Eng. A* **2010**, *527*, 6638–6648. [CrossRef]
18. Chan, W.L.; Fu, M.W. Experimental studies and numerical modeling of the specimen and grain size effects on the flow stress of sheet metal in microforming. *Mater. Sci. Eng. A* **2011**, *528*, 7674–7683. [CrossRef]
19. Engel, U.; Eckstein, R. Microforming—From basic research to its realization. *J. Mater. Process. Technol.* **2002**, *125*, 35–44. [CrossRef]
20. Saotome, Y.; Yasuda, K.; Kaga, H. Microdeep drawability of very thin sheet steels. *J. Mater. Process. Technol.* **2001**, *113*, 641–647. [CrossRef]
21. Ma, Z.W.; Cao, Z.Y.; Lu, J.B.; Zhang, Y.J.; Liu, W. Size effects on springback behavior of H80 foils. *Rare Met.* **2018**, *37*, 1082–1090. [CrossRef]
22. Lu, H.N.; Wei, D.B.; Jiang, Z.Y.; Liu, X.H. Modelling of size effects in microforming process with consideration of grained heterogeneity. *Comp. Mater. Sci.* **2013**, *77*, 44–52. [CrossRef]
23. Fang, Z.; Jiang, Z.Y.; Wei, D.B.; Liu, X.H. Study on springback in micro V-bending with consideration of grain heterogeneity. *Int. J. Adv. Manuf. Technol.* **2015**, *78*, 1075–1085. [CrossRef]
24. Ha, N.S.; Le, V.T.; Goo, N.S. Investigation of punch resistance of the Allomyrira dichtoloma beetle forewing. *J. Bionic Eng.* **2018**, *15*, 57–68. [CrossRef]
25. Ha, N.S.; Le, V.T.; Goo, N.S.; Kim, J.Y. Thermal Strain Measurement of Austin Stainless Steel (SS304) during a Heating-cooling Process. *Int. J. Aeronaut. Space Sci.* **2017**, *18*, 206–214. [CrossRef]
26. Chin, G.Y.; Mammel, W.L. Computer solutions of the Taylor analysis for axisymmetric flow. *Trans. TMS AIME* **1967**, *239*, 1400–1405.
27. Chan, W.L.; Fu, M.W. Experimental and simulation based study on micro-scaled sheet metal deformation behavior in microembossing process. *Mater. Sci. Eng. A* **2012**, *556*, 60–67. [CrossRef]
28. Xue, S.; Wang, C.; Chen, P.; Xu, Z.; Cheng, L.; Guo, B.; Shan, D. Investigation of Electrically-Assisted Rolling Process of Corrugated Surface Microstructure with T2 Copper Foil. *Materials* **2019**, *12*, 4144. [CrossRef]

materials

MDPI

Article

Parametric Effects of Single Point Incremental Forming on Hardness of AA1100 Aluminium Alloy Sheets

Sherwan Mohammed Najm [1,2,*], Imre Paniti [1,3], Tomasz Trzepieciński [4], Sami Ali Nama [5], Zsolt János Viharos [3,6] and Adam Jacso [1]

1 Department of Manufacturing Science and Engineering, Budapest University of Technology and Economics, Műegyetem rkp 3, H-1111 Budapest, Hungary; imre.paniti@sztaki.hu (I.P.); jacso.adam@gpk.bme.hu (A.J.)
2 Kirkuk Technical Institute, Northern Technical University, Kirkuk 41001, Iraq
3 Centre of Excellence in Production Informatics and Control, Institute for Computer Science and Control (SZTAKI), Kende u. 13-17, H-1111 Budapest, Hungary; viharos.zsolt@sztaki.hu
4 Department of Manufacturing and Production Engineering, Faculty of Mechanical Engineering and Aeronautics, Rzeszow University of Technology, Al. Powst. Warszawy 8, 35-959 Rzeszów, Poland; tomtrz@prz.edu.pl
5 Engineering Technical College, Middle Technical University, Baghdad 10074, Iraq; drsami@mtu.edu.iq
6 Department of Management and Business Law, Faculty of Economics and Business, John von Neumann University, Izsáki Str. 10, H-6000 Kecskemét, Hungary
* Correspondence: sherwan.mohammed@gpk.bme.hu

Abstract: When using a unique tool with different controlled path strategies in the absence of a punch and die, the local plastic deformation of a sheet is called Single Point Incremental Forming (SPIF). The lack of available knowledge regarding SPIF parameters and their effects on components has made the industry reluctant to embrace this technology. To make SPIF a significant industrial application and to convince the industry to use this technology, it is important to study mechanical properties and effective parameters prior to and after the forming process. Moreover, in order to produce a SPIF component with sufficient quality without defects, optimal process parameters should be selected. In this context, this paper offers insight into the effects of the forming tool diameter, coolant type, tool speed, and feed rates on the hardness of AA1100 aluminium alloy sheet material. Based on the research parameters, different regression equations were generated to calculate hardness. As opposed to the experimental approach, regression equations enable researchers to estimate hardness values relatively quickly and in a practicable way. The Relative Importance (RI) of SPIF parameters for expected hardness, determined with the partitioning weight method of an Artificial Neural Network (ANN), is also presented in the study. The analysis of the test results showed that hardness noticeably increased when tool speed increased. An increase in feed rate also led to an increase in hardness. In addition, the effects of various greases and coolant oil were studied using the same feed rates; when coolant oil was used, hardness increased, and when grease was applied, hardness decreased.

Keywords: SPIF; single point incremental forming; sheet forming; hardness; ANN; relative importance (RI)

Citation: Najm, S.M.; Paniti, I.; Trzepieciński, T.; Nama, S.A.; Viharos, Z.J.; Jacso, A. Parametric Effects of Single Point Incremental Forming on Hardness of AA1100 Aluminium Alloy Sheets. *Materials* **2021**, *14*, 7263. https://doi.org/10.3390/ma14237263

Academic Editors: Jie Zhou and Valentin Ştefan Oleksik

Received: 14 October 2021
Accepted: 17 November 2021
Published: 27 November 2021

1. Introduction

Incremental Sheet Forming (ISF) is a sheet-forming technique that produces components through a series of small incremental deformations. ISF is a flexible active manufacturing process and is economically feasible for low-volume production due to the absence of a punch and die. SPIF is one of the major types of ISF, and it is known as the simplest process variant of incremental sheet-forming technologies. SPIF is an emerging process that has been identified as suitable for use in small-scale production. Trzepieciński et al. [1] presented a brief overview of state-of-the-art methods of ISF for lightweight materials. The aim of their paper was to guide and inspire researchers by identifying current development trends of valuable contributions in the field of Single Point Incremental Forming (SPIF) of lightweight metallic materials. In SPIF, a rotating tool with a rounded tip at its end is used, and the desired shape is formed from clamped sheet metal [2]. In the literature [3], a review paper on new advances and future opportunities considered single point incremental to be one of the forming technologies of future-proof materials in aerospace applications. Furthermore, there is governmental, academic, and business interest in developing new manufacturing technologies, and there is also interest in ISF's impact on the environment, particularly in what ways and to what extent ISF reduces energy needs [4]. Two exergy analyses of traditional forming and hydroforming of ISF were contrasted by Dittrich et al. in [5]. After analysing the environmental impact of these forming techniques in the supply chain, they concluded that ISF is significantly less harmful to the environment, particularly for prototypes and non-mass production. Sustainability guidelines were developed by Ingarao et al. [6] regarding the advantages and disadvantages of SPIF related to the amount of energy necessary to form sheets and to economic material use in each process. Ingarao et al. proved that SPIF supports saving material with respect to CO_2 emissions, because it supports recycling and facilitates novel ways of preparing raw materials. An overview of the history of ISF was written by Emmens et al. [7], Li et al. [8], and Behera et al. [9]: they discussed the enormous benefits and many advantages of SPIF and particularly referenced the flexibility of the process, which allows SPIF to be used in more applications in industries and processes. Hence, SPIF will be considered an essential process for the industry in the future.

By studying the mechanical properties of components formed through SPIF and standardising effective process parameters, the SPIF process can become a significant industrial application embraced by numerous companies. To set optimal input process parameters of SPIF with multiple pure copper sheets, Raju and Sathiya [10] utilised a hybrid optimisation technique by connecting it with Taguchi grey relational analysis and the methodology of the response surface. They showed that the feed rate is the most significant parameter, followed by step depth and tooltip diameter. On the other hand, lubrication plays a vital role in the successful forming of components during the SPIF process. Lubrication reduces friction at the contact zone of the tool and sheet [11]. Recently, José et al. [12] studied how mineral oil, sunflower, soybean, and corn lubricants influence friction and wear effects on aluminium parts manufactured using the SPIF process. Using a scanning electron microscope, they found the following: the experimental characterisation of the sample's surface showed that the worn surface of the metallic material samples produced using vegetable oils increases surface roughness compared to those produced with mineral oils. With respect to aluminium alloy foils, Najm and Paniti [13,14] noted that a flat tool yielded better outcomes than a hemispherical tool in various conditions of the SPIF process. The best geometric accuracy was achieved when the smallest corner radius flat tool was used because a decrease in spring-back was observed. ISF-as-a-Service was introduced by Paniti [15], who distinguished first-order and second-order bottleneck parameters. He described the main capabilities of an incremental sheet-forming service provider in cloud manufacturing. The hardening and normal anisotropy coefficients are the most influential factors on traditional Forming Limit Diagrams (FLDs), as found by Fratini et al. [16] when studying the relationship between material formability and mechanical properties. On the other hand, Zhang [17] concluded that the most influential factor on formability is forming temperature, followed by vertical step depth, sheet thickness, and tooltip diameter. The above-mentioned results were obtained on Mg alloy sheets formed by using warm incremental sheet forming. Liu et al. [18] asserted that formability and maximum vertical force increase when vertical step-down increases, and they linearly increase when sheet thickness is increased. In their study [19], Li et al. found that mechanical properties and the thinning rate were affected by three parameters of ISF. It was shown that an increase in the tool diameter considerably improved the microhardness of the product's surface. An increase in the tool diameter and a decrease in the step size ultimately increase the tensile strength rate. Li et al. claimed that contrary to step size, sheet thickness significantly affects yield strength. In [20], Manco et al. proved that, due to variations in tool path, tool trajectory could be considered an essential parameter for the optimisation of process design by comparing the smallest thickness of the sheet with predicted thickness using the sine law. Krasowski et al. [21] analysed and discussed experimental investigations of the effects of selected SPIF parameters on the formability of DC04 sheets and the susceptibility to crack formation on truncated cones produced through SPIF, and they found that lubrication conditions clearly affect the formability of DC04 steel sheets.

At present, many methods of artificial intelligence are used in various applications, including the metal forming process. More specifically, ANN generates predictive models for end-milling machining, powder metallurgy, and high-speed machining [22–24]. Moreover, machine learning techniques with controlled manufacturing are used to develop various effective predictive models [25–29]. Trzepieciński et al. [30] presented an analysis of the interaction between SPIF process parameters and the main roughness parameters. They found that predictive models of ANNs for Ra and Rz were characterised by performance measures of R^2 values between 0.657 and 0.979. In other studies, different tool materials and shapes were investigated experimentally to study factors including formability, geometric accuracy [31], and surface roughness [32] on an AlMn1Mg1 sheet formed using SPIF under various forming conditions. The researchers evaluated the performance of an Artificial Neural Network (ANN) and Support Vector Regression (SVR). Two different ANN models were built in the study: an R-squared value with other validation metrics and a feed-forward neural network with a backpropagation algorithm were used. A close correspondence was found between predicted roughness, formability, and geometric accuracy in the experimental results. The researchers derived regression equations to analytically predict surface roughness in terms of Ra and Rz. Baruah et al. [33] claimed that lubrication was the largest contributing factor in the process of ISF in all three directions (rolling, transverse, and angular) when surface roughness in ISF is meant to be reduced. In fact, to date, the applied lubricant and the viscosity of the lubricants on the ISF process have not been optimised or discussed, as attested by [5,34,35]. In addition, Kumar and Gulati [34] claimed that all parameters investigated in their study were significant for forming force except lubricating oil viscosity, and they also noted that surface roughness decreased when viscosity increased [35]. According to the literature, ANN is a helpful tool—before starting new experiments—for predicting and designing predictive models to estimate expected results, behaviour, or direction based on the use of the parameters of the studied process. Using ANN before starting an actual experiment has the essential benefits of selecting the correct parameters, reducing processing time, increasing efficiency, minimising errors, and comparing actual results with predicted ones so as to reach the best values. In addition, ANN is considered one of the most powerful tools for solving engineering problems by predicting experimental data. In addition, ANN can serve as a valuable means to generate and assess different processes and prepare the final details of tools.

Under normal conditions, the hardness behaviour in SPIF is as follows: formed parts achieve higher hardness than an unformed sheet. Using different path strategies and different forming angles, Al-Attaby et al. [36] showed that the tool path affects the hardness and microstructure of the formed sheet. In all cases relating to the forming angle, hardness increased. Regarding the two-point incremental forming process, Mostafanezhad et al. [37] analysed the formability of aluminium 1050: the scholars used the response surface methodology experimentally. They found that wall angle is the most influential factor with respect to the thinning ratio; initial thickness, followed by step-down, has a significant impact on forming force.

The above-detailed issues, the need for well-defined mechanical properties of SPIF components, and the lack of referent analytical models prompted the authors to investigate the effects of SPIF process variables on the hardness of truncated cones formed from AA1100 aluminium alloy sheets. Moreover, as a novelty and aim contributing to the significance of this paper, different regression equations were derived to determine the hardness of the components of a truncated cone using SPIF. In addition, the Relative Importance (RI) of parameters of SPIF on hardness was assessed and classified by utilising the joint partitioning weight of the built neural network. To the best of the authors' knowledge and according to the literature introduced, such an experimental process has not been reported to date. In this research, the influences of feed rate, various kinds of grease and coolant oil, spindle speed, and tool diameter on the hardness of the AA1100 aluminium alloy sheet formed by SPIF were investigated. Conventionally, AA1100 alloy is employed for radiator components [38]. However, as a final general aim, it was posited that the appropriate selection of properties would improve the application of the AA1100 alloy. Furthermore, based on the diver's mechanism of deformation, Song et al. [39] found three different regions of deformation (bending/stretching, shear, and stretch/shear). Based on these research projects, in this study, hardness was measured in three different positions on the inside wall of the cone, and measured data were compared to the primary hardness of the sheet involved in the experiment.

2. Material and Methods

2.1. *Workpiece Material*

Aluminium and aluminium alloys have become attractive materials for application in the aerospace and automotive industries owing to their beneficial properties. In the experiments conducted in this study, single point incremental forming tests were conducted using a blank sheet of AA1100 aluminium alloy were produced by Xuzhou Bozhan Aluminum Technology Co. Ltd, Xuzhou, Jiangsu, China with an initial hardness of 42.87 HV. The initial thickness of the sheet used is 0.6 mm, with an original surface roughness of 0.29 μm. AA1100 aluminium belongs to the 1xxx series with less than 1% alloying elements. The main uses of the 1xxx series aluminium alloys are foil and strip for packaging, chemical equipment, tank car or truck bodies, spun hollowware, and elaborate sheet metal work because of their high corrosion resistance and formability [40]. The 1xxx alloys are essentially characterised by superior corrosion resistance, usefulness for fabricating chemical tanks and piping, or their excellent electrical conductivity, as in bus bar applications. These alloys have relatively poor mechanical properties [41].

If strength is not an essential factor, AA1100 aluminium is selected to create fuel tanks, cowlings, and oil tanks of aircraft due to the corrosion resistance and the economic weight. The before-mentioned grade of aluminium can be utilised to repair aircraft wingtips and tanks because it is weldable [42]. AA1100 commercially pure aluminium is highly resistant to chemical attack and weathering. This low-cost material is characterised by excellent solderability and susceptibility to deep drawing. It is used for high-purity applications such as chemical processing equipment. In addition, examples of common 1xxx series aluminium alloy applications include nameplates, fan blades, flue lining, sheet metal work, spun holloware, and fin stock [43]. It is also used to produce decorative parts, giftware, cooking utensils, rivets, and reflectors. A SPECTROMAXx optical emission spectrometer manufactured by SPECTRO, Kleve, Germany was used to determine the chemical composition of the AA1100 alloy used, and the test of mechanical properties was conducted with a United testing machine according to the ASTM B557M-15 standard test methods for tension check. The results of the tests of the aluminium sheet were compared with the nominal values in the ASM Handbook [44]: these tests show the conformity to the standard composition of the alloy (ISO 19000 standard). The mechanical properties and the chemical composition of the sheet material are shown in Tables 1 and 2, respectively.

Table 1. Mechanical properties of the AA1100 aluminium alloy sheet.

Property	Ultimate Tensile Stress, MPa	Yield Strength, MPa	Elongation, %
Actual	110	95	20
Nominal	110	103	25
Standard Deviation, σ	0	4	2.5

Table 2. Chemical composition of the AA1100 aluminium alloy sheet (in wt.%).

Element	Si	Fe	Cu	Mn	Mg	Cr	Ni	Zn	Ti	Pb	B	Sn	V	Al
Actual	0.110	0.482	0.004	0.005	0.001	0.0005	0.004	0.021	0.021	0.0005	0.003	0.001	0.014	balance
Nominal	0.5	0.5	0.2	0.04	0.01				Other 0.15 max					balance

2.2. Experimental Setup

In this study, a Boxford 300VMCi milling machine built by Boxford in Halifax, UK with 0.01 mm accuracy was used. The ISO format using G and M codes was used to program a cone shape with large and small diameters of 80 and 10 mm, respectively. An inward spiral path strategy was used to deform the cone part, in which case a spiral tool trajectory is advantageous to the successful forming of the same parts [45]. The strategy utilised in this study was developed by Skjoedt et al. [46] to overcome the difficulty of reaching maximum axial loads at each layer (step down) and to prevent the appearance of a line on the inner side of the formed part. A wall angle of 45°, a context contour of 0.5 mm for the step size, and an inward spiral path were applied, as shown in Figure 1.

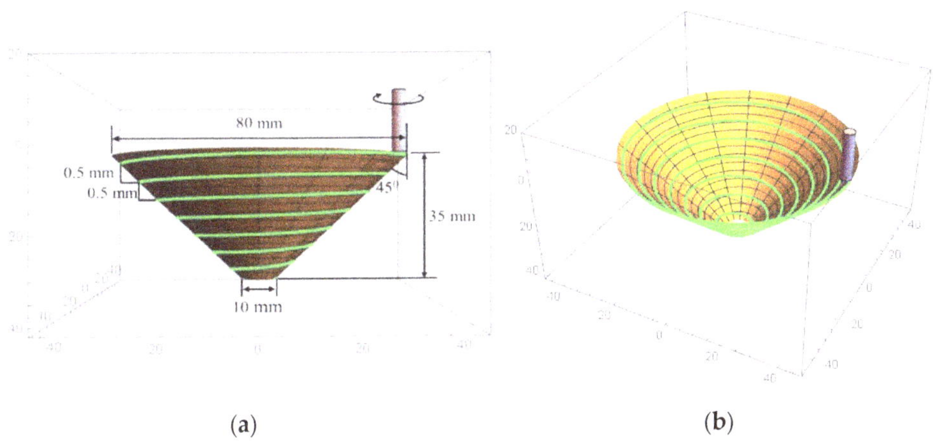

Figure 1. (**a**) CAD geometry and dimensions of the experimental product and (**b**) view of an inward spiral path.

In the SPIF process, only one tool can be used, and the feed rate, spindle speed, lubricant, and forming conditions should be selected beforehand. Furthermore, in the current study, the mentioned parameters were considered parametric values and were changed in the subsequent part forming. A primary step was conducted to select the best values of these parameters to fix them in the subsequent forming process, and only one of these parameters was changed in each forming group. To this end, a matrix of 3 factors with the same levels (feed rate, spindle speed, and tool diameter) was applied in the first step. The best values of feed rate, spindle speed, and tool diameter were chosen depending on the best geometrical accuracy and maximum depth. In the second step, the best values selected from the first step were applied in order to study the coolant type.

Forming tools with different diameters (4, 6, 8, and 10 mm) were used in the experiment, as shown in Figure 2a. The tools used in the experiment are made of carbon steel with a hardness of 30 HRC and are 100 mm in total length. Plain carbon steel was used for manufacturing the clamping rig, which was fixed to the CNC machine table with a simple-to-use fixture system, as shown in Figure 2b.

A digital Vickers microhardness tester supplied by TIME Group Inc., Beijing, China was used to measure the hardness of the component formed using SPIF (see Figure 3a) based on Equation (1). For each set, three products were experimentally formed to study the process parameters of different forming conditions (see Figure 3b). The hardness of each part was measured at three zones: the upper, middle, and lower zones along the inner wall of the formed part. Figure 3c shows a formed part after it was cut to the desired shape in the proper size for the preparation of the test samples and for establishing the three zones of the hardness measurement. The hardness measurement was repeated three times at different points inside each mentioned zone. The hardness value of each zone was calculated as the average value of the selected zone hardness. The average hardness value of the three zones was considered the average value of the measured component. The appropriate piece of the section was mounted by a mounting press device and polished by a Metaserv type 250/RP device manufactured in (Buehler, Lake Bluff, IL, USA) before hardness was measured (see Figure 3d).

$$H_v = 1854.4\,\frac{F}{d^2} \tag{1}$$

where F is penetration force (N), and d is average diagonal distance (d1 + d2)/2.

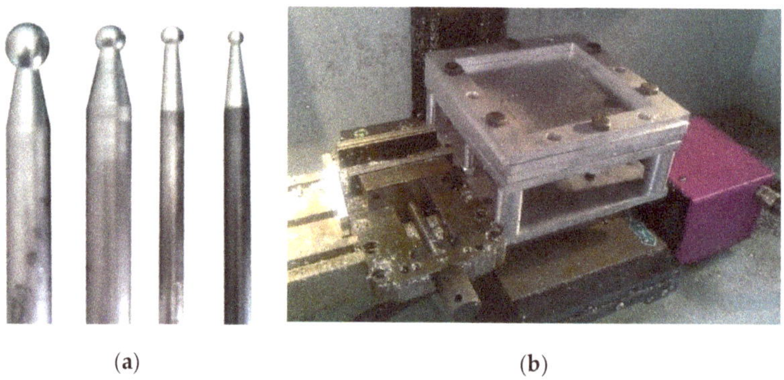

Figure 2. (**a**) Forming tools with different diameters, (**b**) fixture rig on the CNC machine table.

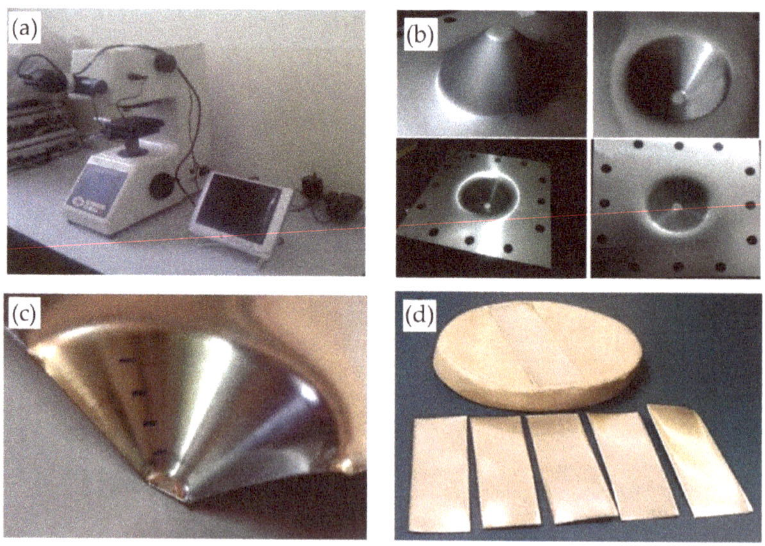

Figure 3. (**a**) Digital microhardness device, (**b**) formed parts, (**c**) three zones of hardness measurement, (**d**) hardness test sample of the formed part.

For hardness measurement, the hardness tester was calibrated before testing using the calibration standard block, and 100 N was applied on the formed part with a Vickers diamond pyramid indenter for 15 s. The results were recorded automatically on a digital screen after the adjustment of the rhomb corner had been triggered by the indenter.

There is no internationally accepted term for the definition of Environmentally Acceptable Lubricants (EALs), and they still lack standardisation. The American Society for Testing and Materials (ASTM) used "environmentally acceptable" as a phrase for defining EALs [47]. There is an overall trend towards using EALs. In the present study, different coolant types (four different grease types, as shown in Figure 4a–d, and one coolant oil) were used to carry out the experiment. Supergrees EP2 and Kaucuklu grease produced by Petrol Ofisi, Istanbul, Turkey, Zinol grease from Universal Lubricants (ZINOL) L.L.C, Sharjah, United Arab Emirates, Gp Grease Calcium type was produced by United Grease & Lubricants Co. LLC based in Ajman, United Arab Emirates, and the coolant oil was also by Petrol Ofisi, Istanbul, Turkey. Table 3 lists grease properties based on their commercial name and standard denominations, and Table 4 presents coolant oil properties. It is worth mentioning that viscosity values of different greases were assumed based on the ISO 3448:1992 standard for viscosity grading systems [48].

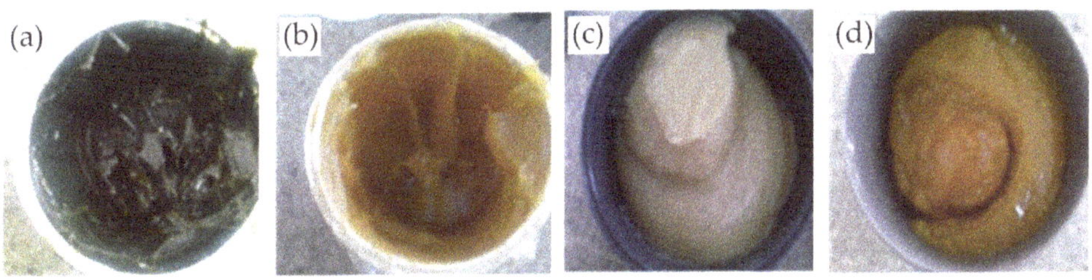

Figure 4. Different types of grease: (**a**) EP2, (**b**) Kaucuklu, (**c**) Zinol, (**d**) Gp Grease Calcium.

Table 3. Selected properties of the greases used.

Grease Type	ISO Viscosity Grade	Average Dropping Point, °C (at 25 °C)	Flash Point, °C	Viscosity at 40 °C, mm^2/s
EP2	ISO VG 15	90	180	15
Kaucuklu	ISO VG 22	88	172	22
Zinol	ISO VG 32	88	170	32
Gp Grease Calcium	ISO VG 46	58	60	46

Table 4. Selected properties of the coolant oil used.

Acidity, pH	Kinematic Viscosity at 29 °C, mm^2/s	Boiling Point, °C
1.086	1.086	95

Lubricants cannot be used in forming processes where high loads are applied, and thus, Syahrullail et al. [49] suggested using an appropriate additive to solve this problem. Consequently, the difference between using coolant oil and grease is that grease forms a mixture with small disintegrating particles (debris) of either the formed sheet or, in rare cases, the tool. Due to heat generation, sometimes the debris repeatedly sticks to the sheet surface or passes between the tool and the formed sheet. Diabb et al. [50] observed aluminium flakes in the used lubricant: this phenomenon was caused by wear adhesion on alloy sheets of SPIF components. In the case of coolant oil, which flows continuously on the sheet, debris can be washed away from the forming zone. However, when grease is used, a smoother surface can be produced compared to the scenario where coolant oil is used due to the flattening and roughening effects exerted by the debris, as stated in [51].

On the other hand, coolant oil continuously flows during the forming process, whereas grease is applied on the sheet surface only once at the beginning of the process. In other words, coolant oil has higher exergy than grease due to the difference in the amount of material used, which means that an increased environmental impact is observable. Figure 5a,b illustrate the processes of using grease and coolant oil.

Figure 5. (**a**) Forming process using grease, (**b**) forming process using coolant oil.

Four different tool rotation speeds (500, 1000, 1500, and 2000 rpm) were used to study hardness behaviour. In addition, in the scope of the current experiments, four different feed rates (200, 400, 600, and 800 mm/min) were implemented to investigate the effects of changes in the feed rate on sheet hardness.

3. Results and Discussion

3.1. Feed Rate

Four different feed rates were used with different lubricants (oil and grease). At the same time, other experimental parameters were fixed: the tool speed was 2000 rpm, the tool diameter was 10 mm, and coolant oil was used. Table 5 and Figure 6a,b show the results of hardness measurements for different feed rates. Changing the lubricant type resulted in inverse values of hardness: it increased when the feed rate was increased and coolant oil was used, and it decreased when grease was used.

An increase in feed rate led to an increase in hardness, and this was inversely proportional to formability. The increase in feed rate caused a decrease in formability, as mentioned in [52]. A decrease in hardness is due to changes in surface asperities because the peaks of the aspirates formed by the generated debris shoot and break. By attaching the debris to the tool and cultivating the sheet surface, new grooves can be created, and the sharp peaks of the asperities can likewise be crashed. Finally, through continuous cultivation and crashing, the contact area between the tool and the formed sheet will increase.

Table 5. Effects of different feed rates on hardness.

Feed Rate, mm/min	Hardness HV							
	Coolant Oil				Grease			
	Top	Middle	Bottom	Standard Deviation, σ	Top	Middle	Bottom	Standard Deviation, σ
200	43.89	40.40	39.02	2.0494	63.30	58.61	50.56	5.2610
400	39.16	43.61	41.78	1.8262	59.80	56.70	49.13	4.4816
600	44.08	46.44	48.22	1.6957	51.29	53.17	44.70	3.6317
800	47.76	53.84	49.54	2.5522	43.99	45.72	45.50	0.7689

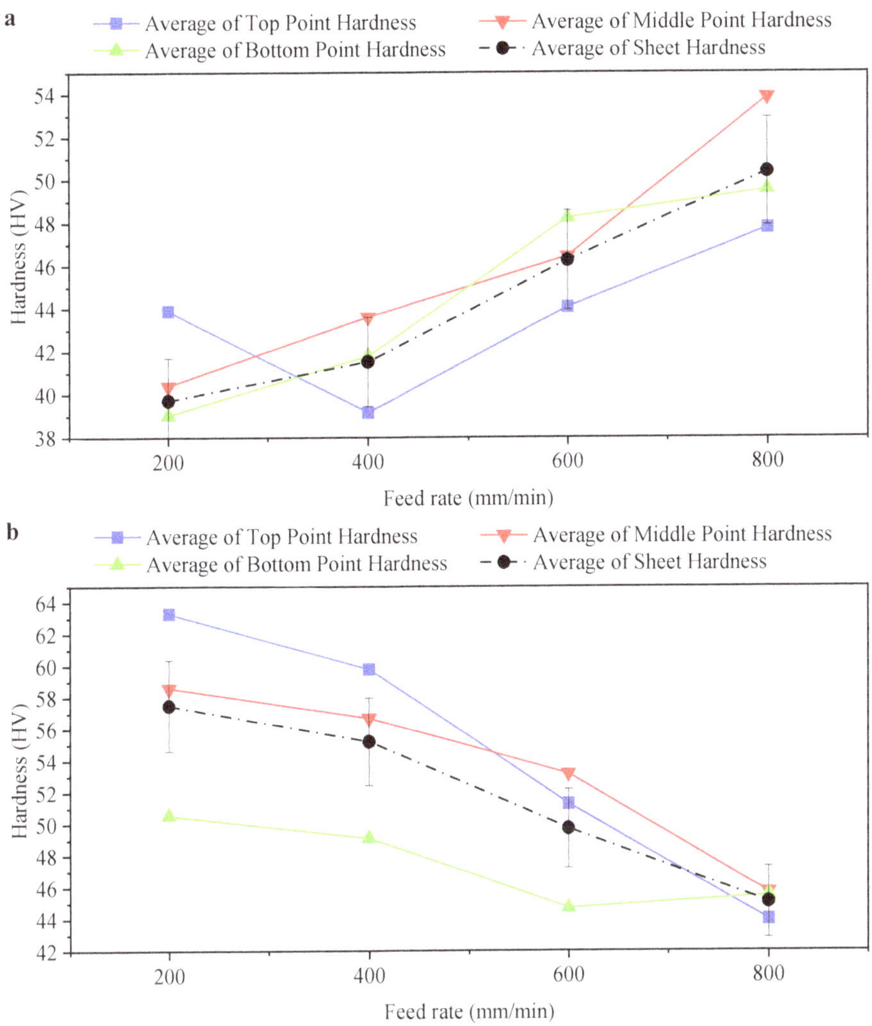

Figure 6. (**a**) Effects of different feed rates on hardness of formed sheet when using coolant oil, (**b**) effects of varied feed rates on hardness of formed sheet when using grease.

Hol et al. [53] mentioned that, in the case of normal forces, the sheet surface asperities are in the plastic condition, and they are further affected by only a little stress in the underlying bulk material. They claim that this stress is perpendicular to the normal force and generates increased plastic deformation of asperities. Finally, because of the enormous strain of the underlying material, this situation leads to an increased contact area, which is recognised as a decrease in effective hardness.

3.2. Tool Speed

Table 6 lists different tool rotation speeds with the experimentally obtained hardness values. Figure 7 shows that an increase in tool speed led to an increase in hardness. High speed causes the resulting particles to impact the surface of the sheet faster than in the case of low speeds, and this results in the hardening of the surface. On the other hand, in the case of high tool speeds, the tool head, in the same contact area, travels on the sheet with more passes than a tool at low speeds. Due to stretching with longitudinal deformation, the sheet material seems to be undergoing cold working conditions. Cold working creates a different type of crystal deformation, such as compressing, twisting, and bending, and this results in comparatively uniform plain crystalline particles. New imperfections created by these movements result in more resistance and, finally, increase hardness.

Table 6. Effect of different speed values on hardness.

Tool Speed, rpm	Hardness HV			
	Top	Middle	Bottom	Standard Deviation, σ
500	45.83	41.78	40.10	2.4050
1000	50.50	47.48	42.75	3.1895
1500	56.83	55.96	45.90	4.9601
2000	61.58	75.51	57.71	7.6439

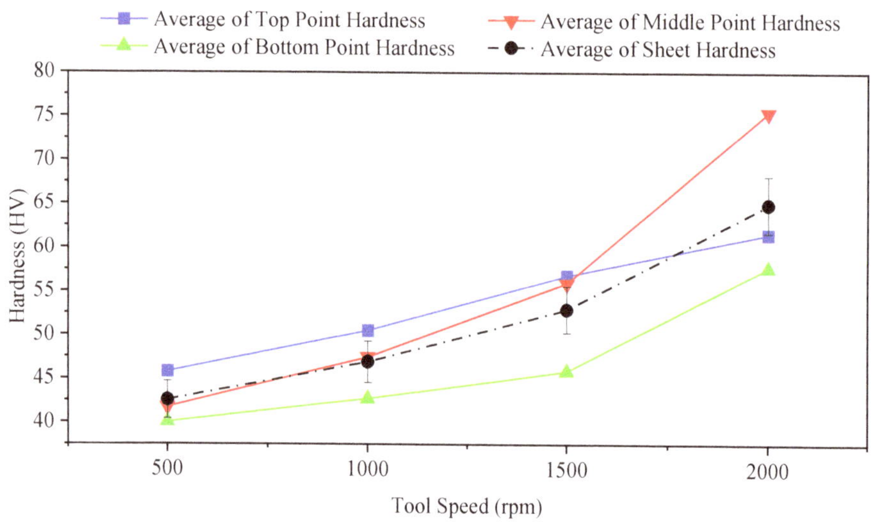

Figure 7. Effect of different tool speeds on hardness of formed sheet (feed rate: 600 mm/min; tool diameter: 10 mm; coolant: oil).

3.3. *Tool Diameter*

The effects of tool diameter on hardness are presented in Table 7 and in Figure 8. Decreases in values are due to increases in tool diameter. McAnulty et al. [54] found different behaviours for the effects of changes in tool diameter on formability. Asgari et al. [55] concluded that a tool diameter of 3 mm results in increased hardness in an aluminium alloy 1100-O sheet relative to 5 or 10 mm tool diameters. A decrease in tool diameter from 10 to 3 mm causes ultimate tensile stress and yield stress to decrease by 7% and 24%, respectively. Furthermore, a reduction in the tool diameter causes a decrease in grain size [55]. Shrivastava and Tandon [56] discussed various parameters of the pre-production sheet and studied the effects of such parameters on the ISF process and on the final properties of products. They claimed that the forces needed to form the sheet in ISF are affected by grain size. Increasing the grain size leads to a decrease in forming forces, yield stress, and hardness [56]. The researchers used different tools for forming, and all of the formed parts showed decreased hardness irrespective of the diameter of the applied tools. The results of this study show that hardness decreased in the case of any diameter of the tool irrespective of the hardness value at each point. A tool diameter of 4 mm showed higher hardness than other diameters at all points. Tool diameters of 10 and 8 mm, on the other hand, produced lower hardness values than tool diameters of 6 and 4 mm. In addition, larger tools, which passed through the formed sheet more times than smaller tools, made the material of the formed part softer (and also caused more heating through increased friction).

Table 7. Effects of different tool diameters on hardness.

Tool Diameter, mm	Hardness HV			
	Top	Middle	Bottom	Standard Deviation, σ
4	75.51	57.71	66.61	7.2668
6	56.33	50.65	53.01	2.3299
8	43.22	40.62	44.26	1.5308
10	41.78	40.10	42.57	1.0300

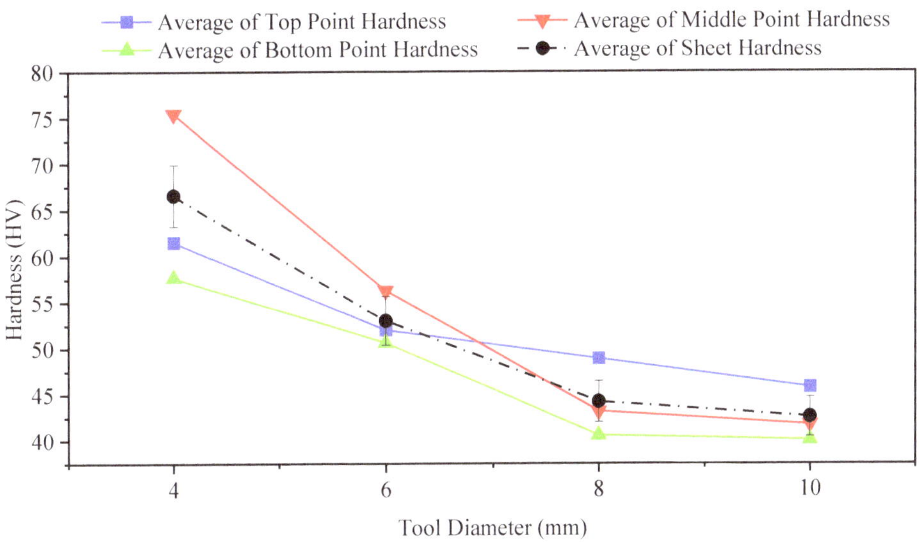

Figure 8. Effect of different tool diameters on hardness of formed sheet (feed rate: 600 mm/min; tool speed: 2000 rpm; coolant: oil).

3.4. Grease Grade

Table 8 and Figure 9 show the impact of different greases on the hardness results. The grease with the highest drop point resulted in the lowest hardness and vice versa. The grease type named "Gp Grease Calcium-ISO VG 46" provided the highest hardness (with the lowest drop point): this was due to the fact that this grease entered between the tool and the formed sheet and cooled the local forming zone faster. Regarding the properties of different greases, it is shown that the use of the grease with a higher flash point resulted in a more stable hardness value. In fact, it can be noted that the use of grease rather than the use of coolant oil produced more homogeneous hardness values at different points of the same sheet, with a slight difference in this conclusion for coolant oil compared to Gp Grease Calcium type grease.

3.5. Regression Equations to Calculate the Hardness of SPIF Components

Regression enables one to find an alternative method to quickly and more economically calculate SPIF components' hardness, rather than having to resort to an experimental process. In view of this, in our experiments, regression equations capable of calculating the hardness of SPIF were used instead of actual measurements. Consequently, different equations were used in this study: Equation (2) Linear Cross-Validation Regression, (3) Linear Cross-Validation with Multiple Regression of Viscosity, (4) Multiple Regression, and (5) Equation Based on Biases and Weight. These equations are as follows:

Table 8. Effects of different grease types on hardness.

Grease Type	ISO Viscosity Grade	Average Dropping Point (at 25 °C)	Flash Point, °C	Hardness HV			
				Top	Middle	Bottom	Standard Deviation, σ
Gp Grease Calcium	ISO VG 15	58	60	46.81	72.00	56.95	10.3487
Zinol	ISO VG 22	88	170	46.34	42.70	40.74	2.3202
Kaucuklu	ISO VG 32	88	172	42.73	38.12	40.62	1.8843
EP2	ISO VG 46	90	180	45.81	43.30	38.18	3.1751

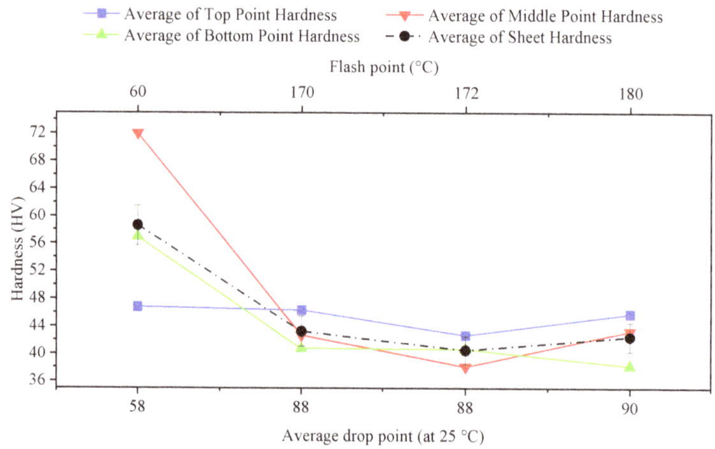

Figure 9. Effects of different grease types on hardness of formed sheet (feed rate: 600 mm/min; tool speed: 2000 rpm; tool diameter: 10 mm).

Linear Cross-Validation Regression:

$$H = \left(Fr_{Coeff} \times Fr + S_{Coeff} \times S + D_{Coeff} \times D + L_{Coeff} \times L_v + C\right)/4$$
$$H = (0.0658 \times Fr + 0.0028 \times S - 9.3993 \times D + 0.2829 \times L_v + 234.7775)/4 \quad (2)$$

Linear Cross-Validation with Multiple Regression of Viscosity:

$$H = \left(Fr_{Coeff} \times Fr + S_{Coeff} \times S + D_{Coeff} \times D + (L_{Coeff1} \times L_v - L_{Coeff2} \times {L_v}^2\right) + C)/4$$
$$H = (0.0749 \times Fr + 0.007482 \times S - 8.2277 \times D + (3.5691 \times L_v - 0.0950 \times {L_v}^2) + 217.7624)/4 \quad (3)$$

Multiple Regression:

$$\begin{aligned} H = &\left(Fr_{Coeff1} \times Fr - Fr_{Coeff2} \times Fr^2\right) + (S_{Coeff1} \times S - S_{Coeff2} \times S^2) \\ &+ (D_{Coeff1} \times D^2 - D_{Coeff2} \times D) + (L_{Coeff1} \times {L_v}^2 \\ &- L_{Coeff2} \times L_v) + C \\ H = &(0.029255 \times Fr - 0.000010 \times Fr^2) + (0.0278 \times S - 0.000009 \times S^2) \\ &- (17.4122 \times D - 1.0414 \times D^2) \\ &- (0.9343 \times L_v - 0.0244 \times {L_v}^2) + 87.4763 \end{aligned} \quad (4)$$

Equation Based on Biases and Weights:

$$H = b_2 + LW \times Exp^{\left(-(b_1 + IW \times x)^2\right)}$$

$$H = 3.9206 + [LW] \times exp\left(-\left(\begin{bmatrix} -5.1413 \\ 1.2616 \\ 0.6642 \\ 2.7010 \\ 1.5268 \\ -2.5527 \\ 0.9677 \\ 0.9160 \\ 2.3191 \\ -4.4413 \end{bmatrix} + \begin{bmatrix} -1.1773 & -0.7032 & -1.7944 & -1.8489 & 1.2188 & 2.0108 & -0.6296 & 1.9369 & 0.7717 & 0.3779 \\ -1.7346 & -1.1106 & 0.1240 & 0.7292 & 0.0289 & -0.3600 & 0.6209 & -2.3675 & -0.6774 & -1.6394 \\ 0.3222 & -0.2106 & -0.1155 & 1.1132 & 2.3911 & -0.5010 & -1.7080 & 0.0188 & 0.4250 & -0.6822 \\ 0.1749 & -0.6833 & -0.8397 & 1.7357 & -1.2086 & -0.6854 & -0.8242 & 1.5373 & -1.4357 & 0.2100 \end{bmatrix} \times \begin{bmatrix} x \\ Fr \\ S \\ D \\ L \end{bmatrix}\right)^2\right) \quad (5)$$

where LW = [6.4552 −2.1276 2.7984 7.8625 4.7908 4.3514 −6.4366 −4.2600 2.3087 4.7381], H is hardness, Fr is feed rate, S is spindle speed, D is tool diameter, L_V is viscosity of the lubricant, C is the intercept, and *Coeff* is a coefficient.

Many different validation metrics are used for assessing and measuring the agreement between a predictive model and physical observations with the aim of selecting the best models or equations, and choosing the proper validation metric can be a crucial point and a challenge for evaluating results. In this study, the equations developed were compared and validated with the validation metrics listed in Table 9. In order to check the equations in question, different validation metrics were used to test performance on the basis of the results of the equations used in the hardness calculation. The criteria of validation consist in minimising error. To this end, Root Mean Square Error (RMSE) and Mean Absolute Error (MAE) were used for validation in this study. RMSE can be more sensitive to the error in case the MAE is more stable. However, RMSE and MAE are more accurate evaluation metrics compared to other metrics [57]. The equation's more reliable performance is guaranteed by a condition where MAE and RMSE values are close to 0. Nevertheless, the large variance between RMSE and MAE values represents significant variations in error distribution. Consequently, Mean Relative Error (MRE) was used to measure the precision of the equations applied.

Table 9. Assessment of best alternative equations with different validation metrics for hardness calculation.

Validation Metric	Linear Cross-Validation Regression	Linear Cross-Validation with Multiple Regression of Viscosity	Multiple Regression	Equation Based on Biases and Weights
Mean Error	0.0000	0.0000	−0.0002	−0.0306
Mean Absolute Error	4.8183	3.6826	2.7811	1.8954
Mean Square Error	40.9419	29.2784	22.0436	20.2431
Root Mean Square Error	6.3986	5.4110	4.6951	4.4992
Mean Relative Error	0.0963	0.0727	0.0555	0.0367
Standard Deviation, σ	6.5955	5.5775	4.8396	4.6376
Standard Error of Mean	1.5996	1.3527	1.1738	1.1248

As can be seen in Table 9, the suggested Equation Based on Biases and Weight shows much greater reliability compared to other equations, and the next most reliable equation is the Multiple Regression equation. Consequently, both could be applied to precisely calculate the hardness of the SPIF component. Figure 10 illustrates the ability of the developed equations to precisely calculate the hardness of SPIF components compared to the real values of hardness. The fluctuating and uneven hardness in Figure 10 is normal because the values are for different components formed in various conditions using SPIF. All of the hardness data were used in predictions models; by sorting these data from low to high or vice versa will affect the random selection of data as training and testing values, which may make it challenging to distinguish the difference between the actual values of hardness and the predicted values by various models in the figure. The HV of points 1–4 is related to the feed rate, 5–8 is related to the spindle speed, 9–12 is related to the tool diameter, and the rest is related to the four types of grease and one coolant oil.

The significant change in the values is because these values were obtained from different components formed in various process conditions. The goal is to employ all of these data to derive different equations despite fluctuations in the data because the mentioned parameter will have parametric values that can be changed in the equations; in the end, the best equation is selected based on the predictive values of hardness that are closest to the actual values with minimum error.

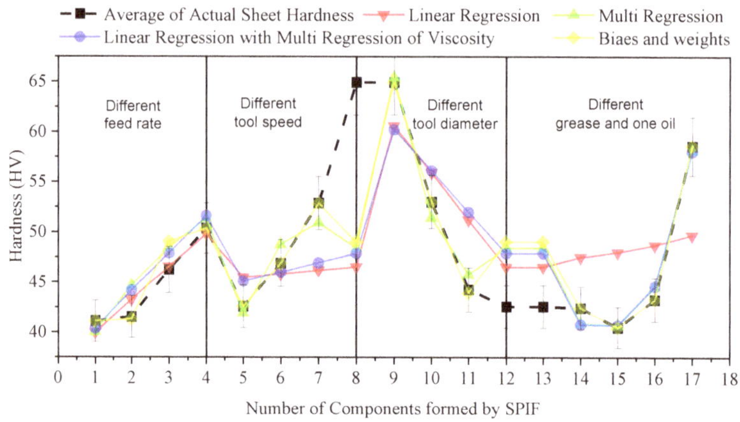

Figure 10. Calculated and real hardness of SPIF components.

The outcomes achieved via the proposed equations versus the real data are presented in Figure 11a–d. The solid line shows a hypothetical exact fit of actual and calculated hardness values, over which data are superimposed. Data dispersion and deviation are based on the ability of the selected equation to predict the hardness values with minimum errors; i.e., a large number of data points that do not match the approximate line (superimposed line) means high error, and a large number of points that match this line means lower error. A satisfactory agreement between the experimental and calculated values was observed for two equations, represented in Figure 11c and d. This figure shows that the equations can appropriately estimate hardness. The connection between the calculated and real data reveals that the calculated values were in agreement with values from the real experiment.

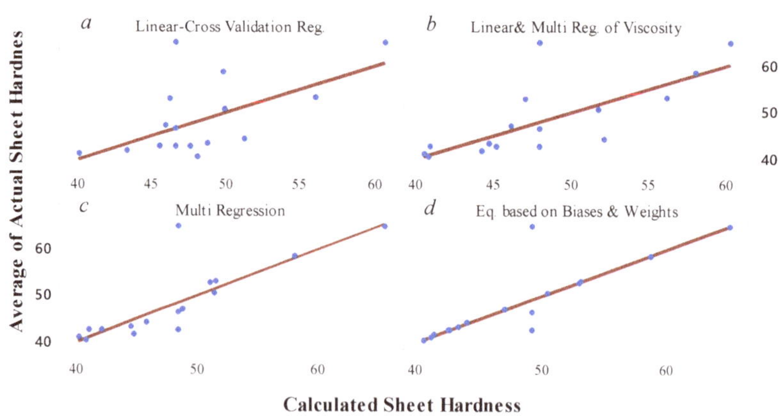

Figure 11. Actual and calculated values obtained with equations of (**a**) Linear Cross-Validation Regression, (**b**) Linear Cross-Validation with Multiple Regression of Viscosity, (**c**) Multiple Regression, and (**d**) Equation Based on Biases and Weights.

3.6. Contribution Analysis of Input Variables

Various methods can be used to evaluate the contribution of each parameter individually and its effective rate on the output. To generate dependent variables, different algorithms for determining the RI of different parameters were employed in this study as a predictor of the predicted output, which is hardness. The algorithms used are Garson's algorithm [58] and the Most-Squares (MS) algorithm, which was proposed by Ibrahim [59]. The equations of the mentioned algorithms are defined in Equations (6) and (7). Both of the algorithms are based on the connection weights of the neurons of the ANN model, which was built for this purpose using MATLAB R2020a [60]. The network structures consist of input, hidden, and output layers, and the number of neurons is (4-10-1). As a mathematical tool for making predictions in machine learning for the purpose of training multilayer networks, a backpropagation learning algorithm was used: this algorithm is called "multilayer perceptron" (MLP), the concept of which was established by Werbos in 1974 and Rumelhart, McClelland, and Hinton in 1986 [61]. The Garson method has also been used in many studies, as presented in [62–67]. Goh [68] applied the Garson algorithm and claimed that RI estimation requires the partitioning of the hidden output weights into elements connected to each neuron in the input layers. Nevertheless, a comparative study of seven different algorithms included the above-mentioned methods to assess relative importance, as outlined in [59]. The author asserted that the Most-Squares method is a better method in comparison to the other methods, and the Most-Squares method seems to outperform all the other methods, as described by the equations below.

$$RI(\%) = \frac{\left[\sum_{j=1}^{n_h} \left(y_{vj} \ / \ \sum_{k=1}^{n_v} y_{kj}\right) hO_j\right]}{\sum_{y=1}^{n_v} \left[\sum_{j=1}^{n_h} \left(y_{vj} \ / \ \sum_{k=1}^{n_v} y_{kj}\right) hO_j\right]} \tag{6}$$

$$RI(\%) = \frac{\sum_{j=1}^{n_v} \left(y^i_{vj} - y^f_{vj}\right)^2}{\sum_{j=1}^{n_v} \sum_{v=1}^{n} \left(y^i_{vj} - y^f_{vj}\right)^2} \tag{7}$$

where n_v is the number of neurons in the input layer, n_h is the number of neurons in the hidden layer, y_j is the absolute value of connection weights between the input and the hidden layers, hO_j is the absolute value of connection weights between the hidden and the output layers, $\sum_{j=1}^{n_v} \left(y^i_{vj} - y^f_{vj}\right)^2$ is the sum squared difference between initial connection weights and final connection weights from the input layer to the hidden layer, and $\sum_{j=1}^{n_v} \sum_{v=1}^{n} \left(y^i_{vj} - y^f_{vj}\right)^2$ is the total of the sum squared difference of all inputs.

The replacement of the input parameters with greater RI significantly influences the outcomes as compared to changes in the parameters with lower RI values [66,67,69]. In the current study, the input parameters are the feed rate, spindle speed, tool diameter and different coolant oils, whereas the outcome is the hardness of the component formed using SPIF.

With regard to the relative importance and weights analysis, the most significant factor affecting hardness, as described in Figure 12a, is the feed rate with an RI of 31%, followed by tool diameter with an RI of 25%. Another interesting observation, based on Figure 12b, is a slight difference in the RI ratios. The tool diameter with an RI of 29% is the most influential factor on hardness, followed by feed rate with an RI of 28%. Relying on the Garson method to assess the contribution, it can be claimed that spindle speed and viscosity have nearly the same RI, with values of 23% and 21%, respectively. When the MS method was used, the relative importance of spindle speed and viscosity was 24% and 19%, respectively, the values of which express the effects on hardness. Based on the above-mentioned facts, it can be asserted that the most significant parameters that jointly influence SPIF hardness are the feed rate and tool diameter. A slight difference is acceptable due to the difference in algorithms of the two methods and thanks to the small amount of data studied. It is worth mentioning that the Garson method partitions hidden output connection weights into components associated with each input neuron and uses absolute values of connection weights. At the same time, connection weights between the hidden and the output layers were not used in the MS method: instead, connection weights between the input and hidden layers were used.

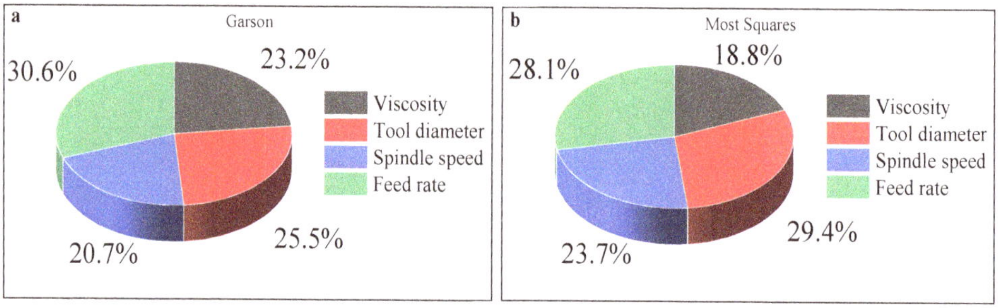

Figure 12. Relative importance of different input variables according to the (**a**) Garson and (**b**) Most-Squares algorithms.

4. Conclusions

In this study, the effects of four process parameters on hardness were investigated experimentally: the influences of feed rate, spindle speed, and tool diameter on hardness were successfully analysed. Regression equations were established using both linear and quadratic functions based on biases and weights generated from the ANN model of influential forming parameters. Two different methods of RI were used to assess the effects of SPIF parameters on outputs. The RI values of the Garson and Most-Squares methods revealed that feed rate and tool diameter are influential factors, impacting hardness by 29.5% and 27% on average, respectively. Based on our study, the effects of four parameters on hardness can be summarised as follows:

- An increase in the feed rate increases hardness when coolant oil is used. Hardness decreases when grease is used (which happens by way of filling the grooves between asperities with debris carried by the grease).
- The hardness of the component increases when tool speed increases.
- Increases in tool diameter result in a decrease in the hardness of components.
- Grease properties are certain to affect hardness values.
- The use of grease instead of coolant oil generates homogeneous hardness values at different points of the same formed sheet.

From the first four findings, it can be concluded that in order to increase the hardness of a SPIF component made of AA1100 aluminium alloy, high feed rates and high tool speed have to be applied, and coolant oil must be used instead of grease if tools smaller than 8 mm in diameter are used. A significant finding of the study is that the Biases and Weights-Based model, jointly used with Multiple Regression, yielded the best calculation of hardness. In the scope of the calculations, the results were assessed using different validation metrics; in the case of the above-mentioned models, the lowest MRE values were 0.0367 and 0.0555, respectively.

5. Recommendations for Further Research

This research presented many issues that require further investigation. Given this, it is worth researching the impact of the studied parameters on sheets of different materials and thicknesses. It is essential to understand whether the mechanical properties of the forming tool affect the hardness of the components or not. A detailed study and evaluation of the different microstructures derived from the different processing conditions of the ISF process should be investigated to examine the impact of the parameters on the grain size, which affects the component hardness. In addition, based on the authors' knowledge, tool shape and tool material are vital parameters, and the hardness of formed components might be closely related to these parameters, which also merits investigation. Furthermore, different cases should be analysed with different parameter values to test and validate the developed regression equations.

Author Contributions: Conceptualisation, S.M.N., I.P. and S.A.N.; methodology, S.M.N., I.P., S.A.N., and Z.J.V.; formal analysis S.M.N., I.P., T.T., S.A.N., Z.J.V., and A.J.; investigation, S.M.N. and S.A.N.; data curation, S.M.N., I.P., and Z.J.V.; writing—original draft preparation, S.M.N.; writing—review and editing, S.M.N., I.P., and T.T.; supervision, I.P. and S.A.N.; project administration, I.P.; funding acquisition, S.M.N., I.P., and Z.J.V. All authors have read and agreed to the published version of the manuscript.

Funding: This research received no external funding.

Institutional Review Board Statement: Not applicable.

Informed Consent Statement: Not applicable.

Data Availability Statement: The data presented in this study are available on request from the corresponding author.

Acknowledgments: The research reported in this paper was supported by the National Laboratory of Artificial Intelligence funded by the NRDIO under the auspices of the Ministry for Innovation and Technology and by the European Commission through the H2020 project EPIC (https://www.centre-epic.eu/) (accessed on 26 November 2021) under grant No. 739592. In addition, this research was supported by Hungary's TEMPUS Public Foundation and Stipendium Hungaricum Scholarship.

Conflicts of Interest: The authors declare no conflict of interest.

References

1. Trzepieciński, T.; Oleksik, V.; Pepelnjak, T.; Najm, S.M.; Paniti, I.; Maji, K. Emerging Trends in Single Point Incremental Sheet Forming of Lightweight Metals. *Metals* **2021**, *11*, 1188. [CrossRef]
2. Szpunar, M.; Ostrowski, R.; Trzepieciński, T.; Kaščák, L'. Central Composite Design Optimisation in Single Point Incremental Forming of Truncated Cones from Commercially Pure Titanium Grade 2 Sheet Metals. *Materials* **2021**, *14*, 3634. [CrossRef]
3. Trzepieciński, T.; Najm, S.M.; Sbayti, M.; Belhadjsalah, H.; Szpunar, M.; Lemu, H.G. New Advances and Future Possibilities in Forming Technology of Hybrid Metal–Polymer Composites Used in Aerospace Applications. *J. Compos. Sci.* **2021**, *5*, 217. [CrossRef]
4. Cooper, D.R.; Gutowski, T.G. Prospective Environmental Analyses of Emerging Technology: A Critique, a Proposed Methodology, and a Case Study on Incremental Sheet Forming. *J. Ind. Ecol.* **2020**, *24*, 38–51. [CrossRef]
5. Dittrich, M.A.; Cao, J.; Roth, J.T.; Xia, Z.C.; Kiridena, V.; Ren, F.; Henning, H.; Gutowski, T.G. Exergy analysis of incremental sheet forming. *Prod. Eng.* **2012**, *6*, 169–177. [CrossRef]
6. Ingarao, G.; Ambrogio, G.; Gagliardi, F.; Di Lorenzo, R. A sustainability point of view on sheet metal forming operations: Material wasting and energy consumption in incremental forming and stamping processes. *J. Clean. Prod.* **2012**, *29–30*, 255–268. [CrossRef]
7. Emmens, W.C.; Sebastiani, G.; van den Boogaard, A.H. The technology of Incremental Sheet Forming—A brief review of the history. *J. Mater. Process. Technol.* **2010**, *210*, 981–997. [CrossRef]
8. Li, Y.; Chen, X.; Liu, Z.; Sun, J.; Li, F.; Li, J.; Zhao, G. A review on the recent development of incremental sheet-forming process. *Int. J. Adv. Manuf. Technol.* **2017**, *92*, 2439–2462. [CrossRef]
9. Behera, A.K.; de Sousa, R.A.; Ingarao, G.; Oleksik, O. Single point incremental forming: An assessment of the progress and technology trends from 2005 to 2015. *J. Manuf. Process.* **2017**, *27*, 37–62. [CrossRef]
10. Raju, C.; Sathiya Narayanan, C. Application of a hybrid optimization technique in a multiple sheet single point incremental forming process. *Measurement* **2016**, *78*, 296–308. [CrossRef]
11. Kumar, A.; Gulati, V.; Kumar, P.; Singh, V.; Kumar, B.; Singh, H. Parametric effects on formability of AA2024-O aluminum alloy sheets in single point incremental forming. *J. Mater. Res. Technol.* **2019**, *8*, 1461–1469. [CrossRef]
12. Zavala, J.M.D.; Martínez-Romero, O.; Elías-Zúñiga, A.; Gutiérrez, H.M.L.; la Vega, A.E.; Taha-Tijerina, J. Study of Friction and Wear Effects in Aluminum Parts Manufactured via Single Point Incremental Forming Process Using Petroleum and Vegetable Oil-Based Lubricants. *Materials* **2021**, *14*, 3973. [CrossRef]
13. Najm, S.M.; Paniti, I. Study on Effecting Parameters of Flat and Hemispherical end Tools in SPIF of Aluminium Foils. *Teh. Vjesn.-Tech. Gaz.* **2020**, *27*, 1844–1849. [CrossRef]
14. Najm, S.M.; Paniti, I. Experimental Investigation on the Single Point Incremental Forming of AlMn1Mg1 Foils using Flat End Tools. *IOP Conf. Ser. Mater. Sci. Eng.* **2018**, *448*, 012032. [CrossRef]

15. Paniti, I. Adaptation of Incremental Sheet Forming into cloud manufacturing. *CIRP J. Manuf. Sci. Technol.* **2014**, *7*, 185–190. [CrossRef]
16. Fratini, L.; Ambrogio, G.; Di Lorenzo, R.; Filice, L.; Micari, F. Influence of mechanical properties of the sheet material on formability in single point incremental forming. *CIRP Ann.* **2004**, *53*, 207–210. [CrossRef]
17. Zhang, S.; Tang, G.H.; Wang, W.; Jiang, X. Evaluation and optimization on the formability of an AZ31B Mg alloy during warm incremental sheet forming assisted with oil bath heating. *Measurement* **2020**, *157*, 107673. [CrossRef]
18. Liu, Z.; Li, Y.; Meehan, P.A. Experimental investigation of mechanical properties, formability and force measurement for AA7075-O aluminum alloy sheets formed by incremental forming. *Int. J. Precis. Eng. Manuf.* **2013**, *14*, 1891–1899. [CrossRef]
19. Li, Y.; Chen, X.; Zhai, W.; Wang, L.; Li, J.; Guoqun, Z. Effects of process parameters on thickness thinning and mechanical properties of the formed parts in incremental sheet forming. *Int. J. Adv. Manuf. Technol.* **2018**, *98*, 3071–3080. [CrossRef]
20. Manco, L.; Filice, L.; Ambrogio, G. Analysis of the thickness distribution varying tool trajectory in single-point incremental forming. *Proc. Inst. Mech. Eng. Part B J. Eng. Manuf.* **2011**, *225*, 348–356. [CrossRef]
21. Krasowski, B.; Kubit, A.; Trzepieciński, T.; Slota, J. Experimental Analysis of Single Point Incremental Forming of Truncated Cones in DC04 Steel Sheet. *Adv. Mater. Sci.* **2020**, *20*, 5–15. [CrossRef]
22. Zain, A.M.; Haron, H.; Sharif, S. Prediction of surface roughness in the end milling machining using Artificial Neural Network. *Expert Syst. Appl.* **2010**, *37*, 1755–1768. [CrossRef]
23. Amirjan, M.; Khorsand, H.; Siadati, M.H.; Eslami Farsani, R. Artificial Neural Network prediction of Cu–Al2O3 composite properties prepared by powder metallurgy method. *J. Mater. Res. Technol.* **2013**, *2*, 351–355. [CrossRef]
24. Ezugwu, E.O.; Fadare, D.A.; Bonney, J.; Da Silva, R.B.; Sales, W.F. Modelling the correlation between cutting and process parameters in high-speed machining of Inconel 718 alloy using an artificial neural network. *Int. J. Mach. Tools Manuf.* **2005**, *45*, 1375–1385. [CrossRef]
25. Li, E. Reduction of springback by intelligent sampling-based LSSVR metamodel-based optimization. *Int. J. Mater. Form.* **2013**, *6*, 103–114. [CrossRef]
26. Marouani, H.; Aguir, H. Identification of material parameters of the Gurson–Tvergaard–Needleman damage law by combined experimental, numerical sheet metal blanking techniques and artificial neural networks approach. *Int. J. Mater. Form.* **2012**, *5*, 147–155. [CrossRef]
27. Lela, B.; Bajić, D.; Jozić, S. Regression analysis, support vector machines, and Bayesian neural network approaches to modeling surface roughness in face milling. *Int. J. Adv. Manuf. Technol.* **2009**, *42*, 1082–1088. [CrossRef]
28. Hussaini, S.M.; Singh, S.K.; Gupta, A.K. Experimental and numerical investigation of formability for austenitic stainless steel 316 at elevated temperatures. *J. Mater. Res. Technol.* **2014**, *3*, 17–24. [CrossRef]
29. Kondayya, D.; Gopala Krishna, A. An integrated evolutionary approach for modelling and optimization of laser beam cutting process. *Int. J. Adv. Manuf. Technol.* **2013**, *65*, 259–274. [CrossRef]
30. Trzepieciński, T.; Kubit, A.; Dzierwa, A.; Krasowski, B.; Jurczak, W. Surface Finish Analysis in Single Point Incremental Sheet Forming of Rib-Stiffened 2024-T3 and 7075-T6 Alclad Aluminium Alloy Panels. *Materials* **2021**, *14*, 1640. [CrossRef] [PubMed]
31. Najm, S.M.; Paniti, I. Artificial neural network for modeling and investigating the effects of forming tool characteristics on the accuracy and formability of thin aluminum alloy blanks when using SPIF. *Int. J. Adv. Manuf. Technol.* **2021**, *114*, 2591–2615. [CrossRef]
32. Najm, S.M.; Paniti, I. Predict the Effects of Forming Tool Characteristics on Surface Roughness of Aluminum Foil Components Formed by SPIF Using ANN and SVR. *Int. J. Precis. Eng. Manuf.* **2021**, *22*, 13–26. [CrossRef]
33. Baruah, A.; Pandivelan, C.; Jeevanantham, A.K. Optimization of AA5052 in incremental sheet forming using grey relational analysis. *Measurement* **2017**, *106*, 95–100. [CrossRef]
34. Kumar, A.; Gulati, V. Experimental investigations and optimization of forming force in incremental sheet forming. *Sādhanā* **2018**, *43*, 159. [CrossRef]
35. Kumar, A.; Gulati, V. Experimental investigation and optimization of surface roughness in negative incremental forming. *Measurement* **2019**, *131*, 419–430. [CrossRef]
36. Al-Attaby, Q.M.D.; Abaas, T.F.; Bedan, A.S. The Effect of Tool Path Strategy on Mechanical Properties of Brass (65–35) in Single Point Incremental Sheet Metal Forming (SPIF). *J. Eng.* **2013**, *19*, 629–637.
37. Mostafanezhad, H.; Menghari, H.G.; Esmaeili, S.; Shirkharkolaee, E.M. Optimization of two-point incremental forming process of AA1050 through response surface methodology. *Measurement* **2018**, *127*, 21–28. [CrossRef]
38. Premkumar, R.; Vignesh, R.V.; Padmanaban, R.; Govindaraju, M.; Santhi, R. Investigation on the microstructure, microhardness, and tribological behavior of AA1100-hBN surface composite. *Koroze A Ochr. Mater.* **2021**, *65*, 1–11. [CrossRef]
39. Song, X.; Zhang, J.; Zhai, W.; Taureza, M.; Castagne, S.; Danno, A. Numerical and experimental investigation on the deformation mechanism of micro single point incremental forming process. *J. Manuf. Process.* **2018**, *36*, 248–254. [CrossRef]
40. Kaufman, J.G. Applications for aluminum alloys and tempers. In *Introduction to Aluminum Alloys and Tempers*; ASM International: Materials Park, OH, USA, 2000; Volume 1100, p. 242.
41. Alcotec.com. Understanding the Alloys of Aluminum. 2021. Available online: http://www.alcotec.com/us/en/education/knowledge/techknowledge/understanding-the-alloys-of-aluminum.cfm (accessed on 19 November 2021).
42. Training, A.; Maintenance, B.; Schools, A.; Books, A.; Jar, T.; Repair, A. Aluminum Alloys. Flight-Mechanic.com. 2021. Available online: https://www.flight-mechanic.com/aluminum-alloys/ (accessed on 19 November 2021).

43. Davis, J.R. *Aluminum and Aluminum Alloys*; ASM Int.: Materials Park, OH, USA, 2001; pp. 351–416. [CrossRef]
44. Committee, A.I.H. *ASM Handbook Volume 2: Properties and Selection: Nonferrous Alloys and Special-Purpose Materials*; ASM International: Materials Park, OH, USA, 1990.
45. Kumar, A.; Gulati, V. Optimization and investigation of process parameters in single point incremental forming. *Indian J. Eng. Mater. Sci.* **2020**, *27*, 246–255.
46. Skjoedt, M.; Hancock, M.H.; Bay, N. Creating Helical Tool Paths for Single Point Incremental Forming. *Key Eng. Mater.* **2007**, *344*, 583–590. [CrossRef]
47. United States Environmental Protection Agency; Office of Wastewater Management Washington. Environmentally Acceptable Lubricants. In *Encyclopedia of Lubricants and Lubrication*; Springer: Berlin/Heidelberg, Germany, 2014; p. 526.
48. IS0 3448:1992(E). *Industrial Liquid Lubricants—IS0 Viscosity Classification*; International Organization for Standardization: Geneva, Switzerland, 1992.
49. Syahrullail, S.; Kamitani, S.; Shakirin, A. Performance of Vegetable Oil as Lubricant in Extreme Pressure Condition. *Procedia Eng.* **2013**, *68*, 172–177. [CrossRef]
50. Diabb, J.; Rodríguez, C.A.; Mamidi, N.; Sandoval, J.A.; Taha-Tijerina, J.; Martínez-Romero, O.; Elías-Zúñiga, A. Study of lubrication and wear in single point incremental sheet forming (SPIF) process using vegetable oil nanolubricants. *Wear* **2017**, *376–377*, 777–785. [CrossRef]
51. Nama, S.A.; Namer, N.S.M.; Najm, S.M. The Effect of using Grease on the Surface Roughness of Aluminum 1100 Sheet during the Single Point Incremental Forming Process. *J. Trends Mach. Des.* **2014**, *1*, 53–56.
52. Kim, Y.; Park, J. Effect of process parameters on formability in incremental forming of sheet metal. *J. Mater. Process. Technol.* **2002**, *130*, 42–46. [CrossRef]
53. Hol, J.; Meinders, V.T.; Geijselaers, H.J.M.; Van Den Boogaard, A.H. Multi-scale friction modeling for sheet metal forming: The mixed lubrication regime. *Tribol. Int.* **2015**, *85*, 10–25. [CrossRef]
54. McAnulty, T.; Jeswiet, J.; Doolan, M. Formability in single point incremental forming: A comparative analysis of the state of the art. *CIRP J. Manuf. Sci. Technol.* **2017**, *16*, 43–54. [CrossRef]
55. Asgari, A.; Sedighi, M.; Riahi, M. Investigation of Punching Parameters Effect on Mechanical Properties of Al-1100-O in Incremental Sheet Metal Hammering Process. *Strength Mater.* **2015**, *47*, 882–889. [CrossRef]
56. Shrivastava, P.; Tandon, P. Investigation of the Effect of Grain Size on Forming Forces in Single Point Incremental Sheet Forming. *Procedia Manuf.* **2015**, *2*, 41–45. [CrossRef]
57. Misra, S.; He, J. Stacked neural network architecture to model the multifrequency conductivity/permittivity responses of subsurface shale formations. In *Machine Learning for Subsurface Characterization*, 1st ed.; Misra, S., Li, H., He, J., Eds.; Elsevier Inc.: Amsterdam, The Netherlands, 2020; pp. 103–127.
58. Garson, G.D. Interpreting Neural-Network Connection Weights. *AI Expert* **1991**, *6*, 46–51.
59. Ibrahim, O.M. A comparison of methods for assessing the relative importance of input variables in artificial neural networks. *J. Appl. Sci. Res.* **2013**, *9*, 5692–5700.
60. Beale, M.H.; Hagan, M.; Demuth, H. Deep Learning Toolbox Getting Started Guide. In *Deep Learn. Toolbox*; The MathWorks Inc.: Natick, MA, USA, 2019. [CrossRef]
61. Riedmiller, P.M. Machine Learning: Multi Layer Perceptrons. Albert-Ludwigs-University Freibg. AG Maschinelles Lernen. Available online: http://ml.informatik.uni-freiburg.de/_media/documents/teaching/ss09/ml/mlps.pdf (accessed on 10 October 2021).
62. Zarei, M.J.; Ansari, H.R.; Keshavarz, P.; Zerafat, M.M. Prediction of pool boiling heat transfer coefficient for various nano-refrigerants utilizing artificial neural networks. *J. Therm. Anal. Calorim.* **2020**, *139*, 3757–3768. [CrossRef]
63. Ding, H.; Luo, W.; Yu, Y.; Chen, B. Construction of a Robust Cofactor Self-Sufficient Bienzyme Biocatalytic System for Dye Decolorization and its Mathematical Modeling. *Int. J. Mol. Sci.* **2019**, *20*, 6104. [CrossRef] [PubMed]
64. Zhou, B.; Vogt, R.D.; Lu, X.; Xu, C.; Zhu, L.; Shao, X.; Liu, H.; Xing, M. Relative Importance Analysis of a Refined Multi-parameter Phosphorus Index Employed in a Strongly Agriculturally Influenced Watershed. *Water Air Soil Pollut.* **2015**, *226*, 25. [CrossRef]
65. Shabanzadeh, P.; Yusof, R.; Shameli, K. Artificial neural network for modeling the size of silver nanoparticles' prepared in montmorillonite/starch bionanocomposites. *J. Ind. Eng. Chem.* **2015**, *24*, 42–50. [CrossRef]
66. Vatankhah, E.; Semnani, D.; Prabhakaran, M.P.; Tadayon, M.; Razavi, S.; Ramakrishna, S. Artificial neural network for modeling the elastic modulus of electrospun polycaprolactone/gelatin scaffolds. *Acta Biomater.* **2014**, *10*, 709–721. [CrossRef]
67. Rezakazemi, M.; Razavi, S.; Mohammadi, T.; Nazari, A.G. Simulation and determination of optimum conditions of pervaporative dehydration of isopropanol process using synthesized PVA–APTEOS/TEOS nanocomposite membranes by means of expert systems. *J. Memb. Sci.* **2011**, *379*, 224–232. [CrossRef]
68. Goh, A.T.C. Back-propagation neural networks for modeling complex systems. *Artif. Intell. Eng.* **1995**, *9*, 143–151. [CrossRef]
69. Nabipour, M.; Keshavarz, P. Modeling surface tension of pure refrigerants using feed-forward back-propagation neural networks. *Int. J. Refrig.* **2017**, *75*, 217–227. [CrossRef]

materials

MDPI

Article

Optimization Method of Tool Parameters and Cutting Parameters Considering Dynamic Change of Performance Indicators

Daxun Yue, Anshan Zhang *, Caixu Yue, Xianli Liu, Mingxing Li and Desheng Hu

Key Laboratory of Advanced Manufacturing and Intelligent Technology, Ministry of Education, Harbin University of Science and Technology, Harbin 150080, China; 1920100002@stu.hrbust.edu.cn (D.Y.); yuecaixu@hrbust.edu.cn (C.Y.); Xianli.liu@hrbust.edu.cn (X.L.); 1910100005@stu.hrbust.edu.cn (M.L.); 1920110094@stu.hrbust.edu.cn (D.H.)

* Correspondence: aszhang@hrbust.edu.cn

Abstract: In the process of metal cutting, the cutting performance of cutting tools varies with different parameter combinations, so the results of the performance indicators studied are also different. So in order to achieve the best performance indicator it is necessary to get the best parameter matching combination. In addition, in the process of metal cutting, the value of the performance index is different at each stage of the processing process. In order to consider the cutting process more comprehensively, it is necessary to use a comprehensive evaluation method that can evaluate the dynamic process of performance indicators. This paper uses a dynamic evaluation method that considers the dynamic change of performance indicators in each stage of the cutting process to comprehensively evaluate the tool parameters and cutting parameters at each level. For the purpose of high processing efficiency and long tool life, tool wear rate and material removal rate are used as performance indicators. In the case of specified rake angle, cutting speed and cutting width, titanium alloy is studied by end milling cutter side milling. The tool parameters and cutting parameters in milling process are optimized by using a dynamic comprehensive evaluation method based on gain horizontal excitation. Finally, the parameter matching combination that can make the performance indicator reach the best is obtained. The results show that when the rake angle is 8°, the cutting speed is 37.68 m/min, and the cutting width is 0.2 mm, the tool wear rate and material removal rate are the best when the clearance angle is 9°, the helix angle is 30°, the feed per tooth is 0.15 mm/z, and the cutting depth is 2.5 mm.

Keywords: cutting performance; matching combination; dynamic change process of performance indicator; dynamic evaluation method; comprehensive evaluation

Citation: Yue, D.; Zhang, A.; Yue, C.; Liu, X.; Li, M.; Hu, D. Optimization Method of Tool Parameters and Cutting Parameters Considering Dynamic Change of Performance Indicators. *Materials* **2021**, *14*, 6181. https://doi.org/10.3390/ma14206181

Academic Editors: Tomasz Trzepieciński and Valentin Ştefan Oleksik

Received: 14 September 2021
Accepted: 13 October 2021
Published: 18 October 2021

1. Introduction

Metal cutting is one of the most widely used metal parts manufacturing methods [1]. However, in the process of machining, the cutting performance of the tool is different under different parameter combinations, and the required performance indicator results are also different. Therefore, in order to make the performance indicator reach the optimal value, it is necessary to study the matching between each parameter in the cutting process, and get the optimal parameter matching combination of each performance indicator.

At present, many scholars have studied the optimal matching combination among parameters. Zhang et al. [2] conducted the orthogonal test of high speed milling of aviation aluminum alloy with end mills, and obtained the optimal combination of milling parameters through range analysis. Kubilay et al. [3] taking the turning of Ti6Al4V with indexable turning tool as the research object. The average roughness height, maximum roughness height, and material removal rate were taken as the performance indicators. The parameters are cutting speed, feed speed, and cutting depth. The surface response function of three performance indicators was obtained by using the surface response method, and the optimal parameter combination was obtained by multi-objective optimization algorithm. Tamal et al. [4] proposes a bayesian regularization neural network for agents and the beetle antenna search algorithm to optimize the algorithm of data driven auxiliary agent optimization algorithm, the algorithm we get the material removal rate, surface roughness and cutting force on cutter diameter, spindle speed, feed speed, and cutting depth function and to get the optimized parameters combination. Juan et al. [5] used the orthogonal cutting simulation method to obtain the relevant performance indicator values as optimization parameters, and proposed a multi-objective particle swarm optimization algorithm to optimize the performance indicator values. Zhang et al. [6] used the Markov chain Monte Carlo method to solve the reliability model of tool life, and then used the multi-objective optimization algorithm combining grey correlation analysis, radial basis neural network and particle swarm optimization algorithm to search for the optimal processing parameters of the whole blade-disc tunnel processing. Mohammed et al. [7] took cutting force and surface roughness as performance indicators and combined gray correlation method (GRA) and expectation function analysis (DFA) to optimize the milling parameters in the milling process of epoxy glass fiber to obtain the best combination of milling parameters. Fang et al. [8] respectively established the correlation function model of the above performance indicators on milling times and milling parameters for the energy consumption, processing cost, and processing time of CNC machine tools in the process of multi-pass milling. On this basis, an improved adaptive simulated annealing particle swarm optimization algorithm was proposed to solve the optimal solution of milling parameters, and the optimal combination of milling parameters was obtained. Vimal et al. [9] took surface roughness, tool wear, and cutting force as performance indicators and used grey-fuzzy evaluation method to optimize the cutting speed, feed speed, and cutting depth in the turning process of glass fiber reinforced plastics, and obtained the optimal parameter combination of cutting speed, feed speed, and cutting depth. Viswanathan et al. [10] took PVD coated carbide turning tools dry turning magnesium alloy as the research object, taking cutting speed, feed per revolution, and cutting depth as the optimized parameters, and taking cutting force, material removal rate, flank face wear, and surface roughness as the performance indicators, then carried out a parameter optimization test. Principal component analysis (PCA) and grey correlation analysis (GRA) were used to optimize the parameters and get the best combination of parameters. Suresh et al. [11] taking surface roughness and material removal rate as performance indicators, the grey-fuzzy comprehensive evaluation method was used to optimize the cutting speed, feed per revolution, and mass fraction of SiC-Gr in the turning process of Al-SiC-Gr composites. Gnanavelbabu et al. [12] controlled turning of aluminum-based boron carbide composites numerically, and the cutting force, tool wear, and other performance indicators in the turning process are measured. The optimum parameter combination among spindle speed, feed per revolution, cutting depth, and B_4C mass fraction was obtained by grey-fuzzy comprehensive evaluation method.

Metal cutting is a dynamic process, so the performance indicator values are constantly changing with the movement of the cutting tool in the cutting process. Based on the above reference studies, it can be seen that the research methods of parameter matching include range analysis method, optimization algorithm, and evaluation method to obtain the best parameter level combination, but these methods do not consider the data dynamic change process of performance indicator in different stages of cutting process. Therefore, it is important to use an optimization method that takes into account the data of all performance indicators and the data dynamic change process of performance indicators in different stages.

Ti6Al4V is a difficult material to process and is widely used in many industrial fields because of its good thermodynamic properties [13,14]. However, when cutting titanium alloy, there will be serious friction between the tool and the workpiece, so there is a large cutting force and high temperature in the cutting contact area, these problems lead to accelerated tool wear speed, tool life is low, and ultimately lead to low processing efficiency, affecting the application of Ti6Al4V [15,16]. In addition, milling is one of the most commonly used machining processes because of its ability to produce complex geometric shapes [17,18]. Therefore, this paper chooses milling as the processing method to be studied. In this paper, for the purpose of high efficiency and long tool life, this paper takes tool wear rate and material removal rate as performance indicators, and titanium alloy is milled by end milling cutter side as the research object. According to the dynamic change of performance indicator, the dynamic evaluation method based on gain level excitation in reference [19] was used to comprehensively evaluate each level of tool parameters and cutting parameters, and finally the optimal level of each parameter on tool wear rate and material removal rate was optimized.

2. Establishment and Verification of Finite Element Simulation of Milling Process

Finite element method has the characteristics of a simple and clear physical concept, is easy to grasp, has a simple description, is easy to popularize, is a superior method, and has a wide application range. In this paper, the tool wear rate and material removal rate are obtained by finite element method and analytical method under different tool parameter and cutting parameter combinations at different stages.

2.1. Establishment of Finite Element Simulation Model for Milling Process

2.1.1. Finite Element Simulation Model Establishment Process

Finite element simulation is an effective tool to evaluate metal cutting process. In recent years, many researchers have carried out finite element simulation analysis on metal cutting process. Finite element method can be used for chip forming simulation, cutting force simulation, wear simulation, etc [20]. At present, many software have been used in metal cutting process simulation research [21]. Deform-3D (Ohio, USA) is a powerful software for simulation and analysis of metal cutting processes [22]. Finite element simulation of cutting process using Deform-3D has advantages such as reducing workpiece cost and machine energy consumption [23]. Deform-3D simulation system is composed of three main modules, which are the pre-processing module, simulation setting module, and post-processing module [24]. According to the above three modules, Figure 1 shows the process of establishing the cutting simulation model of Deform-3D software.

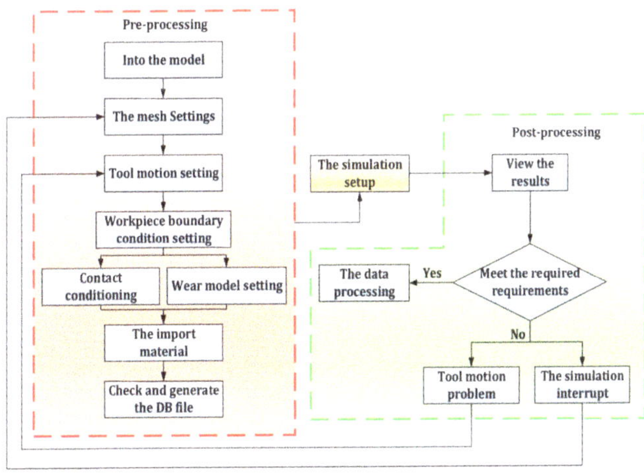

Figure 1. Cutting simulation model establishment process of DEFORM_3D software.

2.1.2. Material Constitutive Model

The constitutive equation of titanium alloy includes stress, strain, temperature, and other parameters [25]. Johnson–Cook material model is simple in form and widely used. It is an ideal model of elastic-plastic strengthening. The model is suitable for the temperature range from room temperature to the melting point of the material [26]. So, the J-C constitutive reinforcement model was selected. The expression form of J-C constitutive model equation is shown in Equation (1).

$$\overline{\sigma} = \left[A + B(\overline{\varepsilon})^n\right] \cdot \left[1 + C \ln \frac{\overline{\varepsilon}}{\overline{\varepsilon}_0}\right] \cdot \left[1 - \left(\frac{T - T_r}{T_m - T_r}\right)^m\right] \tag{1}$$

where σ, ε and ε_0 are equivalent flow stress, equivalent plastic strain rate, and reference plastic strain rate, respectively. T, T_r, and T_m denote the absolute temperature, ambient temperature and melting temperature of the workpiece material, respectively. A, B, C, m, and n are the yield strength, hardening modulus, strain rate sensitivity coefficient, heat softening coefficient, and strain hardening index, respectively. J-C parameters of the Ti6Al4V constitutive model are presented in Table 1.

Table 1. J-C parameters for Ti-6Al-4V alloy [27].

A (MPa)	B (MPa)	C	m	n	$\overline{\varepsilon}_0$ (s^{-1})	T_m (°C)	T_r (°C)
875	793	0.01	0.71	0.386	1	1560	20

2.1.3. Material Parameters

In the pre-processing of finite element simulation, the material attributes of the tool and the workpiece need to be set, and the setting of material parameters plays a crucial role in the accuracy of finite element simulation. The cutter material is carbide, model is YG6, and the tool is not coated. The workpiece material is Ti6Al4V. Table 2 shows the material parameters of tool material YG6 and workpiece material Ti6Al4V.

Table 2. Material parameter [28].

Material Parameter	YG6	Ti6Al4V
Density (g/cm^3)	14.85	4.44
Young's modulus (GPa)	640	112
Poission's Ratio	0.22	0.34
Expansion (/°C)	4.7×10^{-6}	9.4×10^{-6}
Conductivity (W/m·K)	79.6	6.8
Specific heat (J/(kg·°C))	176	565

2.1.4. D Model Establishment, Import and Grid Division

UG is used to conduct 3D modeling of the tool and workpiece, Figure 2a is the model of integral end milling cutter. Figure 2b is the model of square block workpiece, the length is 100 mm. Table 3 shows the key parameters of the integral end milling cutter. In order to speed up the finite element simulation, the 3D model of tool and workpiece is simplified. Because the research object is the end milling cutter side milling block titanium alloy, the main study is of the integral end milling cutter side edge, cut end milling cutter only with a part of the edge, and cut block workpiece in the cutting area of the nearest part of the workpiece. The height of the cut part of the workpiece is determined by the cutting depth. After the model is assembled in UG, all parts are exported in the format of stl file. After all parts are imported into DEFORM-3D software, the original assembly form can still be obtained. Figure 2c shows the tool model and workpiece model after simplifying and importing Deform-3D software.

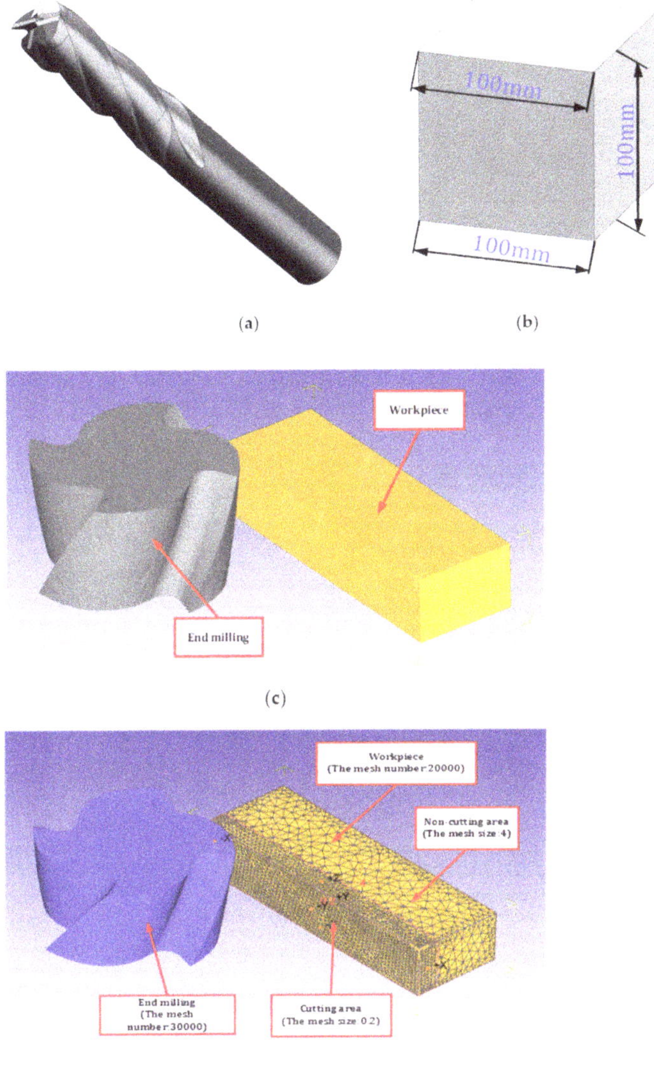

Figure 2. Simulation model of the side milling process: (**a**) End milling cutter model, (**b**) Workpiece model, (**c**) 3D models in simulation software and (**d**) The simulation model.

After importing the 3D model into the DEFORM-3D software, the tool model and the workpiece model are set as rigid body and plastic body respectively. After that, the parts are meshed, and the mesh number of the tool model is 30,000. The number of mesh of the workpiece varies according to the cutting depth. In Figure 2c, the thickness of the workpiece is 2 mm, so the number of mesh of the workpiece is 20,000. When the cutting depth is 3 mm, the mesh number of workpiece is 30,000. When the workpiece is meshing, the workpiece is divided into cutting zone and non-cutting zone. The mesh size of the cutting zone is 0.2, and the mesh size of the non-cutting zone is 4. The resulting figure after mesh division is shown in Figure 2d.

Table 3. The key parameters of the integral end milling cutter.

Num.	Parameter	Value
1	The blade number	4
2	The cutter diameter	10 (mm)
3	The rake angle	8 (°)
4	The width of rake face	1.0 (mm)
5	The first clearance angle	12 (°)
6	The second clearance angle	23 (°)
7	The width of the first flank face	0.7 (mm)
8	The width of the second flank face	0.8 (mm)
9	The helix angle	35 (°)
10	The core diameter	6.2 (mm)

2.1.5. Setting of Tool Wear Model in Finite Element Simulation Software

In this paper, the tool wear rate is selected as one of the performance indicators. Therefore, it is very important to set the tool wear rate in the pre-processing of finite element simulation software. The tool wear rate in Deform-3D is set in the "Tool Wear" window of the "Inter-Object" window. Usui wear model and other wear models can be set in this column. Figure 3 shows the tool wear model setting window.

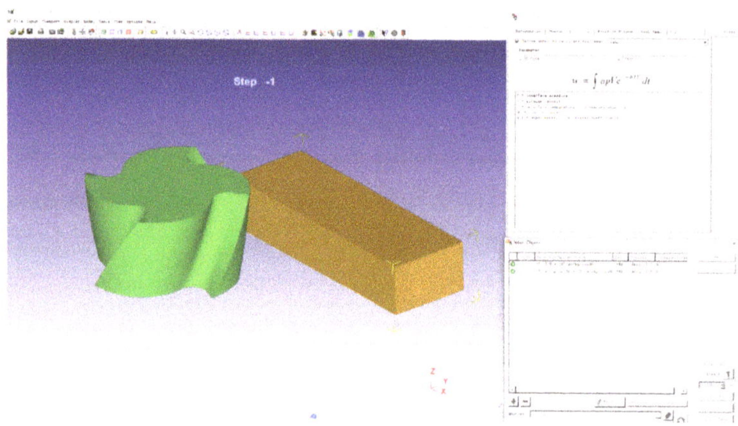

Figure 3. Tool wear model setting.

2.2. *Simulation Parameter Selection and Performance Indicator Setting*

In this paper, the tool wear rate and material removal rate as performance indicators. In this chapter, the data values of tool wear rate and material removal rate are obtained through finite element simulation and analytical method.

2.2.1. Simulation Parameter Selection

According to the comprehensive evaluation study on the importance of parameters to performance indicators in the process of titanium alloy side milling by end milling cutter in the early stage, the clearance angle and helix angle are selected as important tool parameters, and the feed per tooth and cutting depth are important cutting parameters. In this paper, the clearance angle, helix Angle, feed per tooth, and cutting depth are the research objects, and the optimal parameter combination is studied. The values of other parameters not involved in the study are shown in Table 4.

Table 4. Table of fixed parameter values.

Rake Angle (°)	Cutting Speed (m/min)	Cutting Width (mm)
8	37.68	0.2

According to the selected four parameters with the greatest importance, they are taken as the horizontal factor of orthogonal test to carry out test planning. Without considering the interaction between parameters, SPSS software was used to design the orthogonal test of 4 factors and 5 levels, and the parameter level table of $L_{25}(5^4)$ orthogonal test was obtained as shown in Tables 5 and 6 was the orthogonal simulation test table corresponding to Table 5.

Table 5. Variable parameter levels.

	Clearance Angle (°)	Helix Angle (°)	Feed Per Tooth (mm/z)	Cutting Depth (mm)
1	8.00	30.00	0.05	1.00
2	9.00	32.00	0.10	1.50
3	10.00	33.00	0.15	2.00
4	11.00	34.00	0.20	2.50
5	12.00	35.00	0.25	3.00

Table 6. Orthogonal test table.

	Clearance Angle (°)	Helix Angle (°)	Feed Per Tooth (mm/z)	Cutting Depth (mm)
1	9.00	32.00	0.20	1.50
2	12.00	30.00	0.10	1.50
3	11.00	30.00	0.15	2.00
4	8.00	32.00	0.25	2.00
5	10.00	30.00	0.20	2.50
6	9.00	33.00	0.15	2.50
7	11.00	32.00	0.10	3.00
8	12.00	32.00	0.05	2.50
9	12.00	34.00	0.20	2.00
10	11.00	34.00	0.25	2.50
11	10.00	34.00	0.05	3.00
12	9.00	34.00	0.10	1.00
13	9.00	30.00	0.25	3.00
14	11.00	33.00	0.05	1.50
15	8.00	33.00	0.20	3.00
16	8.00	30.00	0.05	1.00
17	10.00	32.00	0.15	1.00
18	12.00	35.00	0.15	3.00
19	8.00	34.00	0.15	1.50
20	10.00	35.00	0.25	1.50
21	9.00	35.00	0.05	2.00
22	11.00	35.00	0.20	1.00
23	12.00	33.00	0.25	1.00
24	8.00	35.00	0.10	2.50
25	10.00	33.00	0.10	2.00

2.2.2. Setting Performance Indicators

One of the performance indicators in this paper is material removal rate, and the research object is milling block titanium alloy with end milling cutter. Therefore, the formula of material removal rate V within a certain time is shown in Equation (2).

$$V = v_f \cdot a_p \cdot a_e = \frac{50 \cdot v_c \cdot f_z \cdot z}{3 \cdot \pi \cdot d} \cdot a_p \cdot a_e \quad (2)$$

where v_f is feed speed, v_c is cutting speed, f_z is feed per tooth, z is the number of teeth, d is cutter diameter, a_p is cutting depth, a_e is cutting width.

Many scholars have done a lot of research on extending tool life by controlling tool wear rate [29], so tool wear rate is chosen as the performance indicator. In the metal cutting process, tool wear is defined as the material loss or deformation of the contact surface caused by friction between the cutting tool and the workpiece. Generally, the tool wear can be mainly divided into adhesive wear, abrasive wear, and so on [30]. The tool wear rate is chosen as one of the performance indicators, so it is extremely important to choose the right wear model. In the process of carbide cutting tools processing Ti6Al4V, the tool wear process under the action of cutting force and cutting heat is very complicated. But adhesive wear will occur regardless of the temperature [31]. Therefore, the adhesive wear model is selected as the research object for FEM, the adhesive wear rate was calculated using the adhesive wear model of Usui et al. [32], its expression is shown in Equation (3).

$$\frac{dW_{\text{Adhesion wear}}}{dt} = A_w \cdot \sigma_n \cdot v_c \cdot e^{\left(\frac{-B_w}{273+T}\right)} \quad (3)$$

where σ_n is the positive pressure, v_c is the chip slip velocity, T is the celsius temperature, and A_w and B_w are the wear characteristic constants, which can be obtained by tool wear test. According to reference [33] and reference [34], A_w = 0.0004, B_w = 7000, in which the values of A_w and B_w need to be input into Deform-3D software.

2.3. *Finite Element Simulation Results*

After Deform-3D post-processing, the tool wear rate when machining to 10 mm was obtained. The simulation results are shown in Figure 4, and the dynamic change diagram of milling simulation is shown in Figure 5. Among them, the cutting length of 10 mm is divided into 4 sections, respectively marked l_1 phase, l_2 phase, l_3 phase, l_4 phase. According to Equation (2), the material removal rate of titanium alloy processed by end milling cutter is obtained. Table 7 shows the data table of tool wear rate and material removal rate under different combinations of tool parameters and cutting parameters in each stage.

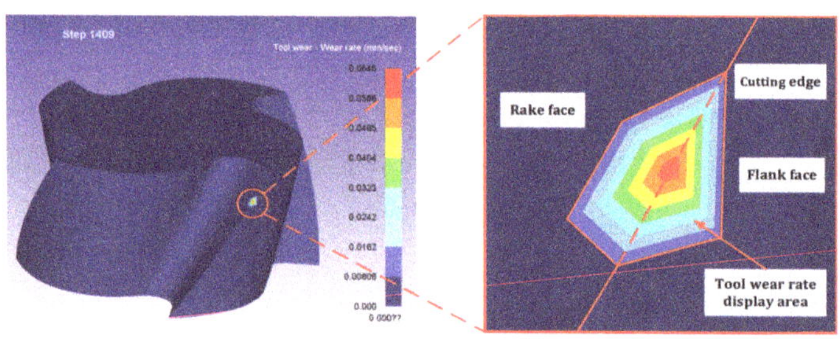

Figure 4. Simulation results of tool wear rate.

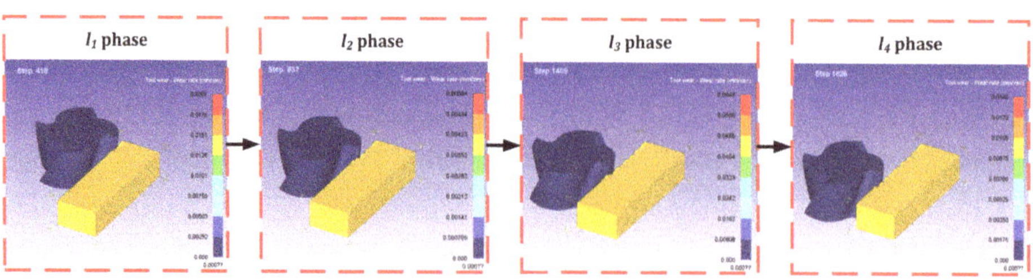

Figure 5. Dynamic change diagram of milling simulation.

Table 7. Performance indicator data table.

	The First Stage (l_1)		The Second Stage (l_2)		The Third Stage (l_3)		The Fourth Stage (l_4)	
	Wear Rate (mm/s)	V (mm^3/s)	Wear Rate (mm/s)	V (mm^3/s)	Wear Rate (mm/s)	V (mm^3/s)	Wear Rate (mm/s)	V (mm^3/s)
1	0.00934	4.8	0.00547	4.8	0.00156	4.8	0.02260	4.8
2	0.07030	2.4	0.04810	2.4	0.06770	2.4	0.13100	2.4
3	0.01680	4.8	0.01240	4.8	0.00945	4.8	0.02470	4.8
4	0.00623	8.0	0.01920	8.0	0.01110	8.0	0.13000	8.0
5	0.03900	8.0	0.14100	8.0	0.03130	8.0	0.02760	8.0
6	0.00718	6.0	0.03550	6.0	0.00378	6.0	0.01050	6.0
7	0.00168	4.8	0.00718	4.8	0.11400	4.8	0.01590	4.8
8	0.02320	2.0	0.01390	2.0	0.01760	2.0	0.01180	2.0
9	0.00508	6.4	0.03820	6.4	0.04800	6.4	0.10500	6.4
10	0.01030	10.0	0.02150	10.0	0.00221	10.0	0.00834	10.0
11	0.00338	2.4	0.22500	2.4	0.00802	2.4	0.05170	2.4
12	0.06360	1.6	0.11400	1.6	0.04850	1.6	0.05840	1.6
13	0.00747	12.0	0.05300	12.0	0.09910	12.0	0.06970	12.0
14	0.01660	1.2	0.04650	1.2	0.01180	1.2	0.07540	1.2
15	0.05810	9.6	0.02610	9.6	0.05140	9.6	0.01940	9.6
16	0.01460	0.8	0.03630	0.8	0.07520	0.8	0.04940	0.8
17	0.00944	2.4	0.03190	2.4	0.07750	2.4	0.00726	2.4
18	0.02010	6.4	0.00564	6.4	0.06460	6.4	0.01400	6.4
19	0.00544	3.2	0.00826	3.2	0.00757	3.2	0.01910	3.2
20	0.03630	6.0	0.09990	6.0	0.03680	6.0	0.03020	6.0
21	0.01550	1.6	0.00282	1.6	0.01850	1.6	0.00993	1.6
22	0.01930	3.2	0.01340	3.2	0.02070	3.2	0.01630	3.2
23	0.01380	4.0	0.03740	4.0	0.01010	4.0	0.00943	4.0
24	0.04980	4.0	0.01570	4.0	0.02950	4.0	0.03220	4.0
25	0.01970	3.2	0.01060	3.2	0.04600	3.2	0.02340	3.2

3. Dynamic Evaluation Method Based on Gain Horizontal Excitation

In different stages under different cutting tool parameters and cutting parameters combination of tool wear rate and material removal rate for data analysis, using the dynamic evaluation method based on gain level motivation for evaluating the various levels, the first parameter is selected by evaluating the parameters of the numerical biggest level, and so on, to get optimal levels of other parameters, eventually, become the best parameter combination.

3.1. Dynamic Evaluation Method Based on Gain Level Excitation

The dynamic comprehensive evaluation method based on gain horizontal excitation is a comprehensive evaluation method that considers the dynamic numerical changes of performance indicators at different stages. This method is based on the gain of each evaluated object in different periods, and then calculates the final total dynamic comprehensive evaluation value of each evaluated object through this dynamic evaluation method. Among them, gain represents the change of comprehensive evaluation value in the upper and lower stages. Level represents the comprehensive evaluation value of each stage; Motivation represents the guidance from one stage to the next. Figure 6 shows the flow chart of dynamic comprehensive evaluation method based on gain horizontal excitation.

As shown in Figure 7, the figure shows that the values of performance indicators of the evaluated object are different in different time periods. The dynamic comprehensive evaluation method not only takes into account all performance indicators, but also takes into account the characteristics that performance indicators may have different values in different stages. In this method, all performance indicators of all evaluated objects are comprehensively evaluated at time t_1 to obtain the comprehensive evaluation value of each evaluated object at that time, and the same method is applied at other times. Then, the evaluation values of each evaluation object at different times are unified together, and the dynamic evaluation values of each evaluated object are obtained according to the dynamic evaluation method.

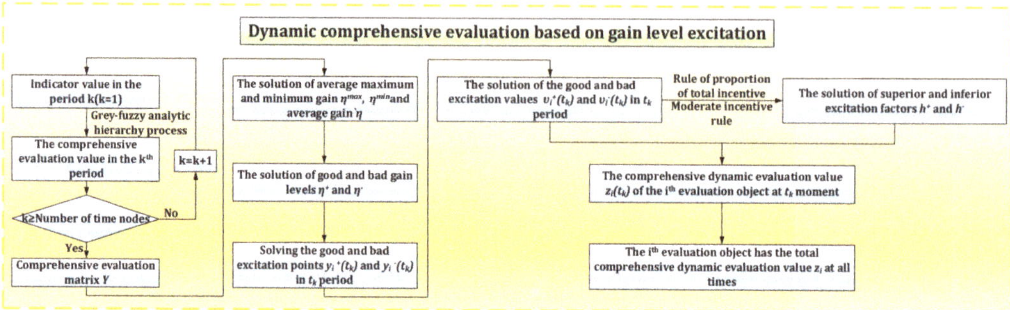

Figure 6. Flow chart of dynamic comprehensive evaluation method based on gain horizontal excitation.

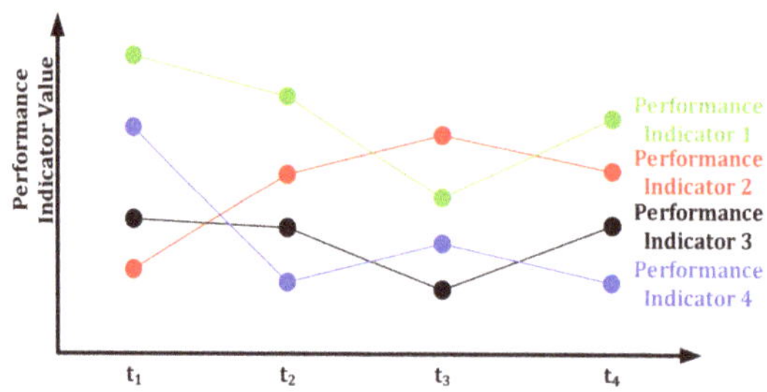

Figure 7. Dynamic change diagram of performance indicators.

3.2. Comprehensive Evaluation of Each Stage Based on Grey-Fuzzy Analytic Hierarchy Process

The purpose of this paper is to optimize the parameter level, so the sequential stereoscopic data table of each level of the studied parameter is shown in Table 8. There are n evaluated objects, that is, each parameter has n levels, s_i is the ith level of this parameter, m performance indicators, $x_{ij}(t_k)$ is the ith ($i = 1, 2, \ldots, n$) evaluated objects at t_k ($k = 1, 2, \ldots, T$) about the indicator x_j ($j = 1, 2, \ldots, m$) observed value.

Table 8. Time series stereoscopic data table.

	t_1				t_2				...	t_T			
	x_1	x_2	...	x_m	x_1	x_2	...	x_m	...	x_1	x_2	...	x_m
s_1	$x_{11}(t_1)$	$x_{12}(t_1)$	...	$x_{1m}(t_1)$	$x_{11}(t_2)$	$x_{12}(t_2)$	...	$x_{1m}(t_2)$	...	$x_{11}(t_T)$	$x_{12}(t_T)$	...	$x_{1m}(t_T)$
s_2	$x_{21}(t_1)$	$x_{22}(t_1)$	...	$x_{2m}(t_1)$	$x_{21}(t_2)$	$x_{22}(t_2)$	...	$x_{2m}(t_2)$	...	$x_{21}(t_T)$	$x_{22}(t_T)$	...	$x_{2m}(t_T)$
...	...	...	...	...	...	...	...	...	...	...	...	...	...
s_n	$x_{n1}(t_1)$	$x_{n2}(t_1)$	...	$x_{nm}(t_1)$	$x_{n1}(T_2)$	$x_{n2}(t_2)$	...	$x_{nm}(t_2)$	...	$x_{n1}(t_T)$	$x_{n2}(t_T)$	...	$x_{nm}(t_T)$

In the process of metal cutting, there exist the influence of parameters on the performance indicator, and there are also the mutual influence between the performance indicator, and the influence degree of these influences is unknown, so the metal cutting process is a fuzzy process. In order to save cost, the number of experiments and data amount in scientific research are limited. Therefore, in order to solve the problem of fuzziness in cutting process and insufficient data due to the limited number of tests, fuzzy theory, and grey theory are needed.

Grey fuzzy evaluation method is an evaluation method that has the advantages of both grey theory and fuzzy theory [35]. The combination of analytic hierarchy process and grey fuzzy evaluation method can better solve the problem of multi-objective comprehensive evaluation. This method is used to comprehensively evaluate the performance indicator data of different evaluated objects at each time to obtain the comprehensive evaluation value of each evaluated object at each time, which is finally combined into the comprehensive evaluation matrix Y, whose expression form is shown in Section 3.3. The calculation formula and related parameters of analytic hierarchy process and grey fuzzy evaluation method can be obtained from reference [36] and reference [37]. Through analytic hierarchy process and grey fuzzy evaluation, the comprehensive evaluation matrix B_k of the evaluated object at the t_k moment is obtained, and its expression is shown in Equation (4).

$$B_k = \begin{bmatrix} y_1(t_k) & y_2(t_k) & \ldots & y_n(t_k) \end{bmatrix} \tag{4}$$

where, $y_i(t_k)$ ($y_i(t_k) \in [0,1]$) is the comprehensive evaluation value of the ith evaluated object at the t_k moment.

3.3. Parameter Level Optimization Based on Dynamic Evaluation Method

After the evaluation matrix of each stage is obtained, the dynamic comprehensive evaluation of the evaluated object is carried out according to the dynamic evaluation method in reference [19], and the dynamic comprehensive evaluation value of the evaluated object is finally obtained.

The comprehensive evaluation matrix B_k of the evaluated object at the t_k moment is obtained through the grey-fuzzy analytic hierarchy process, and all the comprehensive evaluation matrix B_k is combined into the comprehensive evaluation matrix Y, as shown in Equation (5).

$$Y = \begin{bmatrix} y_1(t_1) & y_1(t_2) & \cdots & y_1(t_T) \\ y_2(t_1) & y_2(t_2) & \cdots & y_2(t_T) \\ \vdots & \vdots & \vdots & \vdots \\ y_n(t_1) & y_n(t_2) & \cdots & y_n(t_T) \end{bmatrix} \tag{5}$$

where $y_i(t_k)$ is the comprehensive evaluation value of the ith evaluated object at the t_k moment.

Definition 1. *Mean maximum gain, mean minimum gain and mean gain of η^{max}, η^{min}, and $\overline{\eta}$ evaluated object, respectively. Its calculation formula is shown in Equation (6).*

$$\begin{aligned} \eta^{\max} &= \max_i \left(\frac{1}{T-1} \sum_{k=1}^{T-1} (y_i(t_{k+1}) - y_i(t_k)) \right) \\ \eta^{\min} &= \min_i \left(\frac{1}{T-1} \sum_{k=1}^{T-1} (y_i(t_{k+1}) - y_i(t_k)) \right) \\ \overline{\eta} &= \frac{1}{n(T-1)} \sum_{i=1}^{n} \sum_{k=1}^{T-1} (y_i(t_{k+1}) - y_i(t_k)) \end{aligned} \tag{6}$$

Definition 2. *Respectively, η^+ and η^- are the good and bad gain levels of the evaluated object, and their calculation formula is shown in Equation (7).*

$$\begin{aligned} \eta^+ &= \overline{\eta} + k^+(\eta^{\max} - \overline{\eta}) \\ \eta^- &= \overline{\eta} - k^-(\overline{\eta} - \eta^{\min}) \end{aligned} \tag{7}$$

where k^+ and k^- are corresponding floating coefficients, k^+ and $k^- \in (0,1]$. Floating coefficients k^+ and k^- are used to describe the decision maker's psychological expectation of the overall development of the evaluated object.

After the good and bad gain levels η^+ and η^- are obtained, they are substituted into the following Equation (8),

$$\begin{aligned} \eta^+ &= y_i^+(t_k) - y_i(t_{k-1}), (k = 2, 3, \ldots, T) \\ \eta^- &= y_i^-(t_k) - y_i(t_{k-1}), (k = 2, 3, \ldots, T) \end{aligned} \tag{8}$$

At this point, the excellent and bad excitation points $y_i^+(t_k)$ and $y_i^-(t_k)$ of the ith evaluated object at the t_k moment are obtained by Equation (8).

The excellent and bad excitation points $y_i^+(t_k)$ and $y_i^-(t_k)$ are substituted into Equation (9).

$$\begin{array}{l} v_i^+(t_k) = y_i(t_k) - y_i^+(t_k), y_i(t_k) > y_i^+(t_k) \\ v_i^-(t_k) = y_i^-(t_k) - y_i(t_k), y_i^-(t_k) > y_i(t_k) \end{array} \tag{9}$$

where $v_i^+(t_k)$ and $v_i^-(t_k)$ are the excellent and bad excitation quantities obtained by the ith evaluated object at the t_k stage, respectively. In addition, in the case outside the value range, the superior and inferior excitation quantities are 0. $v_i^+(t_k) = v_i^-(t_k) = 0$ is set at the initial t_k stage without any excitation. Figure 8 is the geometric visual representation of the excellent and bad excitation points. In Figure 8, t_k, t_{k+1} and t_{k+2} stages respectively represent the three situations in which the evaluated object obtains the excellent excitation, does not obtain the bad excitation and obtains the good excitation.

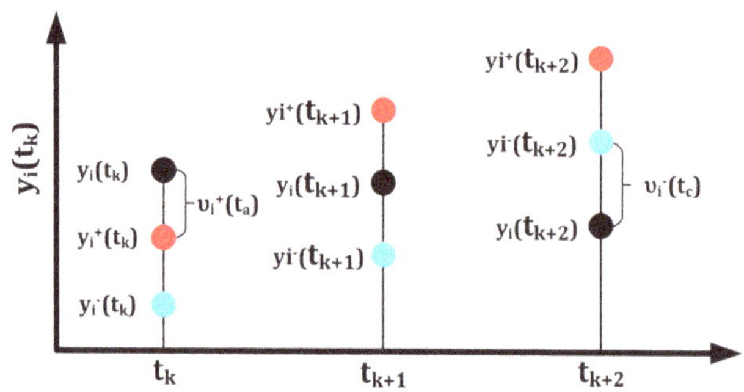

Figure 8. Geometric representation of excellent and bad excitation points and excitation quantities.

After obtaining the excellent and bad incentive points, the dynamic comprehensive evaluation value should also consider appropriate rewards and punishments for the parts above and below the excellent and bad incentive points. Let $z_i(t_k)$ be the dynamic comprehensive evaluation value of the ith evaluated object in the t_k stage, then the calculation formula of $z_i(t_k)$ is shown in Equation (10).

$$z_i(t_k) = h^+ v_i^+(t_k) + y_i(t_k) - h^- v_i^-(t_k) \tag{10}$$

where h^+, $h^-(h^+, h^- > 0)$ are superior and inferior excitation factors respectively; $h^+ v_i^+(t_k)$ and $h^- v_i^-(t_k)$ are the optimal and the inferior excitation values respectively. In addition, according to Equation (9), any $t_k(k = 1, 2, \ldots, T)$ in the moment, $v_i^+(t_k) \times v_i^-(t_k) = 0$, that is, $h^+ v_i^+(t_k)$, $h^- v_i^-(t_k)$ cannot be obtained at the same time.

The proportion rule of the total amount of incentives. For the n evaluated objects, Equation (11) shows that the total amount of good and bad excitation is proportional.

$$r = \frac{h^+ \sum_{i=1}^{n} \sum_{k=1}^{T} v_i^+(t_k)}{h^- \sum_{i=1}^{n} \sum_{k=1}^{T} v_i^-(t_k)} \tag{11}$$

where r $(r \in R^+)$ is the proportion relation between the total amount of excellent incentives and the total amount of bad incentives, which is a reflection of the decision intention of the evaluator. When $r > 1$, indicates that the total amount of excellent excitation is greater than the total amount of bad excitation; When $r < 1$, indicates that the total amount of excellent excitation is less than the total amount of bad excitation; When $r = 1$, it means that the total amount of excellent excitation is equal to the total amount of bad excitation.

Moderate incentive rules. According to Equation (12), the sum of superior and inferior excitation factors h^+ and h^- is 1.

$$h^+ + h^- = 1 \quad (12)$$

When the ratio r between the total amount of optimal excitation and the total amount of inferior excitation is determined, the values of h^+ and h^- can be obtained through Equations (11) and (12).

The dynamic comprehensive evaluation value $z_i(t_k)$ of the ith evaluated object at the t_k stage is obtained through the above steps. Then the total dynamic comprehensive evaluation value z_i of the ith evaluated object at all times is shown in Equation (13).

$$z_i = \sum_{k=1}^{T} \tau_k z_i(t_k) \quad (13)$$

where τ_k is the time factor, $\{\tau_k\}$ is usually taken as a series of increasing type. If there is no specific requirement and time preference can be ignored, $\tau_k = 1$.

3.4. *Comprehensive Evaluation of Parameter Level in Each Stage*

Firstly, the comprehensive evaluation of each level of the clearance angle was carried out, and the grey fuzzy analytic hierarchy process in reference [36] and reference [37] was used to carry out the comprehensive evaluation, and the comprehensive evaluation value of each level of the clearance angle in each stage was obtained.

Use analytic hierarchy process to get the weight of each performance indicator. According to the requirement, the indicator set is $\{x_1, x_2\}$. Among them, x_1 is tool wear rate, and x_2 is material removal rate. First of all, weight distribution was carried out for each performance indicator. In order to highlight the importance of tool life, it was necessary to increase the weight value of tool wear rate, as shown in Table 9. Then, according to reference [36], the weight value of tool wear rate is 0.6, the weight value of material removal rate is 0.4, and the indicator weight matrix P = [0.6, 0.4].

Table 9. Significance comparison scale values among indicators.

x_j/x_h	x_1/x_2	x_2/x_2
U_{jh}	1.5	1.0

According to the weight matrix of performance indicators and the grey fuzzy evaluation method in reference [36], the comprehensive evaluation of the evaluated object in each stage is carried out. As the object is evaluated for each level of parameters. Therefore, all levels of the clearance angle are evaluated comprehensively. According to the performance indicators selected in this paper, phase l_k was used instead of moment t_k to represent the dynamic nodes in the cutting process. The values shown in Table 10 are the average values of the sum of the performance indicator values of the clearance angle at the same level obtained in Table 7 at different stages. And according to the weight matrix of performance indicators and the grey-fuzzy evaluation method, the comprehensive evaluation matrix B_1, B_2, B_3, and B_4 of four stages are obtained, among which the coefficients required by the grey-fuzzy evaluation method can be obtained from reference [37].

$$B_1 = \begin{bmatrix} 0.77701 & 0.88889 & 0.75000 & 0.90000 & 0.57008 \end{bmatrix}$$
$$B_2 = \begin{bmatrix} 0.89557 & 0.78564 & 0.75000 & 0.90000 & 0.61180 \end{bmatrix}$$
$$B_3 = \begin{bmatrix} 0.87574 & 1.00000 & 0.43924 & 0.50000 & 0.38385 \end{bmatrix}$$
$$B_4 = \begin{bmatrix} 0.82976 & 0.91783 & 0.69123 & 0.90000 & 0.33333 \end{bmatrix}$$

Table 10. Stage sequence stereoscopic data table of the clearance angle.

	The First Stage (l_1)		The Second Stage (l_2)		The Third Stage (l_3)		The Fourth Stage (l_4)	
	Wear Rate (mm/s)	V (mm^3/s)	Wear Rate (mm/s)	V (mm^3/s)	Wear Rate (mm/s)	V (mm^3/s)	Wear Rate (mm/s)	V (mm^3/s)
1	0.02683	5.12	0.02111	5.12	0.03495	5.12	0.05002	5.12
2	0.02062	5.20	0.04216	5.20	0.03429	5.20	0.03427	5.20
3	0.09176	4.40	0.10168	4.40	0.03992	4.40	0.02803	4.40
4	0.01150	5.04	0.01728	5.04	0.04477	5.04	0.01450	5.04
5	0.02650	4.24	0.02865	4.24	0.04160	4.24	0.26369	4.24

3.5. Dynamic Evaluation of Parameter Level

After obtaining the comprehensive evaluation value of each level of the clearance angle in each stage, the dynamic evaluation method is used to carry out comprehensive evaluation, and the final comprehensive evaluation value of each level of the clearance angle is obtained.

The comprehensive evaluation matrices B_1, B_2, B_3, and B_4 of each stage obtained in 3.4 were transformed and combined to obtain the comprehensive evaluation matrix Y of five levels about the clearance angle.

$$Y = \begin{bmatrix} 0.77701 & 0.89557 & 0.87574 & 0.80976 \\ 0.88889 & 0.78564 & 1.00000 & 0.91783 \\ 0.75000 & 0.75000 & 0.43924 & 0.69123 \\ 0.90000 & 0.90000 & 0.50000 & 0.90000 \\ 0.57008 & 0.61180 & 0.38385 & 0.33333 \end{bmatrix}$$

According to Equation (6), the average maximum gain $\eta^{\max} = 0.01092$, average minimum gain $\eta^{\min} = -0.07892$, average gain $\overline{\eta} = -0.015588$ were obtained for the comprehensive evaluation matrix Y.

Taking the floating coefficients k^+ and k^- as 0.3, the optimal gain level $\eta^+ = -0.00764$ and the inferior gain level $\eta^- = -0.03459$ of the comprehensive evaluation matrix Y were obtained according to Equation (7).

According to Equation (8), the optimal excitation points $y_i^+(t_k)$ and the inferior excitation points $y_i^-(t_k)$ of each level of the clearance angle in Table 11 at different stages were obtained.

Table 11. The excellent and bad excitation points of each level of the clearance angle at different stages.

	l_1	l_2		l_3		l_4	
	$y_i(l_1)$	$y_i^+(l_2)$	$y_i^-(l_2)$	$y_i^+(l_3)$	$y_i^-(l_3)$	$y_i^+(l_4)$	$y_i^-(l_4)$
1	0.77701	0.76937	0.74242	0.88793	0.86098	0.86810	0.84115
2	0.88889	0.88125	0.85430	0.77800	0.75105	0.99236	0.96541
3	0.75000	0.74236	0.71541	0.74236	0.71541	0.43160	0.40465
4	0.90000	0.89236	0.86541	0.89236	0.86541	0.49236	0.46541
5	0.57008	0.56244	0.53549	0.60416	0.57721	0.37621	0.34926

According to Equation (9), the optimal excitation quantities $v_i^+(l_k)$ and the inferior excitation quantities $v_i^-(l_k)$ at different stages of each level of the clearance angle in Table 12 were obtained.

Table 12. Good and bad excitation scales for each level of clearance angle at different stages.

	l_1		l_2		l_3		l_4	
	$v_i^+(l_1)$	$v_i^-(l_1)$	$v_i^+(l_2)$	$v_i^-(l_2)$	$v_i^+(l_3)$	$v_i^-(l_3)$	$v_i^+(l_4)$	$v_i^-(l_4)$
1	0	0	0.12640	0	0	0	0	0.03139
2	0	0	0	0.06866	0.22800	0	0	0.04758
3	0	0	0.00764	0	0	0.27617	0.25963	0
4	0	0	0.00764	0	0	0.36541	0.40764	0
5	0	0	0.04936	0	0	0.19336	0	0.01593

According to Equations (11) and (12) and the ratio $r = 1$ between the total amount of excellent excitation and the total amount of bad excitation, the optimal excitation factor $h^+ = 0.47894$ and the inferior excitation factor $h^- = 0.52106$ can be obtained.

According to Equation (10), the dynamic comprehensive evaluation value $z(t_k)$ of each level of the clearance angle in Table 13 at different stages is obtained. Set $\tau_k = 1$ and obtain the total dynamic comprehensive evaluation value z for each level of the rear Angle according to Equation (13). According to the total dynamic comprehensive evaluation value z of each level of the clearance angle in the last column in Table 13, 9° clearance angle is selected as the best level.

Table 13. The dynamic comprehensive evaluation table of each level of the clearance angle and the total dynamic comprehensive evaluation table.

	$z(l_1)$	$z(l_2)$	$z(l_3)$	$z(l_4)$	z
1	0.77701	0.95611	0.87574	0.79340	3.40295
2	0.88889	0.74986	1.10920	0.89304	3.65099
3	0.75000	0.75366	0.29534	0.70366	2.50266
4	0.9000	0.90366	0.30960	1.09524	3.20850
5	0.57008	0.63544	0.28310	0.32503	1.81365

Similarly, the total dynamic comprehensive evaluation value of each level of spiral Angle, feed per tooth and cutting depth is shown in Table 14. According to this table, the optimal level of spiral Angle, feed per tooth and cutting depth is determined to be 30°, 0.15 mm/z, and 2.5 mm, respectively.

Table 14. Total dynamic composite evaluation values for each level of the remaining parameters.

	$z_{Helix\ angle}$	$z_{Feed\ per\ tooth}$	$z_{Cutting\ depth}$
1	2.44285	1.99773	1.75018
2	2.20450	1.62204	1.77147
3	2.31967	3.05105	2.48153
4	2.04067	2.39639	2.93874
5	2.06289	2.95389	2.67173

To sum up, when the rake angle is 8°, the cutting speed is 37.68 m/min, and the cutting width is 0.2 mm, the optimal combination of the four parameters studied is 9° clearance angle, 30° helix angle, 0.15 mm/z feed per tooth and 2.5 mm cutting depth.

3.6. Comparison between Parameter Combinations

After the optimal combination of tool parameters and cutting parameters was obtained by the dynamic evaluation method, the finite element method was used to DEFORM-3D finite element simulation of the cutting process of the group of parameters, and the tool wear rate values of the four stages under the group of parameters were obtained. According to Equation (2), the material removal rate of this group of parameters is obtained. Figure 9 is the simulation diagram of wear rate at each stage of parameter combination obtained by dynamic comprehensive evaluation method. Table 15 is the simulation value of tool wear rate at each stage and the numerical table of material removal rate.

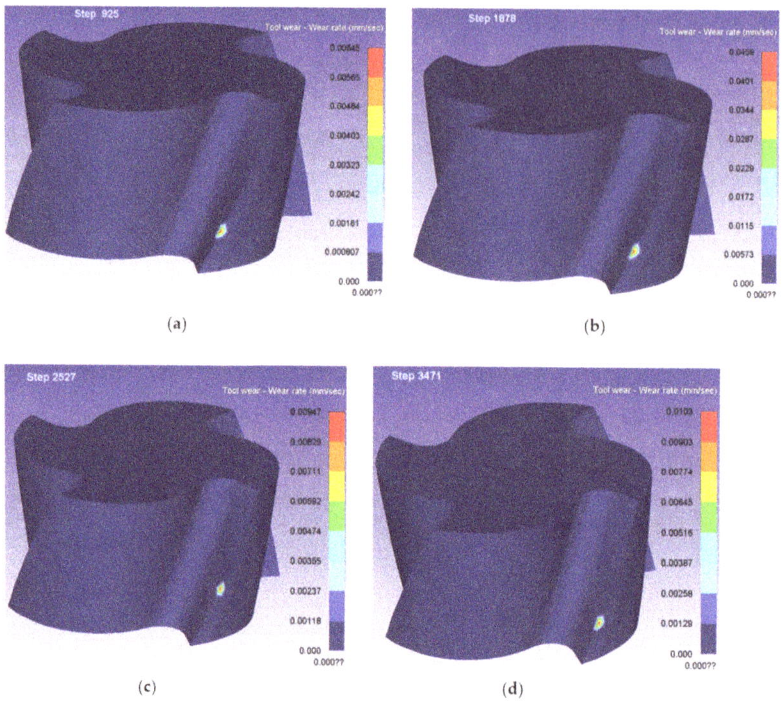

Figure 9. Simulation diagram of tool wear rate at each stage: (**a**) the first-stage wear rate value (**b**) the second-stage wear rate value (**c**) the third-stage wear rate value and (**d**) the fourth-stage wear rate value.

Table 15. The numerical table of tool wear rate and material removal rate corresponding to the optimal parameters.

Wear Rate (mm/s)				V (mm^3/s)
l_1	l_2	l_3	l_4	
0.00645	0.0459	0.00947	0.0103	6.0

By comparing Table 15 with Table 7, the tool wear rate values at each phase in Table 15 are better than most of the tool wear rate values at the same stage in Table 7, and the material removal rate in Table 15 is better than most of the material removal rate in Table 7. Therefore, the parameter combination obtained by dynamic evaluation method has higher comprehensive performance.

4. Validation of the Finite Element Model

In this chapter, the cutting force in three directions is taken as the performance indicator, and the cutting speed is taken as the research object. The accuracy of the simulation model is verified by comparing the simulation value with the experimental value. The reliability of the above research results can be illustrated by verifying the accuracy of the simulation model.

4.1. Setting of Experimental Parameters

Cutting speed is one of the important parameters affecting the cutting force [38]. Therefore, taking the cutting speed as the studied parameter, and taking the cutting forces in three directions as performance indicators, the accuracy of the finite element simulation model was verified by comparing the maximum values of the simulation values and the experimental values. Table 16 shows the milling parameters test table.

Table 16. Settings of milling cutting parameters.

Cutting Speed (m/min)	Feed Speed (mm/min)	Cutting Depth (mm)	Cutting Width (mm)
31.40	400	3	0.8
37.68	400	3	0.8
47.10	400	3	0.8

4.2. The Experimental Device

Figure 10 shows the end milling cutter, titanium alloy workpiece and instrument used in the experiment. Among them, Figure 10a is a CNC milling machine (VDL-1000E, Dalian Machine Tool Group, Dalian, China). Figure 10b is a rotary dynamometer (9171A, Kistler, Winterthur, Switzerland), an end mill and a titanium workpiece for measuring cutting forces in the x-, y-, and z-directions. The cutting force components along x-, y-, and z-directions are recorded by Dynoware signal analyzer software (Kistler, Winterthur, Switzerland).

(a)

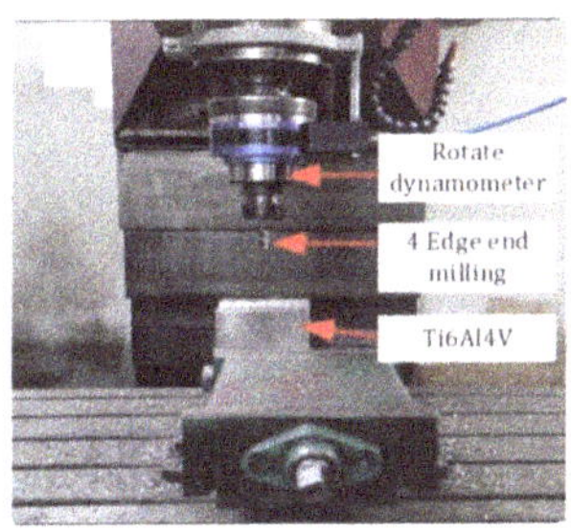

(b)

Figure 10. Experimental instruments and equipment used in the experiment (**a**) CNC milling machine and (**b**) cutting force measuring equipment.

4.3. Reliability Verification of Simulation Model

As shown in Table 17, this table is the cutting force value table under simulation and experiment. The experimental value is the cutting force value within a line, and the simulation value selects the cutting force data within a certain distance and deletes the unreasonable data. Figure 11 shows the comparison between the simulated and experimental values of the maximum cutting forces in three directions. According to the Figure 11, the variation trend of the simulation value and the experimental value with the cutting speed is basically the same, with the maximum error of 24.822% and the minimum error of 14.036%, so the simulation value is in good agreement with the experimental value, so the simulation model of the finite element model is reliable, the data obtained by finite element method can be used directly. Therefore, the reliability of the finite element simulation model shows that the dynamic evaluation results are also accurate and reliable.

Table 17. Obtained cutting forces for different cutting speeds.

	Cutting Speed (m/min)	Simulation (N)	Experiment (N)
F_x	31.40	313.613	266.791
	37.68	634.243	510.750
	47.10	577.552	462.902
F_y	31.40	358.606	289.408
	37.68	532.094	466.600
	47.10	503.908	415.483
F_z	31.40	234.752	188.069
	37.68	168.884	135.653
	47.10	191.216	153.299

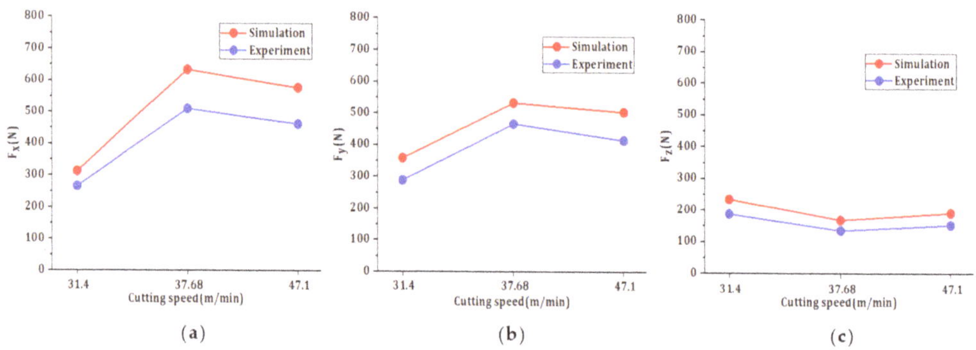

Figure 11. The simulation value of cutting force is compared with the experimental value: (**a**) x-direction (**b**) y-direction and (**c**) z-direction.

5. Conclusions and Prospects

In order to solve the problems of short tool life and low machining efficiency in titanium alloy cutting process, it is necessary to study the optimal matching between tool parameters and cutting parameters. In this paper, the optimization of tool parameters and cutting parameters is studied. Taking titanium alloy side milling with end milling cutter as the research object, and taking tool wear rate and material removal rate as performance indicators. Clearance angle, helix angle, feed per tooth, and cutting depth are optimized parameters. The dynamic comprehensive evaluation method based on gain horizontal excitation is used to comprehensively evaluate each level of tool parameters and cutting parameters, and the optimal level of each parameter is obtained and combined to become the optimal parameter combination. After that, by comparing the optimized parameter combination with the previous parameter combination by finite element method, the comprehensive performance of the optimized parameter combination is determined to be higher. Finally, the reliability of the simulation model is verified by comparing simulation and experiment, so as to illustrate the reliability of the dynamic evaluation results. According to the above research, the following conclusions are drawn:

1. In this paper, the dynamic evaluation method based on gain horizontal excitation was used to optimize the tool parameters and cutting parameters in the process of milling titanium alloy with milling cutter side, and the optimal matching combination of tool parameters and cutting parameters on the tool wear rate and material removal rate was obtained.
2. When the rake angle is 8°, the cutting speed is 37.68 m/min, and the cutting width is 0.2 mm, the machining effect of the clearance angle is 9°, the helix angle is 30°, the feed per tooth is 0.15 mm/z, and the cutting depth is 2.5 mm achieves the best, which can simultaneously meet the requirements of long tool life and high machining efficiency.

In addition, the reliability of simulation model is verified, and the optimization results are also reliable.
3. The comparison between the optimized parameters by finite element method and the parameter combination in Table 6 shows that the optimized parameter combination has higher comprehensive performance.
4. In this paper, the performance indicator value is obtained by simulation, but there is some error between simulation value and experimental value. Therefore, in the future, under the condition of sufficient time and funding, the required numerical value of tool wear rate and material removal rate will be obtained through experiments to make the optimization results more accurate.

Author Contributions: Conceptualization, D.Y.; methodology, D.Y.; validation, D.Y. and D.H.; writing—original draft preparation, D.Y.; writing—review and editing, A.Z., C.Y., X.L., and M.L.; project administration, X.L.; funding acquisition, X.L. All authors have read and agreed to the published version of the manuscript.

Funding: This research is funded by Projects of National Key Research and Development Project (2019YFB1704800).

Institutional Review Board Statement: Not applicable.

Informed Consent Statement: Not applicable.

Data Availability Statement: Not applicable.

Acknowledgments: We thank LetPub (www.letpub.com) for its linguistic assistance during the preparation of this manuscript. And thanks to the English editing service of MDPI.

Conflicts of Interest: The authors declare no conflict of interest.

Abbreviations

Notation	
CNC	Computerized Numerical Control
PVD	Physical Vapor Deposition
Symbol	
σ	The equivalent flow stress
ε	The equivalent plastic strain rate
ε_0	The reference plastic strain rate
T	The absolute temperature
T_r	The ambient temperature
T_m	The melting temperature
A	the yield strength
B	The hardening modulus
C	The strain rate sensitivity coefficient
m	The heat softening coefficient
n	The strain hardening index
V	The material removal rate
v_f	The feed speed
d	The cutter diameter
a_p	The cutting depth
a_e	The cutting width
f_z	The feed per tooth
f_z	The feed per tooth
z	The number of teeth
$W_{\text{Adhesion wear}}$	The adhesion wear
σ_n	The positive pressure
v_c	The chip slip speed
A_w	The wear characteristic constant
B_w	The wear characteristic constant

T	The celsius
l_k	The kth phase
t_k	The period k
n	The number of evaluated objects
s_i	The ith object to be evaluated
m	The number of performance indicators
T	The number of time periods
x_j	The jth performance indicator
$x_{ij}(t_k)$	The value of the ith evaluated object about the jth indicator at the time t_k
B_k	The static evaluation matrix of the kth period
$y_i(t_k)$	The static evaluation value of the ith evaluated object in the kth period
Y	The static comprehensive evaluation matrix
$\eta^{\max}$	The mean maximum gain
$\eta^{\min}$	The mean minimum gain
$\overline{\eta}$	The average gain
η^+	The optimal gain level
η^-	The inferior gain level
$y_i^+(t_k)$	The optimal excitation point of the ith evaluated object at time t_k
$y_i^-(t_k)$	The inferior excitation point of the ith evaluated object at time t_k
$v_i^+(t_k)$	The optimal excitation quantity obtained by the ith evaluated object at time t_k
$v_i^-(t_k)$	The inferior excitation obtained by the ith evaluated object at time t_k
$z_i(t_k)$	The dynamic comprehensive evaluation value of the ith evaluated object at t_k moment
h^+	The optimal excitation factor
h^-	The inferior excitation factor
r	The proportional relationship between the total amount of the optimal excitation quantity and the total amount of bad incentives
τ_k	The time factor
z_i	The total dynamic comprehensive evaluation value of the ith evaluated object
u_{jh}	The weight ratio of the jth performance indicator to the hth performance indicator

References

1. Wang, B.; Liu, Z.; Cai, Y.; Luo, X.; Ma, H.; Song, Q.; Xiong, Z. Advancements in material removal mechanism and surface integrity of high speed metal cutting: A review. *Int. J. Mach. Tools Manuf.* **2021**, *166*, 103744. [CrossRef]
2. Zhang, X.; Ma, J.W.; Jia, Z.Y.; Song, D.N. Machining parameter optimisation for aviation aluminium-alloy thin-walled parts in high-speed milling. *Int. J. Mach. Mach. Mater.* **2018**, *20*, 180–192. [CrossRef]
3. Kubilay, A.; Mohd, D.; Ahmet, H.; Mozammel, M.; Munish, G.; Turnad, G.; Hassan, I. Investigations on surface roughness and tool wear characteristics in micro-turning of Ti-6Al-4V Alloy. *Materials* **2020**, *13*, 2998.
4. Ghosh, T.; Wang, Y.; Martinsen, K.; Wang, K. A surrogate-assisted optimization approach for multi-response end milling of aluminum alloy AA3105. *Int. J. Adv. Manuf. Technol.* **2020**, *111*, 2419–2439. [CrossRef]
5. Juan, C.O.P.; Sepideh, A.; Juan, P.C.R.; Juan, P.C.R. Cutting parameter optimization of Al-6063-O using numerical simulations and particle swarm optimization. *Int. J. Adv. Manuf. Technol.* **2020**, *111*, 2507–2532.
6. Zhang, N.; Wu, D.W.; Jiang, G.J. Optimization of Machining Parameters in blisk processing based on tool reliability. In *IOP Conference Series: Materials Science and Engineering*; IOP: Bristol, UK, 2021.
7. Mohammed, Y.E.K.; Shunmugesh, K. Multi-objective optimization of milling process parameters in glass fibre reinforced polymer via grey relational analysis and desirability function. *Mater. Today. Proc.* **2019**, *11*, 1015–1023.
8. Fang, Y.; Zhao, L.; Lou, P.; Yan, J. Cutting parameter optimization method in multi-pass milling based on improved adaptive PSO and SA. *J. Phys. Conf. Ser.* **2021**, *1848*, 012116. [CrossRef]
9. Vimal, P.; Partha, P.D. Parametric optimization of CNC turning on glass-fibre-reinforced plastic (GFRP) pipes: A grey-fuzzy logic approach. In *IOP Conference Series: Materials Science and Engineering*; IOP: Bristol, UK, 2018.
10. Viswanathan, R.; Ramesh, S.; Maniraj, S.; Subburam, V. Measurement and multi-response optimization of turning parameters for magnesium alloy using hybrid combination of Taguchi-GRA-PCA technique. *Measurement* **2020**, *159*, 107800. [CrossRef]
11. Suresh, P.; Marimuthu, K.; Ranganathan, S.; Rajmohan, T. Optimization of machining parameters in turning of Al-SiC-Gr hybrid metal matrix composites using grey-fuzzy algorithm. *Trans. Nonferrous Met. Soc. China* **2014**, *24*, 2805–2814. [CrossRef]
12. Gnanavelbabu, A.; Arunachalam, V.; Surendran, K.T.S.; Rajkumar, K. Optimization of machining parameters in CNC turning of AA6061-B4C-CNT hybrid composites using Grey-fuzzy method. In *IOP Conference Series: Materials Science and Engineering*; IOP: Bristol, UK, 2020.
13. Swastik, P.; Kalipada, M. Optimization of machining parameter characteristics during turning of Ti-6Al-4V using desirability function analysis. *Mater. Today Proc.* **2018**, *5*, 25740–25749.

materials

MDPI

Article

Statistical Analysis and Optimisation of Data for the Design and Evaluation of the Shear Spinning Process

Sandra Puchlerska [1,*], Krzysztof Żaba [1], Jarosław Pyzik [2], Tomasz Pieja [3] and Tomasz Trzepieciński [4]

1 Department of Metal Working and Physical Metallurgy of Non-Ferrous Metals, Faculty of Non-Ferrous Metals, AGH University of Science and Technology, Al. Mickiewicza 30, 30-059 Kraków, Poland; krzyzaba@agh.edu.pl
2 Sabre Polska Sp. z o.o., Wadowicka 6D, 30-415 Kraków, Poland; jaro.pyzik@gmail.com
3 Pratt&Whitney Rzeszów S.A., Hetmańska 120, 35-001 Rzeszów, Poland; tomasz.pieja@pwrze.utc.com
4 Department of Manufacturing and Production Engineering, Faculty of Mechanical Engineering and Aeronautics, Rzeszów University of Technology, al. Powstańców Warszawy 8, 35-959 Rzeszów, Poland; tomtrz@prz.edu.pl
* Correspondence: spuchler@agh.edu.pl

Abstract: This work proposes a research method that is a scheme that can be universally applied in problems based on the selection of optimal parameters for metal forming processes. For this purpose, statistical data optimisation methods were used. The research was based on the analysis of the shear spinning tests performed in industrial conditions. The process of shear spinning was conducted on the components made of Inconel 625 nickel superalloy. It was necessary to select the appropriate experimental plan, which, by minimising the number of trials, allowed one to draw conclusions on the influence of process parameters on the final quality of the product and was the starting point for their optimisation. The orthogonal design 2^{3-1}_{III} is the only design for three factors at two levels, providing non-trivial and statistically significant information on the main effects and interactions for the four samples. The samples were analysed for shape and dimensions using an Atos Core 200 3D scanner. Three-dimensional scanning data allowed the influence of the technological parameters of the process on quality indicators, and thus on the subsequent optimisation of the process, to be determined. The methods used proved to be effective in the design, evaluation and verification of the process.

Keywords: superalloy; design of experiments; 3D scanning; statistical optimisation

Citation: Puchlerska, S.; Żaba, K.; Pyzik, J.; Pieja, T.; Trzepieciński, T. Statistical Analysis and Optimisation of Data for the Design and Evaluation of the Shear Spinning Process. *Materials* **2021**, *14*, 6099. https://doi.org/10.3390/ma14206099

Academic Editor: Chao Xu

Received: 9 September 2021
Accepted: 11 October 2021
Published: 15 October 2021

1. Introduction

Shear spinning is a method of rotating forming of components on a rotating forming block using a roller. This technique is recommended for small and medium series production of axisymmetric products [1,2]. Plastic deformation of the material in the process of shear spinning occurs by direct action of the pressing rollers on a small contact area with the material. The method is used for the production of products made of carbon and stainless steels, non-ferrous metal alloys and "difficult-to-form" materials [3–5]. The shear formed products have increased strength and hardness, with favourable microstructure directionality. The advantages of shear spinning include: high dimensional accuracy, the possibility of application to large deformations, low tooling costs, and high surface quality [6,7].

There are many methods of analysing and optimising the shear spinning process. The first analytical models of the shear spinning process were developed by Dröge in 1954 [8] to calculate the tangential forces during the formation of cylindrical parts. The author points out that this is a time-consuming method. Another model was developed by Avitzur and Yang [9] for shear forming. They calculated the tangential force assuming a pure shear state for the sheet under the roll. In the model, they focused the feed, roller radius and wall angle. Another model of shear spinning proposed by Kalpakcioglu [10], assumes that the rectilinear deformation with plane deformation allows to estimate the tangential force. The model was verified by Kobayashi [11], who proposed an additional assumption of the bending mechanism. Moreover, the radial and axial force components in the contact area of the material with the roller were estimated. Sortais et al. published a slightly different proposal [12]. They extended the pure shear model to aspects related to flange deformation. Hayama and Murota [13] proposed two models of spinning, assuming axial deformation. In both models, three force components were calculated on the basis of the pressure distribution on the roller, and the unknown friction coefficient was determined by fitting the analytical force calculations to the experimental results. Forces at low feed rates are predicted with high accuracy. In contrast to the axisymmetric models, Wang et al. [14] proposed an asymmetric analytical model for predicting the axial force. Kim et al. [15] obtained a relatively accurate model using experimentally determined material flow curves. Chen et al. [16] calculated all three components of force. The main feature that distinguishes this model from the previous ones is the inclusion of over-spinning as a parameter. Joorabchian and Slater [17] proposed models that adopt axisymmetric and plane strain to estimate the upper limit of the tangential force during shear spinning. Estimates coincide with experimental results. Two analytical models for predicting deformation during conventional spinning are described in the paper of Quigley and Monaghan [18]. In both studies, considerations were made based on changes in the geometry and the constancy of volume. Comparing the authors' suggestions with experiments provides useful insight into the deformation mechanisms in both spinning and shear forming.

The first numerical model of shear spinning was developed by Alberti et al. [19]. Importantly, the ability of the numerical model to predict geometry, strains and loads in the process has been demonstrated, although the authors do not verify their values. A similar approach to the axisymmetric elasto-plastic model was developed by Liu et al. [20]. They investigated the effect of different tool paths on deformation and stresses in the process. A more refined model was proposed by Mori and Nonaka [21], who artificially introduced shear stresses into the oscillometric model. The main advantage of the Mori and Nonaka model compared to the previous proposals is the ability to predict the material gain before the roll and also to use this information to predict failure in the form of cracks. Three-dimensional numerical models are designed to improve accuracy but require longer computational times. The first three-dimensional model was proposed by Quigley and Monaghan [22], who used an intermediate solution algorithm with an elasto-plastic material model. In order to shorten the computation time, the authors attempted both adaptive mesh densification and parallel processing. The authors emphasize the importance of accurate modelling of the distance between the roller and the forming block. They indicate that a slight change in distance has an intense effect on the stresses arising under the tool. Moreover, the calculated forces were shown to be the same as those calculated by Wang et al. [14], with the difference that Wang et al. calculated the forces for the three-step process, while Quigley and Monaghan modelled only part of the first roll pass. Another model of hot shear spinning was developed by Klocke and Wehrmeister [23] to optimise the location of the laser beam and the heated zone. Their analysis was carried out separately, respectively in the Abaqus and Deform simulation programs. The intermediate elastoplastic model for hot process was used by Lu et al. [24] to study over-spinning and under-spinning flange bending. Two elasto-plastic shear spinning models were developed by Kleiner et al. [25] and Klimmek et al. [26]. They used them to study stresses and folds in conventional spinning. Zhan et al. [27] developed a direct elasto-plastic model of the cone spinning process to investigate stress, strain and thickness distributions and predict the effect of feed on tool forces and geometric tolerances of manufactured parts. The authors report that the maximum deviation of the geometry predicted by the model largely coincides with the experiments, with the maximum error of 22%. Gao et al. [28] built a model that investigates the influence of the parameters of the hot spinning process on microstructure, stress state and on damage evolution. The authors concluded that the inner surface of workpiece with the largest voids volume fraction is the place with the greatest potential of fracture. Moreover, the authors suggested an optimal configuration of processing parameters for hot forming of tubes.

To predict process failure caused by cracks or folding in the free surface, the authors propose several experimental approaches. The first is the approach published by Kegg [29], who tried to predict process failure caused by fracture in the rotary crush process. On the basis of experimental observations, he proposed an empirical relation in order to establish the shear spinning of a given material. The Kegg approach was extended by Hayama and Tago [30] to include the occurrence of a failure mode of the process in the form of folds. Additionally, the wall-cracking capacity of the intermediate was discussed by Kegg in [29]. Hayama et al. [31] used strain gauges installed on both sides of the blank, close to the edge of the sheet, to study the deformation and folding of the flange in the process of shear spinning. In publication [32], by capturing the deflection of weld line in flow forming of welded tube, the characteristic and mechanism of circumferential twist in flow forming was investigated. The weld line turned out to be useful for the evaluation of the geometrical characteristics of the circumferential deformation.

One of the methods of assessing the quality of the shear spinning process and products is statistical optimisation, using experimental data. Kawai [33] used a computer database to design shear spinning process. The database consists of approximately one hundred industrial cases that can be searched by dimension, shape and material for both input materials and products. The database was extended by Kawai et al. [2] to also include knowledge of faults and final product quality in the process. A similar approach was presented by Kleiner et al. [25], who applied the DoE (design of experiments) method to determine the influence of process parameters on the initiation of folding in the conventional spinning process. To analyse the forming forces, surface roughness and product geometry, three different approaches based on the DoE methodology were identified. The first of them was proposed by Chen et al. [3], using the DoE in conjunction with regression analysis to establish the empirical relationship between the process parameters and the forming forces and surface roughness in the case of the shear spinning. The second approach, proposed by Kunert et al. [34] uses DoE in conjunction with case-based reasoning to optimise geometry and geometric tolerance in conventional spinning.

In general, the aim of the experimental investigations of shear spinning presented in this paper is to obtain information on the relationship between the input factors (process parameters) and the output quantities (describing, for example, the characteristics of the final product). Such relations are usually presented in the form of a regression function approximating the results of tests on the basis of input factors. A properly conducted experiment, after a correct analysis of the results, allows the significance of the influence of input factors and their possible interactions to be determined giving the researcher the opportunity to select the most important factors and to eliminate the insignificant ones. This information can be used to describe the mechanism of a given phenomenon or to optimise the process parameters. Before planning the experiments, the type and number of dependent variables (responses) and of input variables, their ranges and disturbance factors must be determined. This stage is an important step in the planning of the experiment because the omission of important factors, an excess of them, or the presence of unresolved interrelationships may result in incorrect conclusions from the analysis. Having such test characteristics, an appropriate set of samples is selected. The choice of an experiment design depends on the purpose and specification of the research. The intuitive approach is to select a complete experiment design in which the trials correspond to all possible parameter configurations. Such an approach, however, entails an exponential increase in the number of trials (for a design with n factors on k levels, the number of trials is k^n). Often, an experiment design prepared in this way is impossible to implement due to the duration of the experiments, high costs or the number of available samples. Therefore, statistical planning of experiments is commonly used. Such an approach allows the number of required trials to be significantly reduced while obtaining sufficient information to build an accurate approximation of the dependent variables. The most frequently used designs are factorial, orthogonal and central composite [35]. An orthogonal design minimises the number of trials that are necessary, among which there is any combination of the values of any two factors. A method which uses orthogonal designs is the widely used Taguchi method of process optimisation [36], applied to research on the spinning process by Wang [37], and Sangkharata and Dechjarerena [38]. Analysis of the results obtained following this experimental design enables one to identify the main effects, their significance and interactions between the factors. The main effect is the influence of a specific factor on the dependent variable, independent of other factors. In contrast, in the case of factor interactions, the effect of one factor on the dependent variable varies depending on the value of the other factor. In the analysis of the shear spinning process, Kunert et al. [34] obtained interactions between the working radius of the roll and the distance between the roll and the forming block, as well as interactions between the working radius of the roll and the resulting tool path.

For a complete two-factor experimental design with two levels (labelled as a 2^2 design), a simple way to identify the main effects and interactions is to plot the main effects. Figure 1 shows the results of theoretical experiments T_1 and T_2, for factors A and B changing at two levels A_1 and A_2, and B_1 and B_2, respectively [36].

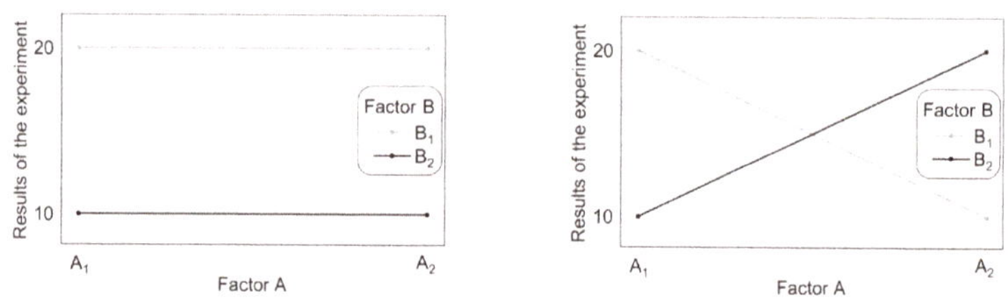

(**a**) (**b**)

Figure 1. Interaction graph for experiment T_1: (**a**) no interaction between factors A and B and (**b**) interaction of factors A and B.

When comparing the marginal means for a specific factor, it is seen that there exists a significant difference between the means related to the individual values of factor B and the fact that there are no differences between the means related to factor A. This indicates that the main effect is the result of factor B. For experiment T_2, there are no differences between the marginal means, however the level of factor A changes the influence of factor B on the results. This means that factors A and B interact. In the graph of effects, the main effect is visible as a parallel line for the results of experiments at individual levels (Figure 1a), while the crossing lines indicate the interaction of factors (Figure 1b).

In the case of orthogonal plans, and in general, fractional factorial designs, attention should be paid to the so-called resolution of the design, describing the occurrence of the interactions involved. Table 1 shows the experimental design for three variable factors at two levels 2_{III}^{3-1} with resolution III. This case involves the effects X_1 i and interaction X_2X_3, effect X_2 and interaction X_1X_3 as well as the effect X_3 i and interaction X_1X_2. This means that it is impossible to distinguish the main effect from the effect of the interaction of the remaining factors. Thus, the design shown in Table 2 does not give a clear answer concerning the main effects and interactions. Further analysis or improvement of the resolution of the plan is required, for example by extending it. However, this is not a problem to analyse when the interactions involved are insignificant. Designs with resolution III are often used to study the overall system response and to pre-select significant factors [39].

Table 1. Results of the experiment T_1 and T_2.

	T_1				T_2		
	A1	A2	Marginal mean		A1	A2	Marginal mean
B1	10	10	10	B1	10	20	15
B2	20	20	20	B2	20	10	15
Marginal mean	15	15		Marginal mean	15	15	

Table 2. Two-level orthogonal design of the experiment for three factors.

No	X1	X2	X3
1	1	1	1
2	1	−1	−1
3	−1	1	1
4	−1	−1	−1

The problem of optimal selection of parameters is a frequent issue when planning production processes and developing technologies. The selection of process parameters that allow a final product that meets the given requirements to be obtained is one of the principal aims of the technologist. Shear spinning parameters are selected based on previous experience, analytical models of the mechanics of the forming process, as well as the results of numerical simulations. For the shear spinning process, many authors, both in theoretical and experimental works [18,40], described the influence of individual process parameters on the occurrence of various types of defects and the quality of the product. Conversely, a problem in the practical application of these results is the need to simultaneously optimise all the quality indicators. The main obstacle is the divergent optima of indicators, where the optimal responses (for example, the desired values of the product features) are obtained for different values of a specific parameter, or there are cases when a change of a parameter increases the value of a certain indicator. The aim of the multidimensional optimisation of parameters is to find values such that the final product quality meets all the requirements within the set tolerance. The approach used in the study, using the concept of the preference function, was originally proposed by Harrington [41] and modified by Derringer and Suich [42]. The last-mentioned optimisation variant was successfully used in the work of Kunert et al. [34] to obtain the optimal set of parameters for the shear spinning process.

This work proposes a research method which is a scheme that can be universally applied in problems involving the selection of parameters for the shear spinning process. For this purpose, statistical data optimisation methods were used. The optimisation of process parameters was based on an analysis of the results of the industrial shear spinning process of drawpieces made of Ni-based Inconel 625 superalloy. It was necessary to select the appropriate type of the experimental plan, which by minimising the number of trials allowed the effect of the process parameters on the final product quality to be determined and was the starting point for their optimisation. The use of 3D scanning and non-contact and non-destructive technology permitted the influence of the technological parameters of the process on quality indicators, and thus on the subsequent optimisation of the shear spinning process, to be determined.

2. Materials and Methods

The material used for the tests consisted of drawpieces (Figure 2a) made of chromium–nickel–molybdenum Inconel 625 superalloy (ATI Flat Rolled Products, Pittsburgh, PA, USA). The test material was subjected to a uniaxial tensile test in room temperature on the Zwick/Roell Z050 machine (Zwick/Roell, Ulm, Germany) in three directions, in relation to the sheet rolling direction: 0°, 45°, and 90°. The stress–strain curve was taken, and the

Lankford coefficient was calculated. Microstructure analysis was also performed using a scanning electron microscope (SEM) Hitachi SU-70 (Hitachi Ltd., Tokyo, Japan). The grain size has been calculated. The drawpieces, with an external diameter of 400 mm, were formed from 1 mm thick sheet metal. The drawpieces were manufactured on a hydraulic press by Schuler Pressen GmbH SMG-260 (Schuler Group, Göppingen, Germany) with a maximum pressing force of 100 T. The preformed drawpieces (Figure 2a) were subjected to the shear spinning process with laser heating in industrial conditions on a Leifeld stand equipped with a semiconductor laser from Laserline GmbH (Laserline GmbH, Mülheim-Kärlich, Germany). The drawpieces were mounted on a spinning block and pressed against a blankholder (Figure 2c). Two symmetrically spaced rollers with a working radius R8 were used for forming. Forming parameters: rotational speed and feed were adjusted. The heating temperature was 600 °C. The laser heating parameters were as follows: the angle of incidence of the laser beam was 135°, the shape of the laser spot was rectangular with dimensions of 10 mm × 40 mm, the laser power was 5 kW. No lubricant was used during forming, only coolant was applied to the spinning block. An element of the combustion chamber of a jet engine was the end product of the forming process (Figure 2b).

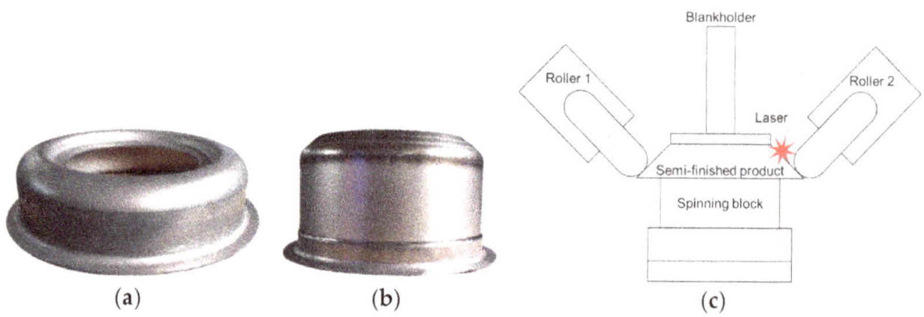

Figure 2. Preformed drawpiece (**a**), finished product (**b**) and stand (**c**).

In order to obtain an axisymmetric product with a varying thickness of the drawpiece wall, experiments were designed to verify the optimal shear spinning parameters. The boundary conditions and key process parameters were determined from the results of extensive cold and laser heated shear spinning studies. It was decided to conduct investigations for three factors at two established levels. The factors selected were the rotational speed of the spinning block (R), the feed rate of the roller (S) and the heating of the material (H).

A 2_{III}^{3-1} experiment was planned. As a starting value for determining the parameter levels, the parameters of the reference test P_0 were adopted as: R_0 = 220 rpm and S_0 = 20 mm/min, H_0 = 0. The experimental design was based on levels increased in relation to the reference values by 15% for the rotational speed, by 20% for the feed rate and by 50% for both rotational speed and feed rate. The differentiated increase for the values of the first level is a result of the need to examine the influence of the feed coefficient (f), defined as the ratio of feed rate to rotational speed. For equal percentage increments of the reference levels, the value of the feed coefficient would be the same in all cases. There are two levels of heating factor: the use of heating (1) or no heating (0). The levels of factors are shown in Table 3. The trial plan and the reference test conditions are shown in Table 4.

Table 3. Levels of factors.

Factor	Level 1	Level 2
S	$S_1 = 1.2S_0 = 36$ (mm/min)	$S_2 = 1.5S_0 = 45$ (mm/min)
R	$R_1 = 1.15R_0 = 253$ (rpm)	$R_1 = 1.5R_0 = 330$ (rpm)
H	1	0

Table 4. The trail plan.

Trial	S (mm/min)	R (rpm)	H	f (S/R) (mm/obr)
P_0	30	220	0	0.136
P_1	36	253	0	0.142
P_2	36	330	1	0.109
P_3	45	253	1	0.177
P_4	45	330	0	0.136
P_5	24	253	0	0.095
P_6	20	253	0	0.080

The 2^{3-1}_{III} experiment, as designed, provides information about the main effects, the possible interactions and their influence on individual continuous quality indicators. Information on the main effects and interactions is obtained from the plot of effects for marginal means versus factor value. Continuous quality indicators were used for the geometrical evaluation of the products. In the Derringer and Suich method [43], for the dependent variables Y_i, which are quality indicators, it is possible to introduce their dependence from k factors (parameters) X_j and an error ϵ_i:

$$\epsilon_i : Y_i = f_i(X_1 \ldots X_k) + \epsilon_i \tag{1}$$

In practice, the accurate value of coefficient f_i is often unknown; thus, the estimator $\hat{Y}_i$ is introduced which is obtained by the regression method. Models of proportional chances for qualitative variables and generalised linear models for quantitative variables are used in this work considering the interaction between factors c_{ij}:

$$\hat{Y}_i = c_0 + c_i \sum_{i=1}^{N} X_i + c_{ij} \sum_{i=1}^{N} \sum_{j=1}^{N} X_i X_j \tag{2}$$

Each estimator $\hat{Y}_i$ has a preference value $d_i \in [0,1]$. The d_i preference determines how much the value of $\hat{Y}_i$ is preferred, making it a value from 1 (most preferred) to 0 (not accepted). Based on the value of d_i, the so-called global preference function G for the variables $\hat{Y}_i$. G function is defined as the weighted geometric mean of the preference value d_i (Equation (3)):

$$G = \left(\prod_{i=1}^{N} d_i^{w_i} \right)^{\frac{1}{k}} \tag{3}$$

where $k = \sum_{i=1}^{N} w_i$ and $w_i > 0$

The G function defined in this way is an indicator of overall quality. If any of the factors d_i is zero, G also becomes zero. In turn, G-values close to 1 indicate that values of d_i are also close to 1 (most preferred). It means that the value of Y_i is close to optimal. Subsequent maximisation of the G-function allows the process parameters that result in the best overall product quality to be found, while the individual indicators are maintained within the tolerance range.

When the value of preference d_i increases (or decreases) with change invariable Y_i, the goal of optimisation is to maximise (or minimise) the response Y_i using a one-way preference transformation. This transform function is monotonous, reaching the value 0 below a certain fixed value T_{min} and value 1 above a certain fixed value T_{max}. In the interpretation, this means that it is most desirable to achieve at least the T_{max} value, while values below T_{min} are not accepted and the preference for them is 0. The one-sided transform function can be expressed by the following function (Figure 3a):

$$G = \left(\prod_{i=1}^{N} d_i^{w_i} \right)^{\frac{1}{k}} \tag{4}$$

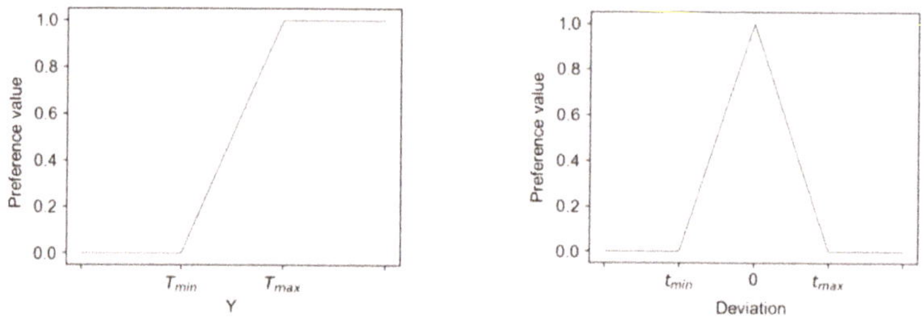

(**a**) (**b**)

Figure 3. One-sided preference function (**a**), two-sided preference function (**b**).

A two-sided transform function is used to obtain a response value close to a certain value in the interval, called the target value. The preference function then reaches 1 for that value and is set to zero outside the specified tolerance range. The rate of growth of the preference rating is controlled by the r and l coefficients. The coefficient r is responsible for the rate of increase of the transform function, and thus the rate of increase in preferences along with the approach to the desired value of S_i. For a certain response Y, its target value (the most preferred) T and the tolerance range $[T - t_{min}, T + t_{max}]$, the transformation of the Y value to the preference value d is determined by the following function (Figure 3b):

$$T(x) = \begin{cases} 0, & x < T - t_{min} \text{ lub } x > T + t_{max} \\ \left(\frac{x-(T-t_{min})}{t_{min}}\right)^{l}, & x \in [T - t_{min}, T) \\ \left(\frac{(T+t_{max})-x}{t_{max}}\right)^{r}, & x \in [T, T + t_{max}] \end{cases} \quad (5)$$

For both a one-sided and two-sided transformation, one can use a spliced transform function. The spliced function is often used to transform qualitative variables.

Thanks to the transformation of preferences and the global preference function, the task of optimising the parameters in terms of the responses of quality indicators comes down to the problem of optimising the scalar function of arguments that are the preferences of the indicators of the specific parameters. The task of optimisation for the global preference function G, indicators Y_i, their transformation T_i, and the set of parameters P is therefore as follows (Equation (6)):

$$\max G(T_1(Y_1(p)), \ldots, T_n(Y_n(p))) \text{ dla } p \in P \quad (6)$$

Among the many methods of solving such a problem, a frequently used method is grid search (GS), which involves searching the parameter space, estimating the value of the objective function, its optimisation cost and validation with the use of a cross test. In the case of a large number of parameters, or a wide range of their values, it is often impossible to exhaustively search for combinations of parameters. The most often used to optimise parameters are, inter alia, gradient methods, heuristic methods, or a randomised procedure for selecting parameter values [43]. In this paper, due to the small number of parameters, an exhaustive search and the cost function determined by the general preference function G (Equation (3)) are used.

After the technological tests, the shape and dimensions of the components were tested using an Atos Core 200 (GOM) optical 3D scanner (GOM, Braunschweig, Germany). STL files were generated from 3D scans. In the next step, the STL files were compared with the product CAD file in order to verify the deviations of the geometry from the nominal shape using GOM Inspect software (GOM, Braunschweig, Germany). Overall comparisons of the 3D surface scans were made with the CAD model and the following surface quality indicators (SURF_G) were generated: maximum positive surface deviation

(MAX), maximum negative surface deviation (MIN), average surface deviation (AVG) and mean square deviation of the surface (SIGMA).

Overall comparisons of the sheet thickness on 3D scans were also made with the CAD model and quality indicators (TH_G) were generated: maximum positive thickness deviation, maximum negative thickness deviation, average thickness deviation, mean square thickness deviation and percentage indicator of out-of-tolerance measurements (TOTALFAIL). In the next step, five component zones (**Z1–Z5**) with different dimensional tolerances for specific nominal thicknesses of component (Figure 4) were designated.

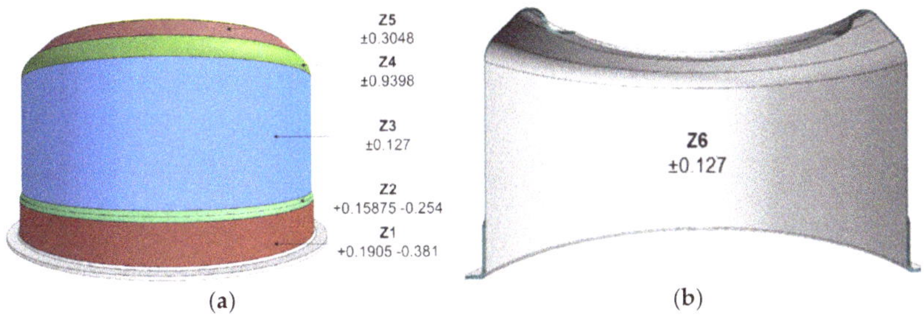

Figure 4. Characteristic zones of the component with specific thicknesses and variations in toler-ance, (**a**) **Z1–Z5**, (**b**) **Z6**.

The tolerances were determined on the basis of the technical drawing of the component. As in the case of overall comparisons, for each Z zone, comparisons of the thickness from the 3D scans were made with the CAD model and the following quality indicators were generated (TH_Zn, where n is the number of the component zone): maximum positive thickness deviation, maximum negative thickness deviation, average thickness deviation, mean square thickness deviation, and percentage indicator of out-of-tolerance measurements. Then, two component zones **Z1** and **Z6** (inner component surface) were designated which had different thickness and diameter tolerances. Then, comparisons of the 3D scan surfaces were also made with the CAD model and quality indicators were generated (SURF_Zn, where n is the number of the component zone): maximum positive surface deviation, maximum negative surface deviation, mean surface deviation, mean square deviation of the surface. All data were interpreted in the 2σ range. For overall indicators, TH_G and SURF_G, a positive (negative) tolerance was determined as the maximum (minimum) values from specific zone tolerances. For the TOTALFAIL indicators, the 50% threshold was adopted as the tolerance.

For the MIN, MAX and AVG indicator family, preference functions were established assuming the value of 1 for the reference value and the value of 0 for the value outside the symmetrical tolerance range. In the case of the MIN, MAX, and AVG indicators, the reference (desired) value is 0 (zero deviation). Preference functions for an established tolerance (t_{min}, t_{max}) take the form (7):

$$\mathrm{PREF}(x) = \begin{cases} 0, & x < t_{min} \text{ lub } x > t_{max} \\ \frac{x + t_{min}}{t_{min}}, & x \in [t_{min}, 0) \\ \frac{t_{max} - x}{t_{max}}, & x \in [0, t_{max}] \end{cases} \quad (7)$$

For the SIGMA and TOTALFAIL indicators, a one-sided preference function was applied, assuming the value 1 for the indicator value 0 and the value 0 for an indicator value above the tolerance limit. In the case of the SIGMA and TOTALFAIL indicators, the reference (desired) value is the value 0 (zero deviations and no measurements outside the tolerance limit). Thus, the preference functions for the established tolerance t take the form (8):

$$\mathrm{PREF}(x) = \begin{cases} \frac{t - x}{t}, & x \in [0, t) \\ 0, & x \geq t \end{cases} \quad (8)$$

The above-defined tolerance functions assume the value 0 outside the tolerance range. In the subsequent process of parameter optimisation, for the values obtained from the estimating models, a situation may occur in which (1) the values of SIGMA and TOTALFAIL are negative, or (2) sets of parameters are non-zero and are pairwise uncoupled, which leads to the zeroing of the global preference function for each set parameters. For this reason, a variation of the preference function was introduced by minimising the preference functions, assuming the value of 0.01 outside the tolerance intervals and the value (0.01, 1) within the tolerance interval.

Functions defined in this way will allow one to obtain a non-zero value of the global preference function in the parameter space, while achieving a low value for parameters resulting in predictions beyond the tolerance range for indicators. The minimising preference functions for the indicators MIN, MAX, AVG (Equation (9)) and SIGMA, TOTALFAIL (Equation (10)), take the form:

$$\mathrm{PREF}(x) = \begin{cases} 0.01, & x < t_{min} \text{ or } x > t_{max} \\ \frac{0.99x + t_{min}}{t_{min}}, & x \in [t_{min}, 0) \\ \frac{t_{max} - 0.99x}{t_{max}}, & x \in [0, t_{max}] \end{cases} \quad (9)$$

and

$$\mathrm{PREF}(x) = \begin{cases} \frac{t - 0.99x}{t}, \; x \in [0,\, t] \\ 0, \; x > t \end{cases} \quad (10)$$

The preference function and minimising preference function for the indicators MIN, MAX, AVG, SIGMA and TOTALFAIL are presented graphically in Figure 5.

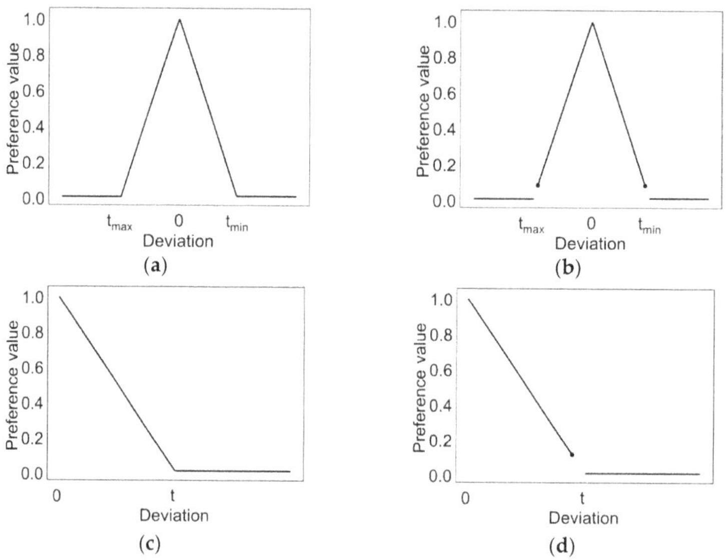

Figure 5. Graphs of (**a**) the preference function for the indicators MIN, MAX, AVG; (**b**) the minimising preference function for the indicators MIN, MAX, AVG; (**c**) the preference function for the SIGMA and TOTALFAIL indicators; (**d**) the graph of the minimising preference function for the SIGMA and TOTALFAIL indicators.

In the process of parameter optimisation, according to the procedure described above, a multivariate optimisation task was built for the quality preference function. This task was then solved using the grid search method. After determining the optimal parameters, the responses for individual indicators were calculated from regression models forecasting the preference function argument for a given indicator. After obtaining the optimal parameters, subsequent trials of the shear spinning process were used to validate the solution obtained and, if necessary, adjust the parameters. Validation was carried out in three technological trials. The parameter sets for each trial were established based on the optimisation results. After the forming tests, the products were subjected to visual quality inspection as well as dimensional and geometry inspection. After industrial validation of the solutions, a measure was introduced on the basis of which the results of the validation tests were compared with the assumptions. To compare the results of the P_i sample, the following measure of the mean relative change in absolute deviation was introduced to the established reference sample P_r for a certain subset of Q indicators (Equation (11)):

$$d_Q(P_i, P_r)\ \frac{1}{|Q|}\sum_{q\in Q}\frac{|q(P_r)| - |q(P_i)|}{|q(P_r)|} \tag{11}$$

where $q(P_i)$ is the value of the indicator q obtained in the trial P_i. A function defined in this way makes it possible for the average and relative improvement of the indicator values for the sample P_i, to be determined. The degree of improvement is considered to be the increase in the value of the indicators towards zero.

Calculations and analyses were performed with the use of the statistical packages available for Python and R languages (statsmodels and rpy2), as well as with the GOM Inspect software.

After the technological trials, the microstructure of the samples was examined. Samples were taken from zones **Z1**, **Z3** and **Z5** (Figure 4).

3. Results and Discussion

Figure 6 shows the results of research on the mechanical properties of the test material.

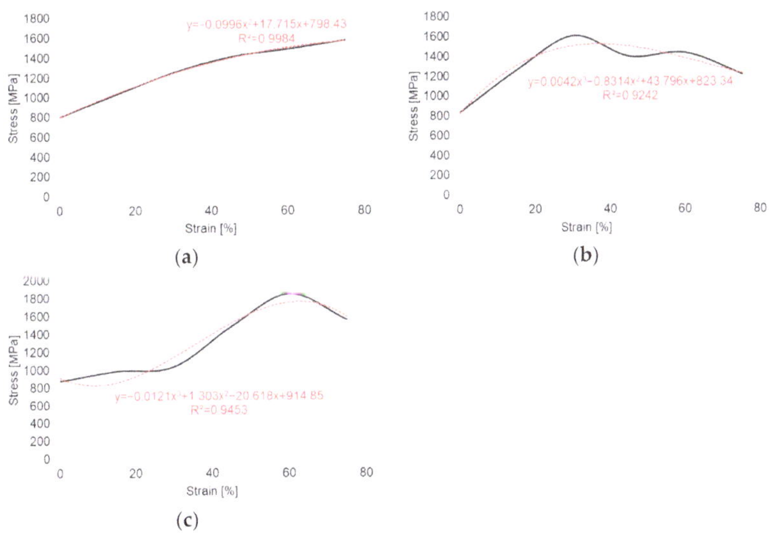

Figure 6. Stress–strain curves (**a**) for 0°, (**b**) 45°, and (**c**) 90°.

Table 5 presents the results of measurements of the test material anisotropy.

The results of the Lankford anisotropy coefficient calculations show that the sheet exhibit a high anisotropy. The R-value is almost twice as large for the sheet in the direction of 45° as for the sheet in the direction of 0°. The mean coefficient of normal anisotropy is $\bar{r} = 1.075$. It should therefore be stated that the sheet has a low ability to be formed by stamping.

Table 5. Lankford coefficient for the test material.

Direction	Lankford Coefficient
0°	0.68
45°	1.25
90°	1.12

Figure 7 shows the results of the batch material microstructure (Figure 7a–b) analysis and after deformation (Figure 7c–e).

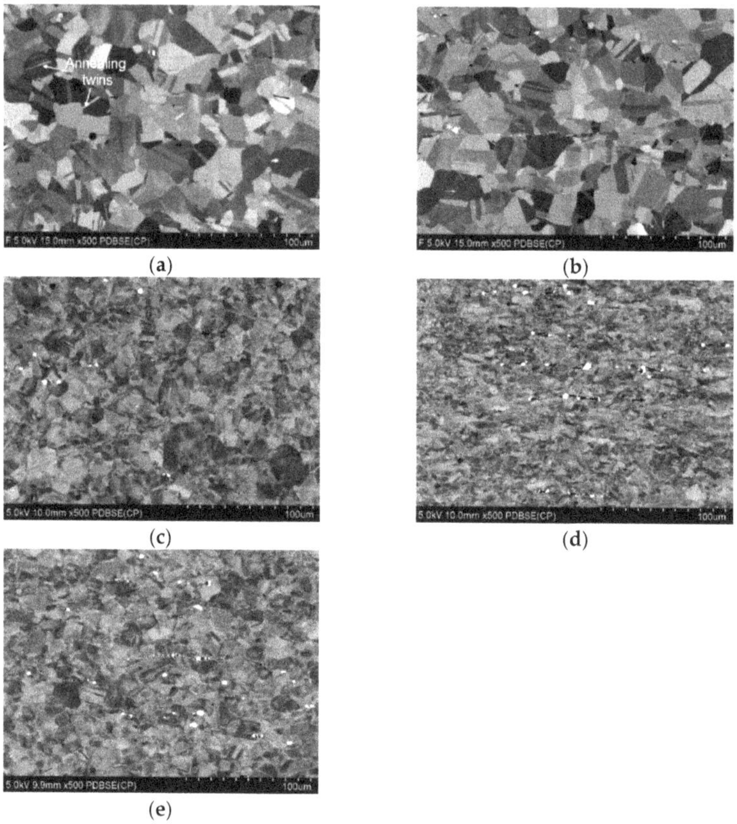

Figure 7. SEM images of the test material microstructure (**a**) cross section, (**b**) longitudinal section. SEM images of the microstructure of the material after deformation taken (**c**) from zone **Z1**, (**d**) zone **Z3**, and (**e**) zone **Z5**.

The mean area of the grains is minimally different in the case of samples taken in the parallel and perpendicular directions. The mean grain size of the samples taken in the perpendicular direction is 1.15% smaller than the mean grain size of the samples taken in the parallel direction. Conversely, the approximate grain diameter in the case of samples taken in the perpendicular direction is 1% larger than in the case of samples taken in the parallel direction. Annealing twins can be distinguished in the structure (Figure 7a). It is a characteristic element of the austenitic alloy structure. They are characterized by a lattice with low stacking-fault energy, which results in a high frequency of stacking errors, and thus resistance to plastic deformation.

The structure of the material after deformation is a structure characteristic of the material after intense deformation. The grains are elongated, the grain boundaries are not clear. The photos show strain localization which plastic deformation occurred. The average grain size for **Z5** (40% deformation) is lower by 46.59% compared to the average grain size of the batch material, for **Z3** (60% deformation) by 73.94%, while for **Z1** (deformation 10%) by 39.47%. The approximate grain diameter for **Z5** (40% deformation) is lower by 24% compared to the average grain diameter of the batch material, for **Z3** (60% deformation) by 44%, and for Z1 (10% deformation) by 19.5%.

Based on a visual assessment of the components, the forming trials were considered successful. All the products were sent for further research. The results of an overall comparison of the components surface and thicknesses with the CAD model and measurement data was generated and analysed according to the experiment plan.

Figure 8 presents maps of the zonal comparison of the thickness of the areas of products that meet the tolerances (green) and those not within the acceptable tolerance (red) together with the measurement data. Compatibility of the sheet thickness of the components within the tolerance limits in five zones (Z1–Z5) and compliance of the surfaces with the tolerance in two zones (Z1, Z6) were analysed. The measurement data for the zonal comparison of the sheet thickness in areas within the tolerance generated from the trials were used to build the models discussed later in this article.

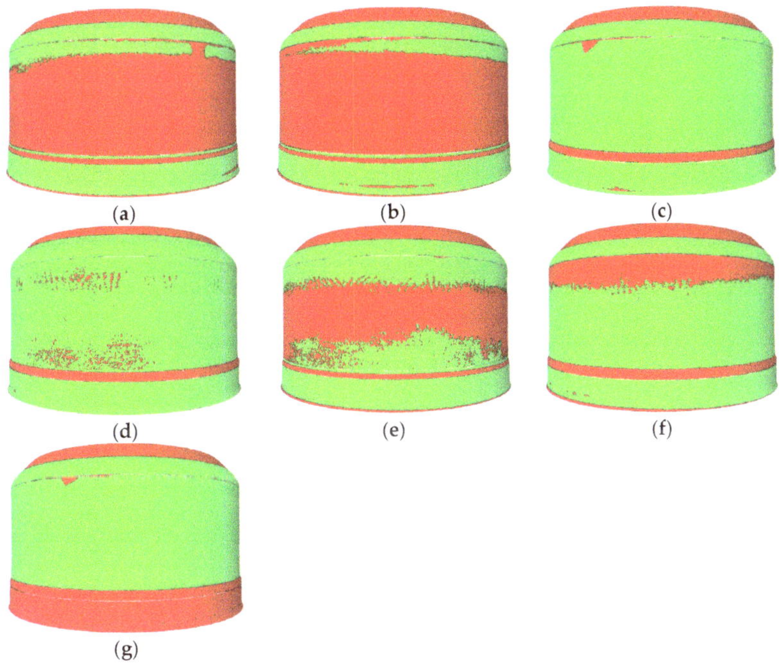

Figure 8. Areas of components after technological tests within given tolerances: P_0 (**a**), P_1 (**b**), P_2 (**c**), P_3 (**d**), P_4 (**e**), P_5 (**f**), and P_6 (**g**).

In the next step, models were built which explain the influence of factors on individual quality criteria. Then an optimisation scheme was launched to return parameters that maximise the global preference function. An attempt was made to determine the influence of S, R and H factors on the quality of the final component (measured using indicators) and their interactions. The potential significant main effects and implicit interactions were determined according to the methodology of analysis of the results of the experimental plan 2_{III}^{3-1}. In the case of the plan considered, these pairs were (S, R:H), (R, S:H) and (H,

S:R). Student's T test was used to determine the significance of differences between the marginal means for the levels of the factors examined.

The dominant interaction was the interaction between the feed rate and the rotational speed of the spinning block S:R, defined in later analysis as the feed factor $F := S/R$. The main effects and interactions identified were used to determine regression models approximating the values of the indicators. S, R and H as well as their mutual interactions were significant factors influencing the values of the indicators. Linear regression models were fitted using the least squares method. The categorical factor H was included in the models using the basic dummy coding, and the coefficient of determination was adopted as a measure of fit. The explanatory variables were standardised.

Initial attempts to obtain correct models with linear S, R, H and F factors and their interactions did not bring the desired results; thus, it was decided to include non-linear factors and apply the following transformations: $f(x) = x^{-1}$, $f(x) = x^{-2}$, $f(x) = x^{-3}$, $f(x) = x^2$, $f(x) = x^3$, $f(x) = \arctan(x)$, $f(x) = \exp(x)$, and $f(x) = \log(x)$. Using such an extended class of factors, satisfactory models were adjusted for most indicators (coefficient of determination 0.7–1.0). Due to the relatively small size of the sample and the number of transformations of factors, a significant number of models that meet the criteria were obtained for many indicators. The selection of the final models was also guided by the evaluation of the trend and the complexity of the model. Models containing strong interactions with H were avoided due to the small number of trials with the value of H = 1. Moreover, models characterised by rapid growth of predicted value outside the approximation range were also avoided.

The regression models obtained, which estimated the indicator values, were used in the multivariate optimisation process which operates by maximising the global preference function. For each of the indicators, after considering the appropriate tolerances, the preference functions were determined in accordance with the relationships (7) and (8). Initial analysis of the models obtained revealed the need to correct the preference function such that the global preference function assumed non-zero values on a non-empty set of parameters. In addition, it was decided to investigate various optimisation variants-apart from the unconditional variant (UV), by zeroing the global preference function for values lying outside the tolerances, the conditional variant (CV) and the minimisation variants: unconditional (UMV) and conditional (CMV), were also analysed.

The conditional variants assume the assignment of a small preference value to the values of indicators outside the tolerance limits, in accordance with Equations (9) and (10). In this way, the possibility of the appearance of indicators with values outside the tolerance limits, when the values of most indicators are close to the preferred value, was allowed for. These variants are particularly important for indicators assuming a value outside the tolerance limits over all the process parameters. In such a situation, the preference function would assume a constant value, ignoring the variability of these indicators and making it impossible to compare their values for different sets of parameters. The minimisation variants solve this problem, allowing one to optimise the values of indicators beyond the given tolerances. In order to obtain minimisation variants, the set tolerances were modified by introducing the so-called minimisation tolerances. Their values were established at a level not lower than the maximum for positive tolerances, and at a level not lower than the minimum for negative tolerances, and the highest/lowest value obtained in the experimental trials $P_1 - P_5$. In the optimisation carried out for such defined tolerances, the indicator values obtained were within the given tolerances or were not worse than those obtained in the experimental tests.

For the Q indicator, the set tolerance t $[t_{min}, t_{max}]$ and the set R of the experimental results, the tolerances of the minimisation variant were defined as follows (Equation (12)):

$$t^{min}\ [\min(t_{min}, \min(R)), \max(t_{max}, \max(R))] \tag{12}$$

The characteristics of the optimisation variants are presented in Table 6.

Table 6. Optimization variants.

	Out-of-Tolerance Value of Preference: 0	Preference Value out of Tolerance: 0.01
Regular tolerances	UV	CV
Minimisation tolerances	UMV	CMV

In addition, considering the regression approximation error, the tolerances were corrected for the deviation of the test results from the predicted values. This deviation was calculated as the root mean square of the residuals.

Basic tolerances were adopted for the area indicators in accordance with the technical specification of the component (Figure 4). For overall indicators, TH_G and SURF_G, a positive (negative) tolerance was determined as the maximum (minimum) of area tolerances. For the TOTALFAIL indicators, the 50% threshold was adopted as the tolerance.

The global preference function—the weighted geometric mean of the value of the preference function—set with relationship (3), takes as a parameter the validity of individual indicators. A varied range of validity allow for emphasis to be given to the optimisation of significant quality indicators, and in the case of similar preference values, parameters are selected for which higher preferences are achieved for indicators with a lower validity. In other words, changes in indicators with a higher validity have a greater impact on the global preference function. Four sets of validity were adopted, favouring the most important families of indicators-quality indicators for thickness measurements, quality indicators for surface measurements and quality indicators for the most validity zones Z1, Z3 and Z6. The validity in individual sets were set at the level of 10 for the preferred indicators, and at the level of 1 for other indicators. Additionally, a set of validity assuming equal significance of all indicators was considered. Defined sets of validity are presented in Table 7.

Table 7. The weight sets used to optimise shear forming parameters.

Set of Weights	Preferred Indicators (Validity 10)	Other Indicators (Validity 1)
W_ALL (all indicators)	TH_G, TH_Z, SURF_G, SURF_Z	
W_TH (thickness indicators)	TH_G, TH_Z	SURF_G, SURF_Z
W_SURF (surface indicators)	SURF_G, SURF_Z	TH_G, TH_Z
W_Z (indicators of zones **Z1, Z3, Z6**)	TH_Z1, TH_Z3, TH_Z6, SURF_Z1, SURF_Z6	TH_G, TH_Z2, TH_Z4, TH_Z5, SURF_G

The Cartesian product $S \times R \times H$ was adopted for the following ranges of process variables $S \in [19, 46]$, $R \in [215, 335]$, $H \in \{0, 1\}$ as the discrete design space of the process, constituting the search space for the grid search scheme. The unit resolution of the ranges was also determined (a step with a value of 1). The adopted ranges are close to the ranges determined by the parameters for which the experiments were performed. The lack of these assumptions could result in an erroneous location of the optimal parameters because the regression models obtained may contain large approximation errors in the values of the factors for which the fit was obtained. For this reason, a condition was also imposed on the value of $F := S/R$ requiring that $F \in [0.07, 0.18]$. Due to the acceptable time complexity of the method for the adopted ranges of parameters, an exhaustive search approach was chosen. For each vector of the design space, the value of the global preference function was calculated, and a set of process parameters was determined that maximised the overall value of preferences. In total, sixteen optimisation sub-processes were carried out (for each optimisation variant and set of validity), allowing the optimal parameters for variously defined priorities and quality requirements to be obtained.

Figure 9 shows the global preference function for selected optimisation variants and sets of validity. In all cases, the local minima and maxima are located in similar areas; the functions tend to adopt maxima for values close to the boundary values of the F-parameter as a result of the strong dependence between regression models and the F-parameter. Different characteristics of the function can be noted. Regular variants have discontinuous preference functions and are reflected in an irregular graph, high variability of functions and jumps in values, while minimisation variants produce in a more regular and smoother course.

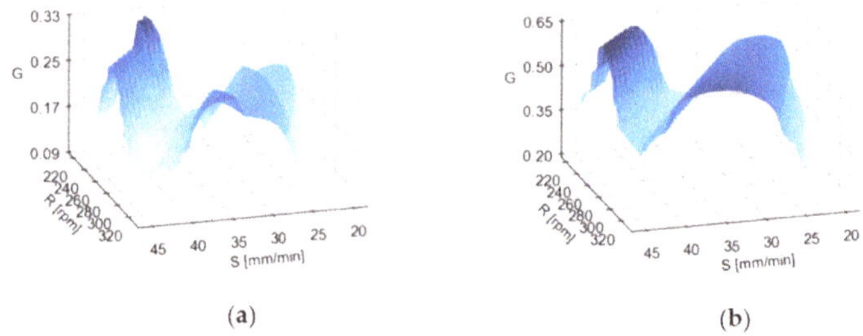

Figure 9. Graphs of the global preference function for (**a**) the variant CV and weights W_ALL and (**b**) CMV and weights W_Z; with H = 1 fixed.

As a result of the high resolution of the design space, the sets of parameters for which the highest values of the global preference function were achieved, were generally spread within a few units for each parameter. It was necessary to aggregate the results to determine the dominant ranges of parameters and, as a result, optimal sets of parameters for each of the tested variants. For this purpose, an examination was made of the frequency of occurrence of particular sets and ranges of parameters in the 5% of the results that were the best (values of the global preference function). In order to illustrate and evaluate the correctness, histograms of the number of parameter sets were prepared (Figure 10).

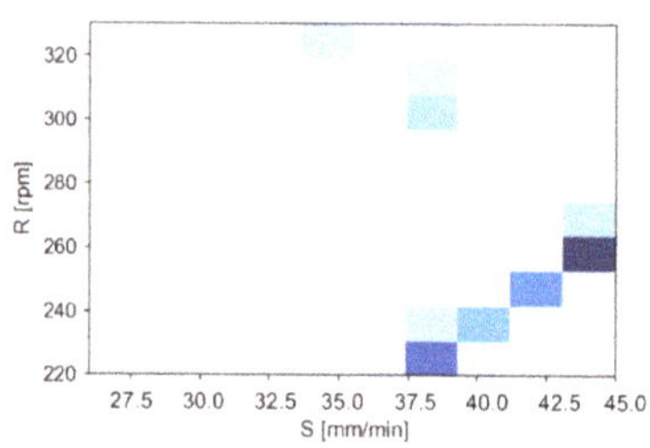

Figure 10. Histogram of the most common sets of parameters in the 5% of the results that were the best (the darker the colour, the higher the frequency of occurrences).

Figure 11 presents the aggregated frequency for all sixteen optimisation variants, revealing the dominant influence of parameters from the ranges: S ∈ [38, 45] and R ∈ [240, 270]. The H parameter was set at level 1, as occurring in most of the tested parameter sets, or having complementary sets (differing only in the H-value) in the 5% of the results that were the best that were analysed.

Table 8 presents the dominant sets of parameters for each optimisation process carried out for various optimisation variants and sets of weights. For the conditional variant, it was not possible to obtain parameters resulting in the values of all indicators within the required tolerances.

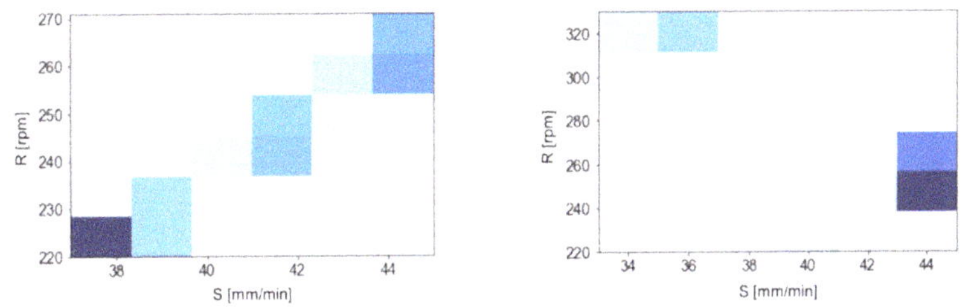

(**a**) (**b**)

Figure 11. Histogram of (**a**) the most common sets of parameters for the best 5% of the CMV results and the W_ALL weight set and of (**b**) the best 5% of the CMV results and the TH weight set (the darker the colour, the higher the frequency of occurrences).

Table 8. Dominant sets of parameters in the optimisation processes performed.

	W_ALL	W_TH	W_SURF	W_Z
UV				
CV	[1, 39, 225]	[1, 39, 225]	[39, 225, 1]	[39, 225, 1]
UMV	[45, 266, 1]	[45, 266, 1]	[45, 266, 1]	[45, 266, 1]
CMV	[37, 330, 1]	[37, 330, 1]	[37, 330, 1]	[37, 330, 1]

Finally, three sets of parameters were selected for which the highest preference values, i.e., the best predicted product quality: V_1 = [45, 259, 1], V_2 = [39, 225, 1] and V_3 = [45, 325, 1], were obtained. The selected parameters predict the improvement of the results measured in the number of indicators whose values are within the given tolerances, and the improvement of the values of individual indicators. The predicted compliance with the tolerance for the quality indicators was evaluated. A comparison with the experiments performed predicts an increase in the number of compatible indicators by 11–53%. The optimal sets of process parameters obtained: V_1 = [45, 259, 1], V_2 = [39, 225, 1], and V_3 = [45, 325, 1], were tested in verifying technological trials. Trial V_3 failed. It was impossible to uninstall a component from the spinning block. Components produced by trials V_1 and V_2 were sent for further research. After 3D scanning the results of an overall comparison of the components surface and thicknesses with the CAD model and measurement data was generated and analysed according to the experiment plan.

Figure 12 presents maps of the zonal comparison of the thickness of the areas of products that meet the tolerances (green) and those not within the acceptable tolerance (red) together with the measurement data. Compatibility of the sheet thickness of the components within the tolerance limits in five zones (**Z1–Z5**) and compliance of the surfaces with the tolerance in two zones (**Z1, Z6**) were analysed.

(**a**) (**b**)

Figure 12. Areas of components after technological tests within given tolerances: V_1 (**a**); V_2 (**b**).

Values of indicators obtained in trials V_1 and V_2 were compared with regression models obtained in the optimisation process. The best compliance for the individual indicators was achieved for the second and third zones, which may result from the characteristics of those areas-they have simple geometry. For some models, the results obtained suggest the required correction of the model had occurred with the need to recognise certain experimental results as outliers.

In order to test the correctness of the approach applied, the results obtained after the validation tests were compared with the theoretical predictions and assumptions of the optimisation process. A comparison was made between the theoretical and resulting values of individual indicators and sets of indicators within the required tolerance ranges. The mean, relative change in the absolute value of the deviation was adopted as a measure of the estimation error (Equation (11)). Good compliance was obtained between the values obtained at the optimisation stage and the actual test results.

Figures 13–15 present comparisons of the results for the V_1 trial with the estimates, (Figure 13a,b), with the best and worst results of the technological tests (Figure 14a,b), and with the absolute deviations of the results for the V_1 trial (Figure 15a,b). There is a good agreement found in the values and directions of the deviations (Figure 13a,b). The results of the validation tests were compared with the results of the P_3 test, for which the best results were found (eighteen indicators compliant with the tolerance limits). The mean, relative improvement of the absolute deviation (Figure 15a,b) for indicators outside the tolerance range compared to the P_3 trial was −10% for the V_1 trial.

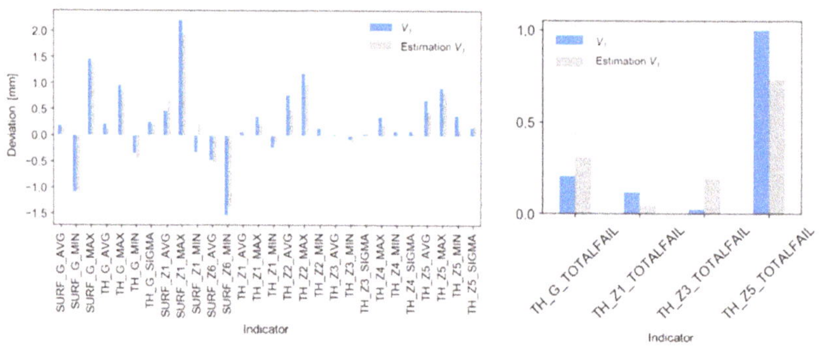

(a) (b)

Figure 13. Comparison of the estimate of the results of the V_1 trial (based on regression models) with the actual results of the trial for the indicators of (**a**) deviations and (**b**) tolerances.

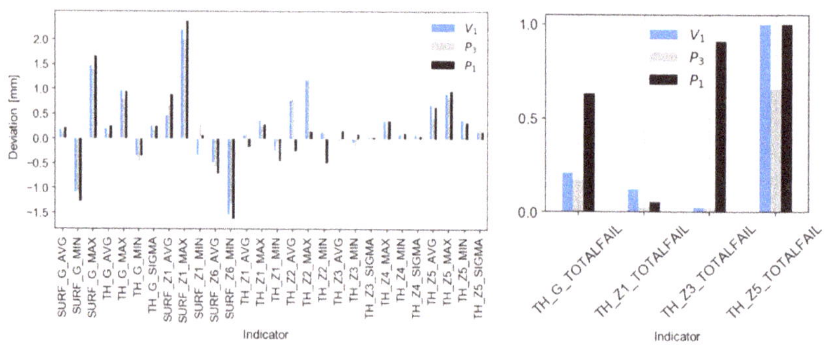

(a) (b)

Figure 14. Comparison of the results of the V_1 trial with the best and worst results of the experiments P_3 and P_1 for the indicators of (**a**) deviation and of (**b**) tolerance.

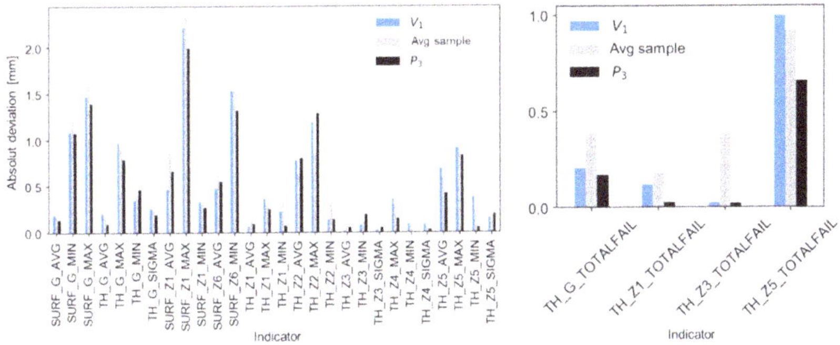

(a) (b)

Figure 15. Comparison of the absolute deviations of the results of the V1 trial with the absolute best and average results of the experiment for the indicators of (**a**) deviation and of (**b**) tolerance.

Figures 16–18 show the comparisons of the results of the V_2 trial with the estimates (Figure 16a,b), with the five best and worst results of the technological tests (Figure 17a,b), and with the absolute deviations of the results of the V_2 trial (Figure 18a,b). There is good agreement in the values and directions of the deviations (Figure 16a,b). The average, relative improvement of the absolute deviation (Figure 18a,b) for indicators outside the tolerance range compared to the P_3 trial was 6% for the V_2 trial.

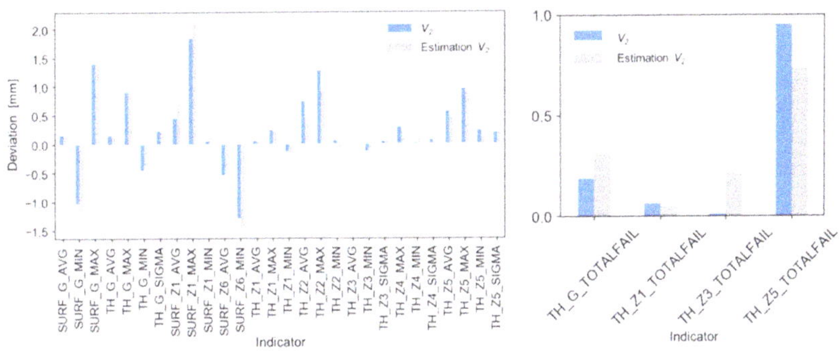

(a) (b)

Figure 16. Comparison of the estimates of the results of the V2 trial (based on regression models) with the actual test results for the indicators of (**a**) deviations and of (**b**) tolerance.

The compliance of the actual values of the indicators was compared with the forecasts obtained from regression models. The values obtained in the trials V_1 and V_2 for the majority of indicators are correctly predicted by the models, keeping the direction and trend of the deviations (Figure 11a,b and Figure 16a,b). For the predicted number of indicators compliant with the tolerance limits, the prediction accuracy for trials V_1 and V_2 was 94% and 97%, respectively. For trials V_1 and V_2, the number of indicators with values within the set tolerance increased by 11% (to twenty indicators within the desired range) for trial V_2 and decreased by 7% for trial V_1 (seventeen indicators within the tolerance range).

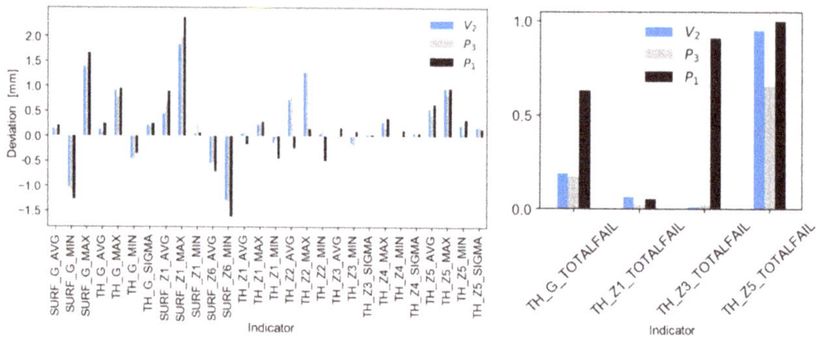

(a) (b)

Figure 17. Comparison of the results of the V2 trial with the best and worst results of the experimental trials P3 and P1 for the indicators of (**a**) deviation and of (**b**) tolerance.

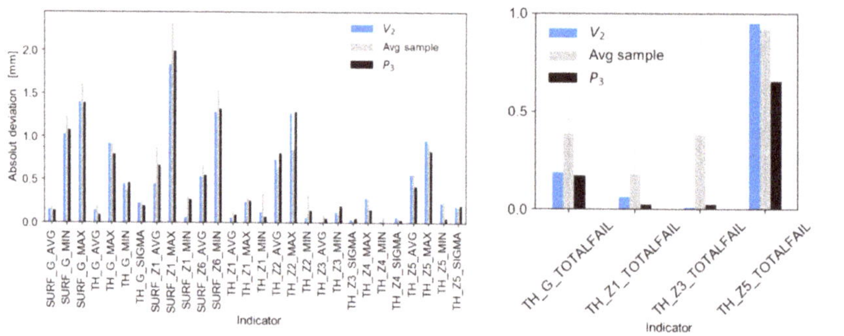

(a) (b)

Figure 18. Comparison of the absolute deviations of the results of the V2 trial with the absolute best and average results of the experimental trials for the indicators of (**a**) deviation and of (**b**) tolerance.

4. Conclusions

Authors should discuss the results and how they can be interpreted from the perspective of previous studies and of the working hypotheses. The findings and their implications should be discussed in the broadest context possible. Future research directions may also be highlighted.

Optimisation schemes for the shear spinning process of Inconel 625 nickel superalloy were analysed in search of a method permitting for effective minimisation of the global response function. The most important criteria in selecting an optimisation scheme are the complexity of the problem, the topology of the process design space and the class of the cost function. Due to the acceptable size of the problem and the discrete parameter space, the grid search method was adopted.

On the basis of multidimensional optimisation, sets of optimal parameters of the shear spinning process with laser heating of Inconel 625 nickel superalloys were selected. For the predicted number of indicators compliant with the tolerance, the prediction accuracies for parameter sets V_1 and V_2 were 94% and 97%, respectively. The values obtained in the trials V_1 and V_2 for most indicators are correctly predicted by the models, keeping the direction and trend of the deviations. Inaccuracies in the results could be the result of the approximation methods used, the quality of model fit, and the measurements carried out by the operator.

The methods used proved to be effective in the design, evaluation and verification of the shear spinning process. They enable excellent control of process variables and the development of the degree of accuracy in predicting the effects of the forming process, along with the number of trials performed. The advantage of the method of process optimisation adopted is the property of self-regulation. As the number of verification tests performed increases and they are included in the input data set, the accuracy of model prediction increases. This results in more precise estimates in future trials until they converge and reach the optimum. The method allows for optimisation without direct access to tools and details of the process, which may be confidential information, unlike FEM, where the user must have a full description, including specifications and models. In addition, the advantage of the adopted method is the lower cost of the creation of such solution, than in the case of FEM. FEM requires appropriate substantive advancement of the personnel, while its implementation usually involves expensive, commercial software. Conversely, the adopted optimisation method used free libraries, and the level of advancement is not as demanding as in the case of FEM.

Author Contributions: Conceptualization, S.P. and J.P.; methodology, S.P. and K.Ż.; software, T.T.; validation, T.P., S.P.; formal analysis, T.T.; investigation, S.P.; data curation, T.P.; writing—original draft preparation, S.P. and K.Ż.; writing—review and editing, T.T. and S.P.; visualization, K.Ż.; supervision, J.P.; All authors have read and agreed to the published version of the manuscript.

Funding: This research received no external funding.

Institutional Review Board Statement: Not applicable.

Informed Consent Statement: Not applicable.

Data Availability Statement: The data presented in this study are available on request from the corresponding author.

Conflicts of Interest: The authors declare no conflict of interest.

References

1. Bache, M.R.; Coleman, C.; Coleman, M.P.; Gray, V.; Boettcher, C. Microstructure evolution in flow formed IN 718 products and subsequent fatigue crack growth properties. *Fatigue Fract. Eng. Mater. Struct.* **2018**, *41*, 2249–2258. [CrossRef]
2. Kawai, K.; Yang, L.N.; Kudo, H. A flexible shear spinning of truncated conical shells with a general-purpose mandrel. *J. Mater. Process. Technol.* **2001**, *113*, 28–33. [CrossRef]
3. Chen, M.D.; Hsu, R.Q.; Fuh, K.H. Forecast of shear spinning force and surface roughness of spun cones by employing regression analysis. *Int. J. Mach. Tools Manuf.* **2001**, *41*, 1721–1734. [CrossRef]
4. Jeong, G.U.; Jin, C.K.; Seo, H.Y.; Kang, C.G. Experimental Investigation on the Deformation Behavior of Inconel 625 Superalloy at High Temperatures. *Metals* **2019**, *9*, 720. [CrossRef]
5. Gustafsson, D.; Moverare, D.; Johansson, S.; Hörnqvist, M.; Simonsson, K.; Sjöström, S.; Sharifimajda, B. Fatigue crack growth behaviour of Inconel 718 with high temperature hold times. *Procedia Eng.* **2010**, *2*, 1095–1104. [CrossRef]
6. Appa Rao, G.; Kumar, M.; Srinivas, M.; Sarma, D.S. Effect of standard heat treatment on the microstructure and mechanical properties of hot isostatically pressed superalloy Inconel 718. *Mater. Sci. Eng. A* **2003**, *355*, 114–125.
7. Plewiński, A.; Drenger, T. Spinning and flow forming hard-to-deform metal alloys. *Arch. Civ. Mech. Eng.* **2009**, *9*, 101–109. [CrossRef]
8. Dröge, K.H. *Forces and Material Flow in Spinning*; Technische Universität Dortmund: Dortmund, Germany, 1954.
9. Avitzur, B.; Yang, C.T. Analysis of power spinning of cones. *J. Eng. Ind.—Trans. ASME* **1960**, *82*, 231–245. [CrossRef]
10. Kalpakcioglu, S. On the mechanics of shear spinning. *J. Eng. Ind.—Trans. ASME* **1961**, *83*, 125–130. [CrossRef]
11. Kobayashi, S.; Hall, I.K.; Thomsen, E.G. A theory of shear spinning of cones. *J. Eng. Ind.—Trans. ASME* **1961**, *83*, 485–495. [CrossRef]
12. Sortais, H.C.; Kobayashi, S.; Thomsen, E.G. Mechanics of conventional spinning. *J. Eng. Ind.—Trans. ASME* **1963**, *85*, 346–350. [CrossRef]
13. Hayama, M.; Murota, T. On the study of metal spinning. *Bull. Fac. Eng.* **1963**, *12*, 53–58.
14. Wang, Q.; Wang, T.; Wang, Z.R. A study of the working force in conventional spinning. In Proceedings of the Fourth International Conference of Rotary Forming, Beijing, China, 17–21 October 1989; pp. 103–108.
15. Kim, C.; Jung, S.Y.; Choi, J.C. A lower upper-bound solution for shear spinning of cones. *Int. J. Mech. Sci.* **2003**, *45*, 1893–1911. [CrossRef]
16. Chen, M.D.; Hsu, R.Q.; Fuh, K.H. An analysis of force distribution in shear spinning of cone. *Int. J. Mech. Sci.* **2005**, *6*, 902–921. [CrossRef]

17. Joorabchian, A.; Slater, R.A.C. Spin forging of sheet metal cones having various cone angles and an upper bound estimate for the tangential component force exerted at the workpiece-roller interfaces. In Proceedings of the 20th International Machine Tool Design and Research Conference, Birmingham, UK, 12–14 September 1979; pp. 512–529.
18. Quigley, E.; Monaghan, J. Metal forming: An analysis of spinning processes. *J. Mater. Process. Technol.* **2000**, *103*, 114–119. [CrossRef]
19. Alberti, N.; Cannizzaro, L.; Lovalvo, E.; Micari, F. Analysis of metal spinning processes by the ADINA code. *Comput. Struct.* **1989**, *32*, 517–525. [CrossRef]
20. Liu, J.H.; Yang, H.; Li, Y.Q. A study of the stress and strain distributions of first pass conventional spinning under different roller-traces. *J. Mater. Process. Technol.* **2002**, *129*, 326–329. [CrossRef]
21. Mori, K.; Nonaka, T. Simplified three-dimensional finite element simulation of shear spinning process based on axisymmetric modeling. *Trans. North Am. Manuf. Res. Inst. SME* **2004**, *32*, 621–627. [CrossRef]
22. Quigley, E.; Monaghan, J. Using a finite element model to study plastic strains in metal spinning. In Proceedings of the 9th International Conference on Sheet Metal, Leuven, Belgium, 2–4 April 2001.
23. Klocke, F.; Wehrmeister, T. Laser-assisted metal spinning of advanced materials. In Proceedings of the 4th Lane Conference, Erlangen, Germany, 21–24 September 2004.
24. Lu, X.Y.; Zhang, S.H.; Hou, H.L.; Li, J.Z.; Zhou, L.X. Three dimensional FE analysis on flange bending for TC4 alloy during shear spinning. *J. Mater. Sci. Technol.* **2006**, *22*, 855–859.
25. Kleiner, M.; Gobel, R.; Kantz, H.; Klimmek, C.; Homberg, W. Combined methods for the prediction of dynamic instabilities in sheet metal spinning. *CIRPAnn. Manuf. Technol.* **2002**, *51*, 209–214. [CrossRef]
26. Klimmek, C.K.; Gobel, R.; Homberg, W.; Kantz, H. Finite element analysis of sheet metal forming by spinning. *J. Jpn. Soc. Technol. Plast.* **2003**, *44*, 372–374.
27. Zhan, M.; Zhanga, J.H.; Xua, Y.L.; Maa, F. Research on variation of stress and strain field and wall thickness during cone spinning. *Mater. Sci. Forum* **2006**, *532*, 149–152. [CrossRef]
28. Gao, P.; Yu, C.; Fu, M.; Xing, L.; Zhan, M.; Guo, J. Formability enhancement in hot spinning of titanium alloy thin-walled tube via prediction and control of ductile fracture. *Chin. J. Aeronaut.* **2021**, *3*, 1–19.
29. Kegg, R.L. A new test method for determination of spinnability of metals. *J. Eng. Ind.—Trans. ASME* **1961**, *83*, 118–125. [CrossRef]
30. Hayama, M.; Tago, A. The fracture of walls on shear spinning—Study on the spinnability of aluminum plates. *Bull. Fac. Eng.* **1968**, *17*, 93–103.
31. Hayama, M.; Murota, T.; Kudo, H. Deformation modes and wrinkling of flange on shear spinning. *Bull. JSME* **1966**, *9*, 423–433. [CrossRef]
32. Gao, P.F.; Li, M.; Zhan, M.; Xing, L.; Ma, F.; Fu, M.W. Circumferential twist in flow forming of tubular parts: Characterization, understanding and control. *J. Manuf. Process.* **2021**, *65*, 144–152, ISSN 1526-6125. [CrossRef]
33. Kawai, K. *A Fact Data Base on Spinning Technology, Rotary Forming*; International Academic Publishers: Oxford, UK, 1989; pp. 59–62.
34. Kunert, J.; Ewers, R.; Kleiner, M.; Henkenjohann, N.; Auer, C. *Optimisation of the Shear Forming Process by Means of Multivariate Statistical Methods*; Technische Universität Dortmund: Dortmund, Germany, 2005.
35. Montgomery, D.C. *Design and Analysis of Experiments*; John Wiley & Sons, Inc.: New York, NY, USA, 1997.
36. Taguchi, G. *System of Experimental Design*; UNIPUB/Kraus International Publications, American Supplier Institute in White Plains: New York, NY, USA, 1987.
37. Wang, L. *Spinning Process*; Durham University: Durham, UK, 2018; Available online: http://etheses.dur.ac.uk/3537/ (accessed on 5 August 2021).
38. Sangkharata, T.; Dechjarerna, S. Spinning Process Design Using Finite Element Analysis and Taguchi Method. *Procedia Eng.* **2017**, *207*, 1713–1718. [CrossRef]
39. Box, G.E.P.; Hunter, S.J.; Hunter, W.G. *Statistics for Experimenters: Design, Innovation, and Discovery*; John Wiley & Sons, Inc.: New York, NY, USA, 2005.
40. Music, O.; Allwood, J.M.; Kawai, K. A review of the mechanics of metal spinning. *J. Mater. Process. Technol.* **2010**, *210*, 3–23. [CrossRef]
41. Harrington, E.C. The Desirability Function. *Ind. Qual. Control* **1965**, *10*, 494–498.
42. Derringer, G.; Suich, R. Simultaneous optimization of several response variables. *J. Qual. Technol.* **1980**, *12*, 214–219. [CrossRef]
43. Kincaid, D.; Cheney, W. *Numerical Analysis*; WNT: Warszawa, Poland, 2006.

materials

MDPI

Review

Single-Point Incremental Forming of Titanium and Titanium Alloy Sheets

Valentin Oleksik [1,*], Tomasz Trzepieciński [2,*], Marcin Szpunar [3], Łukasz Chodoła [4], Daniel Ficek [5] and Ireneusz Szczęsny [5]

1 Faculty of Engineering, Lucian Blaga University of Sibiu, 550024 Sibiu, Romania
2 Department of Manufacturing and Production Engineering, Faculty of Mechanical Engineering and Aerionautics, Rzeszow University of Technology, al. Powst. Warszawy 8, 35-959 Rzeszów, Poland
3 Doctoral School of Engineering and Technical Sciences, Rzeszow University of Technology, al. Powst. Warszawy 12, 35-959 Rzeszów, Poland; d547@stud.prz.edu.pl
4 Department of Integrated Design and Tribology Systems, Faculty of Mechanics and Technology, Rzeszow University of Technology, ul. Kwiatkowskiego 4, 37-450 Stalowa Wola, Poland; l.chodola@prz.edu.pl
5 Department of Aerospace Engineering, Faculty of Mechanical Engineering and Aeronautics, Rzeszow University of Technology, al. Powst. Warszawy 8, 35-959 Rzeszów, Poland; ficekd@prz.edu.pl (D.F.); rm@prz.edu.pl (I.S.)
* Correspondence: valentin.oleksik@ulbsibiu.ro (V.O.); tomtrz@prz.edu.pl (T.T.)

Abstract: Incremental sheet forming of titanium and its alloys has a significant role in modern manufacturing techniques because it allows for the production of high-quality products with complex shapes at low production costs. Stamping processes are a major contributor to plastic working techniques in industries such as automotive, aerospace and medicine. This article reviews the development of the single-point incremental forming (SPIF) technique in titanium and its alloys. Problems of a tribological and microstructural nature that make it difficult to obtain components with the desired geometric and shape accuracy are discussed. Great emphasis is placed on current trends in SPIF of difficult-to-form α-, α + β- and β-type titanium alloys. Potential uses of SPIF for forming products in various industries are also indicated, with a particular focus on medical applications. The conclusions of the review provide a structured guideline for scientists and practitioners working on incremental forming of titanium and titanium alloy sheets. One of the ways to increase the formability and minimize the springback of titanium alloys is to treat them at elevated temperatures. The main approaches developed for introducing temperature into a workpiece are friction heating, electrical heating and laser heating. The selection of an appropriate lubricant is a key aspect of the forming process of titanium and its alloys, which exhibit unfavorable tribological properties such as high adhesion and a tendency to adhesive wear. A review of the literature showed that there are insufficient investigations into the synergistic effect of rotational speed and tool rotation direction on the surface roughness of workpieces.

Keywords: friction; incremental sheet forming; lubrication; microstructure; sheet metal forming; single-point incremental forming

Citation: Oleksik, V.; Trzepieciński, T.; Szpunar, M.; Chodoła, Ł.; Ficek, D.; Szczęsny, I. Single-Point Incremental Forming of Titanium and Titanium Alloy Sheets. *Materials* **2021**, *14*, 6372. https://doi.org/10.3390/ma14216372

Academic Editor: Jan Haubrich

Received: 13 September 2021
Accepted: 22 October 2021
Published: 25 October 2021

Publisher's Note: MDPI stays neutral with regard to jurisdictional claims in published maps and institutional affiliations.

1. Introduction

Titanium is the ninth most abundant element on Earth [1]. At the turn of the 20th and 21st centuries, titanium and its alloys were widely used in various industries such as the automotive, energy, chemical and food industries [2]. Components and entire structures made of titanium and its alloys can be found in particular in aviation and aerospace, where high structural strength and low weight are required [3].

Titanium alloys have a high yield stress of over 1550 MPa for Ti-(α + β) and β-type Ti-based alloys coupled with a relatively low density of 4600 kg/m^3 [4]. The mechanical strength of titanium and its alloys ranges from R_m = 290 MPa for pure titanium Grade 1 to about 1750 MPa for heat treated β-type alloys [4]. The use of titanium alloys allows the strength of the structure to be increased while reducing the weight because titanium alloys have a density that is about 1.7 times lower than high-strength steels [5]. Titanium alloys have the highest specific strength up to a temperature of 600 K when compared to the most commonly used materials, such as steel and aluminum alloys. Titanium and its alloys are covered with a natural layer of titanium oxide TiO_2, which makes these materials resistant to the effects of weather, sea water and many chemicals [5–7]. Titanium has a very low thermal conductivity coefficient, which is 22.08 W/(mK) [8], 13 times lower than that of aluminum.

According to ASTM B265, pure titanium is produced in 7 grades (Grades 1 to 4, 7, 11 and 12) that differing in their degree of contamination. The mechanical properties of titanium depend on its purity. Along with the increasing content of admixtures (Fe, O, N, C, Si and H), the formability decreases and at the same time the strength properties and hardness increase [9]. The mechanical properties of pure titanium can be changed by plastic working [9]. The effects of strain hardening can be eliminated by annealing recrystallisation. The yield stress of pure titanium varies from about 170 MPa (Grade 1) to 480 MPa (Grade 4), and the tensile strength from 240 MPa (Grade 1) to 550 MPa (Grade 4) [10].

Titanium alloys crystallize in two crystalline structures and can be classified into an α-type (hexagonal-closed packed (HCP) crystalline structure stable up to the transformation temperature, near α-type, α + β, a β-type (body-centred cubic (BCC) crystalline structure stable from β-transus to melting point) and near β-type Ti-based alloys [11]. The transformation temperature can be changed by the addition of alloying elements [12]. The α-phase is stabilized by elements such as carbon, nitrogen and aluminum, while chromium, manganese, niobium, molybdenum and vanadium stabilize the β phase. The β-type Ti-based alloys show good plasticity but lower strength than α alloys [13,14]. The low modulus of elasticity of this group of materials ranges between 105 and 120 GPa. Depending on the ratio of the α- and β-type phases, Ti (α + β) alloys are subgrouped into near-α and near-β alloys (Figure 1) [15]. Due to the high content of elements stabilizing the β phase [16], and the slowing down of the aging process, β-type alloys have a low tendency toward work hardening.

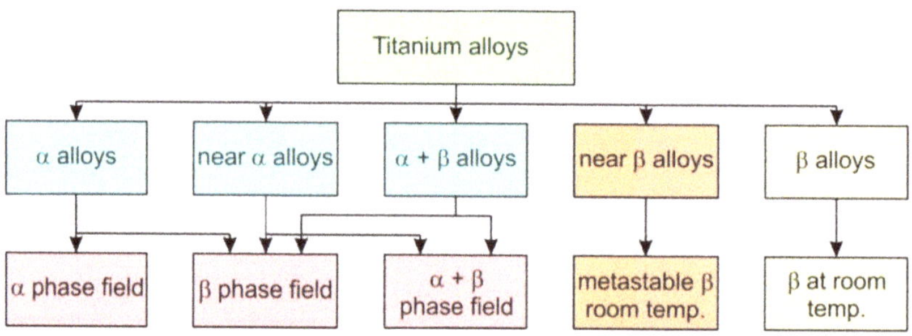

Figure 1. Classification of titanium-based alloys.

The α–β alloy Ti-6Al-4V is the most popular of all titanium alloys, representing more than 50% of the titanium market [17]. However, wide usage of the Ti-6Al-4V sheet is limited by its poor formability at room-temperature [18]. The Ti-6Al-4V alloy can be solution heat-treated, quenched and aged to medium-/high-strength levels and have good formability due to the presence of the β-phase, which has a much higher diffusivity and more slip systems than α and near-α alloys [17]. For these reasons, TI-6Al-4V alloy is also more easily machined than other Ti-based alloys, and that is one of the reasons it is a widely used alloy. The Ti-6Al-4V alloy is designed for a good balance of characteristics, including strength, ductility, fracture toughness, high-temperature strength, creep characteristics, weldability, workability and thermal processability (higher strength is easily obtained by heat treatment) [19]. Based on Ti-6Al-4V, a significant number of titanium alloys have been developed by combining Ti-(4−6)Al with contents between 4% and 5% of β stabilizer elements, which include biomedical Ti-6Al-7Nb, Ti-5Al-2.5Fe, Ti-5Al-2Sn-2Zr-4Mo-4Cr, Ti-6Al-2Sn-4Zr-6Mo, Ti-6Al-6V-2Sn and Ti-7Al-4Mo alloys. New types of titanium, such as gamma alloys (TiAl phases, 46Ti–48Al alloy with Ti_3Al), have been developed [15,20].

Conventional manufacture of metallic products relies on forging, casting and rolling of bulk feedstock materials, followed by subsequent machining to final shapes and dimensions [21]. These traditional manufacturing processes always inevitably result in a large amount of material waste, high manufacturing cost and long leading time [22]. Metal forming involves changing the shape of the material by permanent plastic deformation. Moreover, due to the work hardening phenomenon, the mechanical strength of the material formed increases. The advantages of the sheet metal forming (SMF) processes include no wastage of raw material, better mechanical properties of the product and faster production rate. SMF is used for large batches, which amortize tooling costs, producing large quantities of components during a short time interval. However, the possibility of using conventional stamping processes for small batches or personalized prototypes is very expensive [22]. A variant of the conventional SMF process, incremental sheet-forming (ISF) technology, presents an innovative possibility to decrease the cost of the problem in small volume production. It introduces the use of metallic sheet for small-batch production in an economic way without the need for expensive or dedicated tools [23]. Casting is a manufacturing process in which a liquid material is usually poured into a mold, which contains a hollow cavity of the desired shape, and is then allowed to solidify [24]. The casting products easily have various defects and deficiencies, such as inclusions, pores, shrinkages and cracks inside the casting components. The casting dimensional accuracy is low compared with that of machining components [24]. Machining of the titanium alloys is limited due to of the many machining difficulties occurring as a result of its hardness; chemical reactions that take place between the tool and the machine workpiece; and the high amount of heat generated in the cutting zone [25].

Plastic working of titanium alloy sheets has a significant role in modern techniques for the production of high-quality products with complex shapes [26,27]. Sheet metal forming constitutes a large element in the plastic processing of metals, which is the technique used for the creation of a wide range of components for the automotive, aviation and household appliances industries, as well as for the needs of biomedical engineering [28]. Due to the relatively limited production volume of a series of products made of titanium and its alloys, often for small series, incremental sheet-forming (ISF) methods, i.e., single-point incremental forming (SPIF) and two-point incremental forming (TPIF), are well-suited forming techniques. ISF has generally short forming times and low costs for forming sheet metal components because this process does not require special and costly die tooling.

The main advantages of SPIF and TPIF include [29–31]:

- The possibility of forming elements on a conventional CNC machine;
- Quick and easy consideration of design changes in the components being shaped;
- Significantly lower forming force values compared to conventional stamping;
- A higher value of the deformation limit of the sheet compared to conventional stamping.

The main disadvantage of SPIF is longer forming time compared to traditional methods of sheet metal forming; therefore, this method is economically justified for the production of components in piece and small-lot production [10,29,32]. SPIF and TPIF of titanium alloys have specific properties, the knowledge of which is necessary for the proper management of the process. Recent development trends in sheet metal forming and SPIF of lightweight metals have been provided by Trzepieciński et al. [33,34].

This article provides an overview of the development of incremental forming techniques for titanium and its alloys. In the following chapters, problems linked to the crystal lattice of titanium alloys and its tribological phenomena are considered. SPIF methods used in cold and elevated temperatures and SPIF for biomedical implants are discussed. One of the sections is devoted to the optimization of the forming strategy and the improvement of friction conditions when forming titanium and titanium alloy sheets. The scope of the review covers all applications of ISF methods for product forming, in particular in aircraft, automotive and medical applications.

2. Methods of Review

The Ei Compendex, IngentaConnect, PubMed, Scopus, ScienceDirect and Web of Science database search strategy used in this review was consistent with the principles of a systematic review according to the Preferred Reporting Items for Systematic Reviews and Meta-Analyses (PRISMA) [35] guidelines. The search strategy was limited to articles written in English. To ensure the high quality and timeliness of the review, great emphasis was placed on searching for both review and research papers, review papers, books, chapters and conference proceedings, over the last ten years. Overall, the following numbers of papers were of interest to the authors: 160 research papers in scientific journals, 20 review papers in scientific journals, 10 books, 20 chapters and 30 papers in conference proceedings. On the basis of the titles and abstracts of the bibliographical sources found, an initial selection of articles was carried out, based on the heat-assisted incremental sheet forming method, and on the main purpose of the article (springback analysis, analysis of surface roughness, lubrication conditions, etc.). Articles, the content of which, despite a promising summary, differed from the assumed subject of this review article, were excluded from further analysis. Obviously, duplicate articles found in above-mentioned databases have been removed. In addition, articles by the same authors were reviewed for similar or duplicate content. This approach was often used when publishing extended versions of conference materials in scientific journals.

3. Incremental Sheet Forming Methods

SPIF and TPIF involve the gradual forming of sheet metal using rotating contour tools. Due to the penetration of the tool producing a stretch process in accordance with the assumed trajectory (forming strategy) of the tool movements, the deformed material takes on a final shape through the accumulation of localized plastic deformation of the sheet, in which a hemispherical-head tool controlled by a computer program is used to form the metal [36–38]. Incremental forming is classified into two main varieties:

- Negative (Figure 2a), in which there is an additional element of the tool in the form of a template or a pin supporting the highest area of the detail being formed; forming takes place by appropriate mapping of the template geometry, by "arranging" the sheet on it, with a tool performing programmed movements,
- Positive (Figure 2b), in which the component is formed by the movement of the tool in accordance with the tool trajectory without the supporting template.

The definition of positive and negative is given by the concavity of the part that is shaped, which can be concave up or concave down.

Due to the kinematics of the tools, incremental forming has several varieties that differ in the equipment used. The first type is SPIF; the principle of carrying it out is shown in Figure 3a. The tool is guided by the machine control system along a pre-programmed trajectory, causing incremental deformation of the sheet [39,40]. A variation of SPIF (Figure 4a) is TPIF, which can be performed with the use of a partial die (Figure 3b) or a specific matrix (Figure 3c). The last method increases the geometric accuracy of the formed elements. In two-point forming methods, there is an additional movement of the assembly fixing the edges of the formed sheet, which translates into greater accuracy of the components. In the incremental forming with counter tool (IFWCT) method, an additional mandrel is placed opposite the forming tool and shifted by the thickness of the sheet moves along an appropriately corrected trajectory in relation to the main tool (Figure 3d).

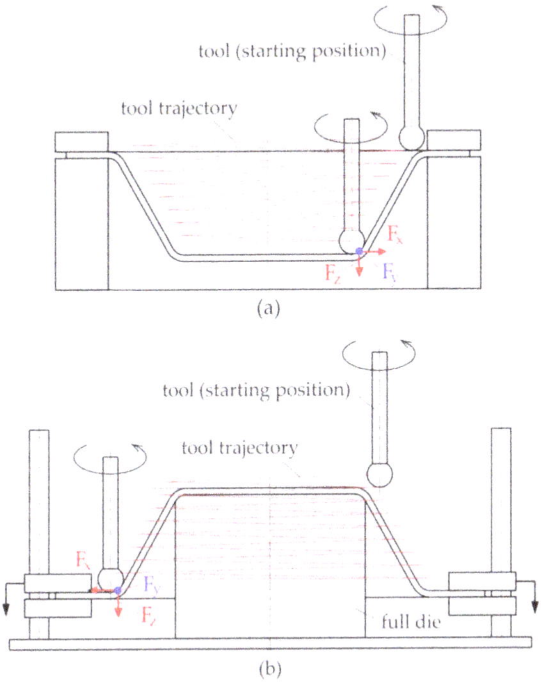

Figure 2. (**a**) Negative and (**b**) positive ISF.

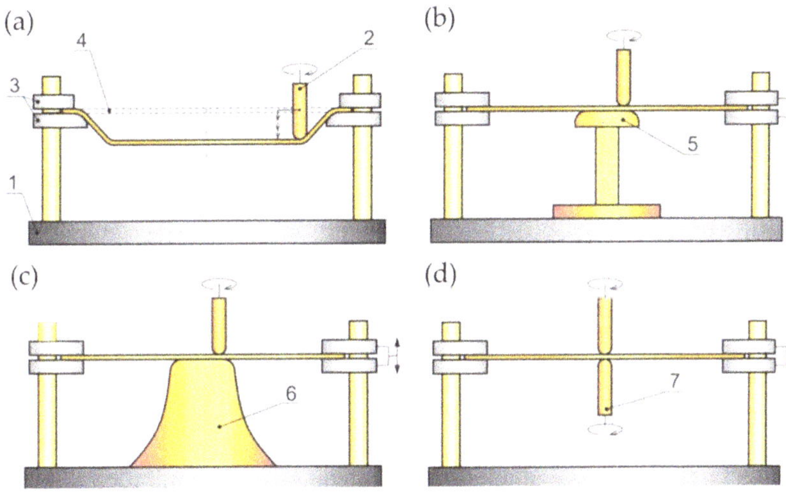

Figure 3. ISF Processes: (**a**) SPIF; (**b**) TPIF with a partial die; (**c**) TPIF a specific die; and (**d**) IFWCT: 1—clamping device, 2—forming tool, 3—plate holder, 4—workpiece (initial position), 5—partial die, 6—specific die and 7—auxiliary forming tool.

Recently, the micro-ISF (μ-ISF) technique developed by Saotome and Okamoto [41] has received increasing attention due to its high operability and flexibility, and it is regarded as a very promising technique for micro-structure manufacturing [42,43]. M-ISF was developed by the use of the scanning electron microscope (SEM) to achieve in situ observation of the forming process. Although the μ-ISF process increases flexibility and reduces installation costs, it presents many problems related to significant shape defects and large decreases in thickness [44].

4. Incremental Sheet Forming towards Biomedical Applications

Titanium is considered to be the most biocompatible of the metals due to its resistance to corrosion in body fluids and bio-inertness [10,45,46]. The strength of titanium is due to a protective oxide film that forms naturally in the presence of oxygen found in body fluids [47]. The oxidized coating adheres strongly to the tissue and is insoluble and chemically impermeable, which prevents it from reacting with the surrounding environment [48]. The biological neutrality of titanium is used in surgery in the form of implants, i.e., spinal implants, dental implants and knee or hip endoprostheses [49,50]. Recent developments in β-type Ti-based alloys for biomedical applications have been provided by Chen et al. [51].

Biocompatibility is defined as the ability of a material to behave properly in contact with biological tissue. Titanium is a metal that additionally exhibits so-called osseointegration, i.e., the ability to create a direct structural connection between its surface and living tissue [52–54]. As a result, no scar tissue or cartilage is formed between bone and the implant, and the connection is very strong. Diffusive oxynitride layers improve the hemocompatibility of titanium and its alloys, including its antithrombogenocity [55,56].

More than 1000 Mg of titanium is implanted in patients worldwide each year [57]. The implant materials must be non-toxic and non-carcinogenic and have similar mechanical properties to the original bones [58,59]. The appropriate geometric profile of the implant and the quality of the surface after forming are guarantees of biocompatibility [60–62]. The implant should be adapted to the authorized requirements of each patient, so the processing of implants is basically a unit production process [63]. Under these conditions, the use of ISF methods seems to be ideal. Since the sheet metal is formed incrementally, the formability and potential of material can be fully explored [64,65]. A number of efforts have been made to manufacture implants and supports for different parts of the human body using ISF, such as the skull [66–69], dental prostheses, knee [70], face [71] and ankle support [72].

Through the years, researchers have investigated and proved several aspects of the process feasibility of using SPIF for manufacturing biomedical implants. Han et al. [73] successfully produced a prosthesis for a skull on a titanium mesh plate, which can increase the contact area and adjust the elasticity modulus of the material. Duflou et al. [66] found that during ISF of a cranio-facial implant, the limit forming angle of pure titanium is 47°. In order to increase this limit, a multi-stage forming strategy was proposed, and the results showed that the limit forming angle was increased to 61°. Duflou et al. made an effort to use ISF to manufacture cranial reconstruction from titanium Grade 2 [66]. This paper discusses the challenges associated with the manufacture of cranio-facial implants with extreme forming angles. Vanhove et al. [74] fabricated titanium Grade 2 thin-shell clavicle implants through SPIF. The fracture-fixation and bone-aligning tasks of these implants call for specific accuracy distribution, while the distinct geometry and post-forming heat treatment influence the production accuracy. The authors have shown the potential to replace bulk generic clavicle implants (Figure 4) by successfully compensating for the forming toolpath in order for it to be brought into a satisfactory accuracy range.

Sbayti et al. [18] evaluated the feasibility of producing a customized acetabular component of a hip prosthesis with an acceptable geometric accuracy using this process and based on the use of Ti-6Al-4V titanium alloy. It was found that the geometric accuracy was significantly improved at $T \geq 400$ °C. Castelan et al. [57] manufactured custom-made cranial implants from a sheet of Grade 2 titanium using 3D printing, CAD/CAM technology and SPIF. Araújo et al. [75] evaluated the SPIF process for the fabrication of customized maxillofacial implants from titanium and demonstrated the role of this technology in manufacturing customized medical parts. Ambrogio et al. [76] manufactured the Ti-6Al-4V and Ti6Al4V-ELI prostheses by SPIF. The authors designed a complete procedure that guides and integrates the jobs of surgeons and engineers during the cranioplasty process. Cytotoxicity tests have shown that SPIF does not affect the biocompatibility of the prostheses.

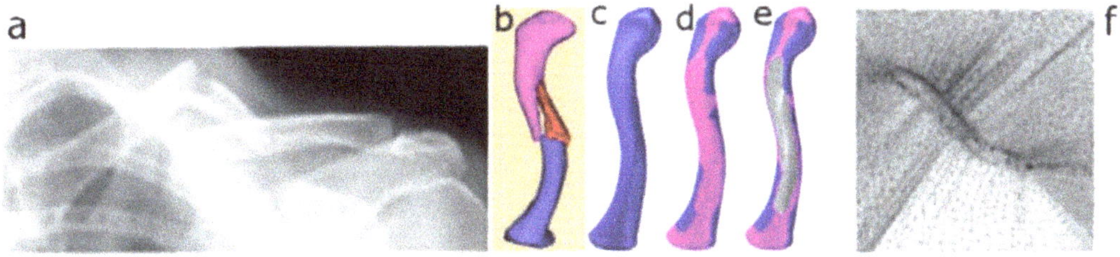

Figure 4. (**a**) Computer tomography scan of a clavicle, (**b**) 3D reconstruction, (**c**) reconstructed clavicle, (**d**) clavicle with predicted muscle attachment sites (blue area), (**e**) clavicle with designed implant and (**f**) implant embedded in extension structure (reproduced with permission from [74]; copyright © 2017 Published by Elsevier Ltd.).

Many solutions have been proposed by Honarpisheh et al. [77] and Mohammadi et al. [78] in order to overcome the formability limitation of titanium alloy in warm SPIF. Results have shown that increasing the working temperature is very effective in decreasing the elastic springback and forming force and improving formability. Saidi et al. [79] developed a reverse engineering approach (Figure 5) associated with a warm SPIF process (Figure 6) in order to produce a Ti-6Al-4V prosthesis of a human skull. The major advantages of this heating system compared to others proposed in the literature are its low cost, great simplicity of installation and limited increase of equipment technology.

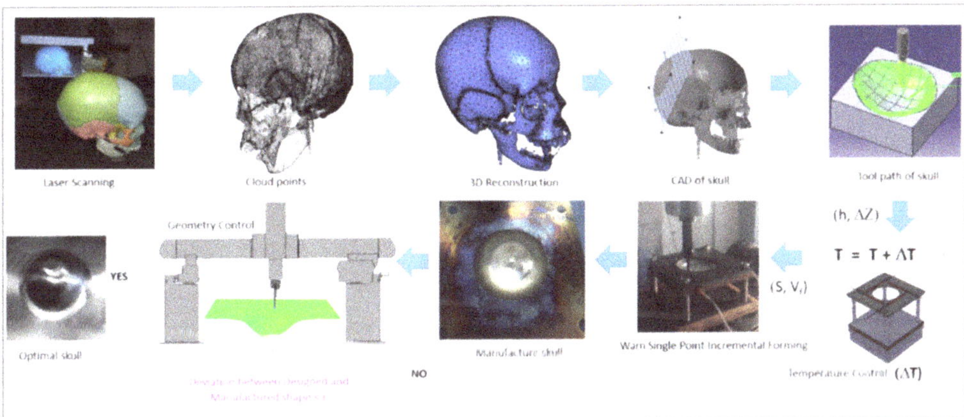

Figure 5. A reverse engineering application to human skull manufacture (reprinted with permission from [79]; copyright © 2018, Springer-Verlag London Ltd., part of Springer Nature).

Castelan et al. [57] physically constructed a CAD model of a skull with 3D Printing, and SPIF was used to manufacture the implant from titanium Grade 2. It was found that dimensional variability can be reduced with changes in the manufacturing process (i.e., forming and cutting) and the heating ramp. Before cutting the final shape of the implant, heat treatment was performed to avoid deformations caused by residual stresses generated during the SPIF.

Palumbo et al. [80] fabricated prototypes of Ti-Gr5 and Ti-Gr23 cranial prostheses via the single-point incremental forming process. The geometry of the prosthesis investigated was obtained by starting from a model of a human skull in poly (methyl methacrylate) material (Figure 7). A defect was created by removing a portion of the skull, and finally the shape of the prosthesis was thus defined by reverse engineering. The comparison between finite element-based simulation numerical data and experimental data revealed that the performance in terms of impact response of the prostheses strongly depends on its thickness distribution, due to the strain hardening phenomena.

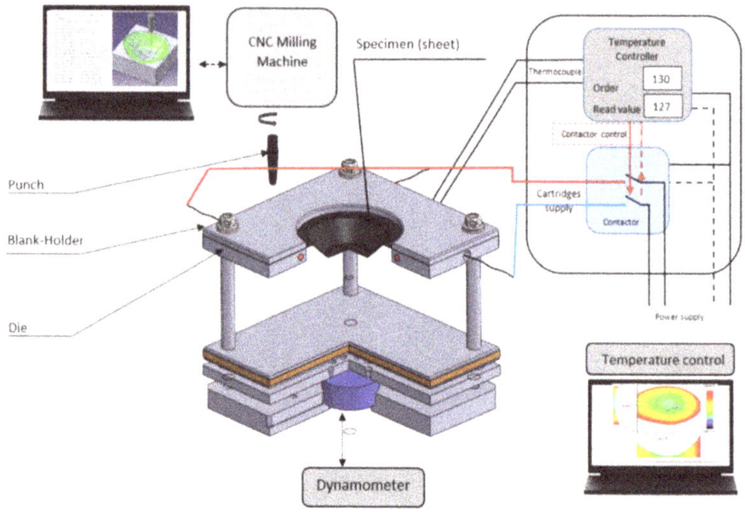

Figure 6. Scheme for warm SPIF (reprinted with permission from [79]; copyright © 2018, Springer-Verlag London Ltd., part of Springer Nature).

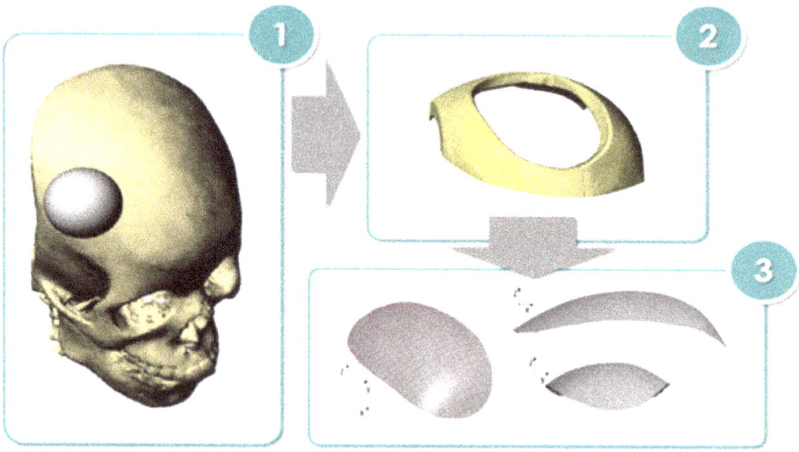

Figure 7. Design process of the prosthesis geometry (reproduced with permission from [80]; copyright © 2020 The Authors; this is an open-access article distributed under the terms of the Creative Commons CC BY license, which permits unrestricted use, distribution and reproduction in any medium, provided the original work is properly cited).

Araújo et al. [71] SPIFed facial implants with titanium Grade 2. The main operating parameters influencing the production of facial implants with SPIF were identified. Investigations demonstrated how circle grid analysis combined with the fracture forming limit diagram (FLD) can be utilized to successfully re-engineer the fabrication of a facial implant. The results show how the overall level of in-plane strains and drawing angles can be controlled in order to obtain a sound facial implant produced by SPIF.

Oleksik et al. [81] presented a study based on experimental research on the surface quality of the medical implants used for the partial resurfacing of the femoral condylar surface of the knee obtained by SPIF. The research examined the punch diameter, the friction conditions and the initial roughness of the punch between the punch and the blank.

5. Thermally-Assisted Incremental Sheet Forming

The hot incremental forming process provides the opportunity to improve the formability of Ti alloys [33]. It is almost impossible to plastic-form Ti alloys at room temperature because of the hexagonal close packed structure of the alpha phase and major plastic deformation by twinning [82]. When increasing the forming temperature, proper lubrication is a key factor affecting the surface roughness obtained, as well as the oxidation process [83]. An elevated process temperature can be provided by different heat sources [84]. In the last two decades, researchers have focused on elevating the temperature of Ti alloy sheets by: direct current (DC) and alternating current (AC), cartridge heaters, induction, laser source, and friction between the tool and the workpiece. Heat sources can be arranged to cover just the precise location [85], the whole sheet or a combination of both. Local heating is energy-efficient, and the sheet oxidation zone is reduced. However, deformation is only enhanced at the tool-workpiece contact zone. Grade 5 seems to be the most favorable for selection for heat-assisted ISF because of its wide application in the aerospace, automotive and medical industries; light weight; strength; and biocompatibility. The method that has been most intensely studied in the last two decades seems to be electrically-assisted ISF due to the low cost of the apparatus needed to upgrade the conventional forming test stand.

In general, heating systems in thermal-assisted ISF processes can be classified as [86]:

- Radiation—a laser light follows the tool trajectory (the workpiece is heated at the actual location of forming) [87–89],
- Convection—hot air blowers heat the entire metal sheet [90],
- Conduction—heater bands are mounted on the external surface of the die to heat the entire workpiece during ISF [91],
- Electricity—an electric current is used to heat the workpiece at the actual location of forming [77,92,93],
- Friction—material is heated due to frictional forces between the rotating tool and the fixed sheet [94–96].

5.1. Electric Heating

Fan et al. [92] were some of the first researchers to apply electric DC current to heat TiAl2Mn1.5 sheet. By connecting the tool to the cathode and the sheet to the anode of the power source, according to Joule's law $Q = I^2 \cdot R \cdot t$ (Q is the heat generated, I is the current, R is the resistance of the conductor and t is the time), electricity heats at the location of the contact point in the deformation zone (Figure 8). It was found that applying a 400 A current with a feed rate of 400 and 800 mm/min, using hemispherical tools with diameters of 8, 10 and 12 mm and step sizes of 0.1 and 0.2 mm, allows one to form a 50° wall angle cone, a distorted tetragonal pyramid and a tetragonal pyramid shape.

In further work, Fan et al. [83] used this method to form Ti Grade 5 with a 400 A current and feed rate of 15 mm/s, achieving a sheet temperature range of 300–500 °C and resulting in a maximum forming angle of 72°. Improved formability due to electric pulses could be affected by a reduction of dislocation density and a boost to its mobility [97]. In addition, a weakened base texture and weakened anisotropy leads to improved formability [98]. Asgar et al. [99] investigated the DC pulse assisted by TPIF process, where the cathode is connected to the upper punch tool and the anode to the bottom, while the tool moves independently. A 0.6-mm-thick titanium alloy sheet can be successfully formed into a pyramidal shape with angles of 30° and 45° using 140 A current. The springback effect was significantly reduced by this method, while it required about twice the forming force compared to the no current variant. As Fan and Gao [100] proved, with a higher feed rate, with a constant step size and wall angle, a higher DC current should be applied to maintain an elevated temperature. Heat level feedback and a proper heat source controller are required to maintain a constant forming temperature. Metallographic structure analysis shows that hot electric-assisted ISF is a process where plastic hardening and annealing occur. To achieve a uniform structure, heat treatment is necessary after forming. Elevated temperatures decrease the tensile strength of Ti Grade 5 due to oxidation, so a protective gas should be applied. Fan and Gao [93] investigated the electric hot ISF of Ti Grade 5 experimentally and compared this to numerical simulation using MSC.Marc software. Simulation revealed three zones of deformation: bending, which takes place at the beginning; shear forming in the middle; and reverse bending at the end. Furthermore, to enhance accuracy and reduce bending during the formation of the first layer, adequate support is required. Fan and Gao [93] also discuss the different regions (bending deformation at the beginning, shear forming in the middle and reverse bending at the end) in assuring the geometric accuracy of the SPIFed components (Figure 9).

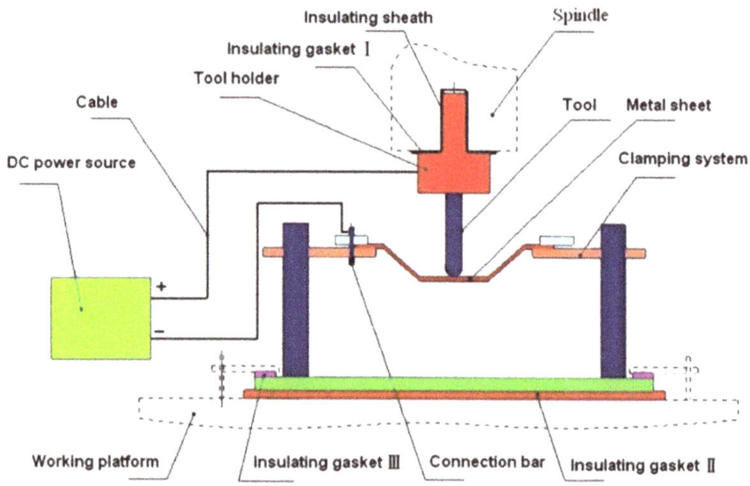

Figure 8. The principle of electric incremental forming (reproduced with permission from [92]; copyright © 2008 Elsevier Ltd. All rights reserved).

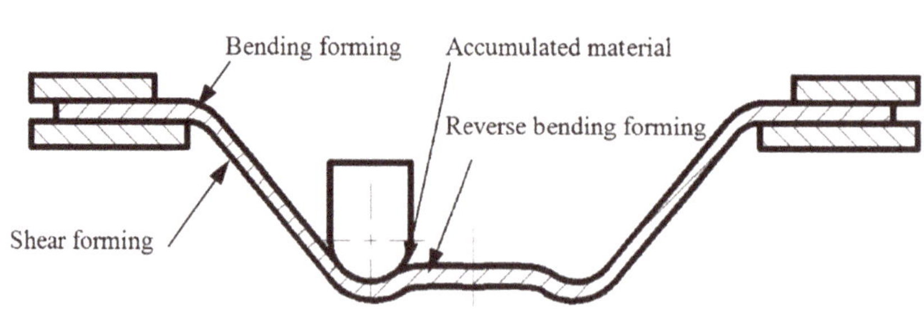

Figure 9. Forming regions in a SPIFed component (reprinted with permission from [93]; copyright © 2014, Springer-Verlag London).

To understand the process a little better, the Johnson–Cook Finite Element Method (FEM) model finds good agreement with experiment during electric hot incremental forming (EHIF) of Ti Grade 5 [77]. Elevated temperature in the contact zone has effects in accelerating the tool wear and on the final surface quality of the part. To reduce the impact of temperature during the EHIF of Grade 5, Liu et al. [101] developed an advanced cooling system integrated with forming tools (Figure 10), equipped additionally with a freely rolling tip. This solution allows the tool temperature to be reduced from over 500 °C to 100 °C and the tip wear from 1 mm to less than 0.1 mm when compared with conventional tools. By this approach, the sheet surface parameter has also been enhanced from Ra = 2.939 μm, Rz = 17.226 μm to Ra = 0.805 μm, Rz = 4.64 μm.

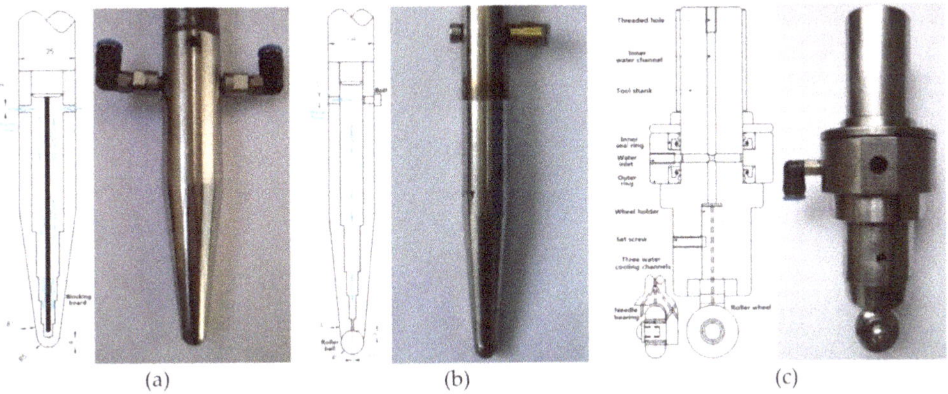

Figure 10. The tools that have been developed for use in E-ISF: (**a**) conventional rigid tool with inner water-cooling channel, (**b**) rolling-ball tool with inner water-cooling channel and (**c**) rolling-wheel tool with inner water-cooling channel (reprinted with permission from [101]; copyright © 2015, Springer-Verlag London).

Najafabady and Ghaei [102] developed a novel approach to the process by using AC (Figure 11a) current instead of DC (Figure 11b) for safety reasons (less sparks and easier to break the circuit). An autotransformer and high-pressure transformer have been employed to adjust the voltage from the 220V electric network to 1–1.5 V and deliver a current of 100–200 A. In their work on Ti Grade 5, 0.62 and 1.05 mm thick constant wall angle conical frustums (CWACF), varying wall angle conical frustums (VWACF) and two types of pyramidal frustum have been experimentally investigated within the 150–200 A range of currents, with a 600 mm/min feed rate, 100 and 600 rpm tool rotational speeds, 0.1 and 0.3 mm step sizes and a target temperature of 600 °C.

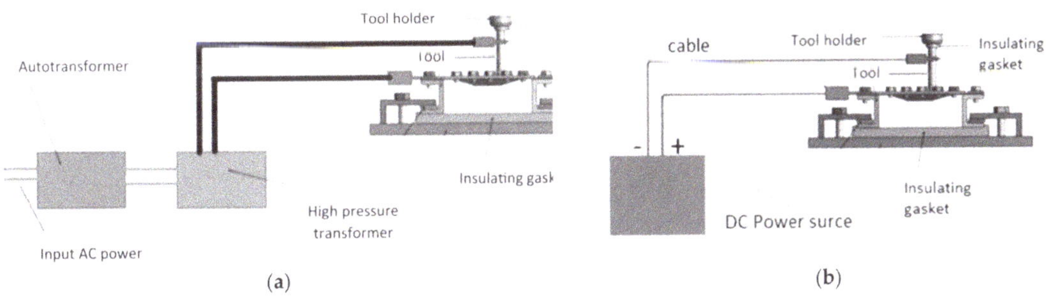

Figure 11. Schematic diagram of the E-ISF process with (**a**) AC and (**b**) DC power (reprinted with permission from [102]; copyright © 2016, Springer-Verlag London).

Increasing the step size from 0.1 to 0.3 mm produced a rise in the inside surface Ra parameter (from 6.8 μm to 11.2 μm). Furthermore, a thicker sheet blank produced an improved outer surface Ra parameter (from 3.2 μm to 0.8 μm). Proper lubrication is a key factor in enhancing surface quality at elevated temperatures. By spraying MoS_2 powder, the Ra parameter has been reduced from above 13 μm under dry conditions to less than

5 μm. A micro hardness test revealed a rise from 291 HV at the flange to 350 HV at the vertex caused by the work-hardening effect. Maximum shape deviation was observed in the workpiece flange with a tendency to decrease towards the vertex—this effect can be significantly reduced in the Electrically-assisted Double Sided Incremental Forming (E-DSIF) process [103]. Figure 12 shows the toolpath strategies utilized.

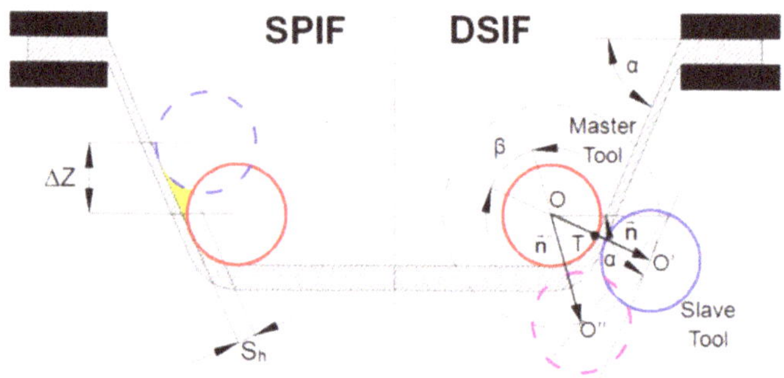

Figure 12. Comparison between tool paths in SPIF and DSIF (reproduced with permission from [103]; copyright © 2015 Elsevier Ltd. All rights reserved).

The general concept of DSIF is shown in Figure 13a, in which the sheet is deformed by the motions of tools on both sides. Conventional DSIF approaches usually rigidly control the displacement of both master and slave tools and may result in loss of contact between the tool and the sheet due to sheet thinning, as shown in Figure 13b. Therefore, Xu et al. [103] proposed a device to ensure stable contact between the tool and the sheet, where the slave tool is supported by an air cylinder, which acts as a spring (Figure 13c).

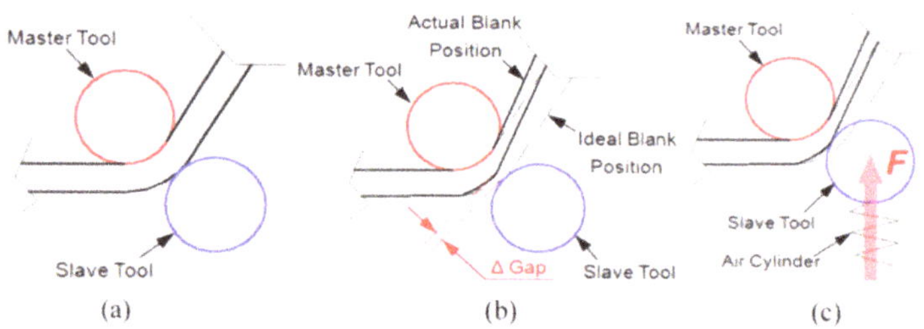

Figure 13. The E-DSIF process: (**a**) ideal position of forming tools, (**b**) loss of contact of slave tool and (**c**) supported slave tool approach to ensure a stable contact (reproduced with permission from [103]; copyright © 2015 Elsevier Ltd. All rights reserved).

Ortiz et al. [104] conducted an experiment where 10kW static heaters were located below the Ti Grade 5 1.6 mm thick sheet. In VWACF, in which the forming angle increases constantly, the angle limit was determined at a furnace temperature of 800 and 900 °C, feed rate of 1000 mm/min, and step size of 0.5 and 1.0 mm. The forming depth limits and angles obtained were 25.4 mm (45°) for room temperature, 44 mm (55°) at 800 °C and 72 mm (70°) at 900 °C. Obviously, VWACF is not sufficient to define the forming limit; subsequently, a constant 70° wall angle conical frustum of 35.3 mm depth was successfully CWACF-formed until breakage. In further work, Ortiz et al. [105] presented the intelligent process model (IPM) to correct the tool path, which improved the model accuracy in response to springback by up to 49%. Khazaali and Fereshteh-Saniee [106] conducted a groove test with the Taguchi method. Three static heaters have been placed below, forming Grade 5 sheet at an elevated initial temperature of up to 190 °C. A tool diameter of 16 mm and a step size of 0.7 mm at 190 °C seem to be the most favorable parameters from the experimental range. In further experiments [107], the authors extended their research on VWACF by conducting 27 experiments (3 levels) using the Taguchi method with three temperature levels (100, 200 and 300 °C), tool diameters (8, 12 and 16 mm), and step sizes (0.4, 0.7 and 1.0 mm). Optimum measurements of the maximum drawing depth—300 °C, Ø8 mm tool and 1mm step size—were determined. Saidi et al. [79] investigated the manufacture of a cranial part from Grade 5 sheet by incremental forming using cartridge heaters as a heat source. In their work, an elevated temperature from 30 °C to 130 °C produced at least a doubling in the reduction in maximum vertical force and of average deviation from the desirable shape. Mohanraj and Elangovan [108] heated up the punching tool to 300 °C in their work using DC resistance. The optimum forming parameters of 300 °C, 0.1 mm step size and 50° wall angle were found for Grade 2. Furthermore, the theoretical thermal model proposed by the authors agrees with the experiment conducted.

5.2. Induction Heating

Ambrogio et al. [86] proposed induction heating 1mm thick nonmagnetic Grade 5 to 600 and 700 °C with cryogenic cooling. The induction coil was located below the sheet and follows the tool path. For cryogenic cooling, liquid nitrogen supplied at 6 bar was sprayed axially onto the forming tool to cause quenching. The cryogenic medium was sprayed while the punch retraced from the last coil. A schematic illustration of the equipment used is presented in Figure 14. During the experiment, cryogenic versus standalone ambient cooling was compared. The induction heating system has an advantage over DC electric heating in terms of sheet surface roughness. The authors obtained an Ra from less than 2.5 μm to over Ra 4 μm in the 1mm step size forming test. Due to cryogenic cooling, the Grade 5 grain growth process was avoided and the micro hardness remained unchanged.

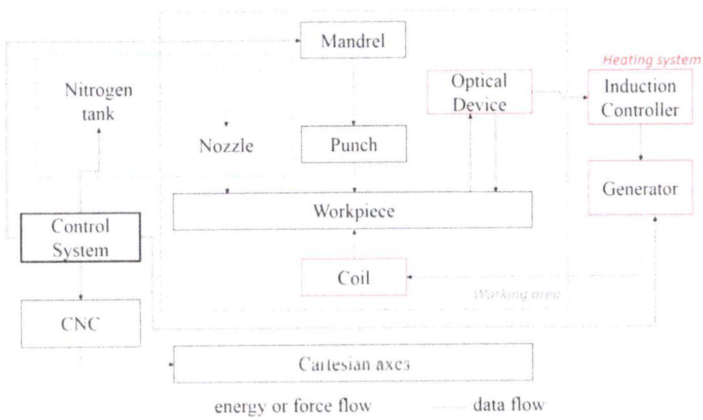

Figure 14. Functional scheme of the experimental equipment (reprinted with permission from [86]; copyright © 2016, Springer-Verlag London).

Gagliardi et al. [91] noted that in order to avoid the coil overheating, a low- to medium-frequency generator with high power is more suitable for non-magnetic sheets, such as Grade 5. Low-medium frequency would be important with regard to process feasibility to ensure that it is not detrimental to the duration of coil working life, while at the same time it increases the heating capacity of the system.

5.3. Laser Heating

Göttmann et al. [87] applied a laser beam to provide localized heat to Grade 2 and Grade 5 during asymmetric incremental sheet forming, in order to compare differences between the heat assisted and the room temperature setup. With experimental parameters of a laser beam with 1700 W power, step size of 0.35 mm, forming velocity of 4000 mm/min and two variants of tool diameter (20 and 30 mm), Grade 5 appears to show significant improvement in forming, although Grade 2 has not improved formability causing thermal failure. To achieve constant temperature in the contact zone, one can place an inside forming tool with a thermocouple, and a PD device could be adopted to control the source power of the laser or the DC electric current. Forming forces are also stabilized by this closed loop steering of temperature [88]. In the work conducted by Mosecker et al. [89], an ISF experiment was extended using a constant laser power source and by adding air cooling through a nozzle behind the tool path. The microstructure changes of Grade 5 and micro hardness differences have also been taken into account. The mainly observed deformation mechanism was the dynamic recovery, but at a temperature of 850 °C, without cooling, dynamic recrystallisation appeared, as well as the deepest deformation obtained. However, at this range of temperatures, Grade 5 micro hardness increases due to the grain size and reduction in the sheet thickness. This effect can be suppressed by applying dynamic recrystallisation cooling.

5.4. Friction Heating

Friction stir incremental forming is a type of localized heat initiation method in which heat is generated by friction between the tool and the workpiece due to the high-speed rotary movement of a punch. In work by Ambrogio et al. [94], titanium Grade 2 and Ti-based Grade 5 alloy were examined by applying high-speed single-point incremental forming. A lathe CNC machine was used as a test stand rather than the generally selected CNC milling machines. This was due to the fact that asymmetrical parts cannot be manufactured using the adopted speed (2500 m/min) because of the feed rate limits of the currently produced CNC milling machines. However, they allow for a quick manufacturing process—a frustum cone of 30° for Grade 2 and 25° for Grade 5 (Ø180 mm upper and Ø42 mm lower diameter) has been successfully formed in less than 1 min. Grün et al. [95] investigated design of experiment (DoE) and One Factor At a Time (OFAT) tests with various tool rotational speeds, feed rates and step sizes during forming VWACF from 1 mm thick Grade 5 using a Ø11 mm ball-ended tool. A major factor affecting formability and the value of the forming forces is tool rpm, while tool feed rate and step size are less significant. Elevated rotation speeds decrease forming forces by increasing contact temperature, although they rapidly increase tool wear and remove material from the sheet surface. The relationship to feed rate rotation direction, climb or 'conventional' (see Figure 15) is also a key parameter, as Uheida et al. [96] investigated in VWACF of Grade 2.

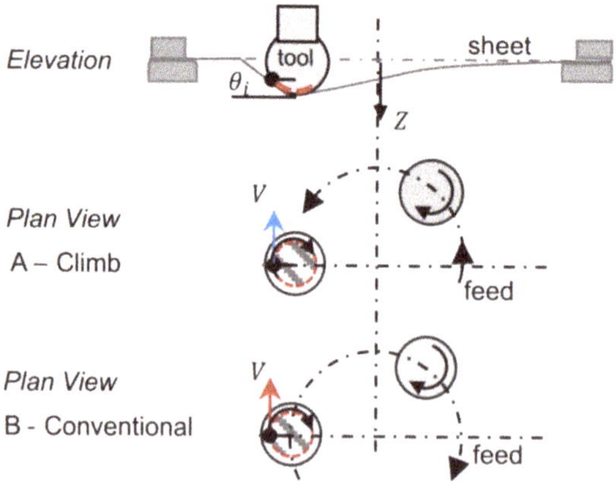

Figure 15. The concept of conventional and climb SPIF (reproduced with permission from [96]; copyright © 2018, The Author(s); this is an open-access article distributed under the terms of the Creative Commons CC BY license, which permits unrestricted use, distribution and reproduction in any medium, provided the original work is properly cited).

Conventional type forming has higher forming temperatures and lower vertical and horizontal forces, leading to better accuracy and a higher forming angle in comparison to climbing. However, the conventional forming direction produced poor surface quality due to increased friction causing the material to overheat.

5.5. Combined Electrical and Friction Heating

Palumbo and Brandizzi [109] investigated the heating of Grade 5 sheet by tool rotation in the range of 800–1600 rpm and additionally replaced the electric static heating with 3000 W electric bands. Electric heating warms up the whole blank, while the tool rotation only heats the location of the tool-blank contact zone. By increasing the tool rotational speed, the sheet surface parameter Ra worsens, but necking is stabilized. Khazaali and Fereshteh-Saniee [110] investigated the groove test as a benchmark of process parameters, in order to determine FLDs, reduce the blank size and reduce the experiment time. In their work, heating elements were placed below the Grade 5 blank from 0.5 mm thick sheet. The sheet temperature needed to be compensated for by a pyrometer and in fact was different from that obtained with the heaters. With constant process parameters of 400 rpm for the tool; 400 mm/min feed rate; and variable step size (0.1, 0.4 and 0.7 mm), temperature (100, 200 and 300 °C) and tool diameter (8, 12 and 16 mm), 27 trials were performed. In conclusion, elevated temperature, larger tool diameter and step size are critical to obtaining improved forming. It was also noted that higher temperature and tool diameter reduces the springback effect and increases shape accuracy—especially between the sheet clamping area and the formed shape (angular difference from 21.2°, 100 °C, tool Ø8 mm to 12.7° for 300 °C, and Ø16 mm tool).

6. Accuracy, Springback Reduction and Toolpath Optimization

Studies on increasing the accuracy of parts produced by SPIF from pure titanium or titanium alloys can be classified in studies that are focused on determining the optimal values of process parameters such as punch diameter, vertical step, friction coefficient, punch speed and tool rotation direction. The studies on springback reduction are focused on finding new technological improvements such as electrically-assisted SPIF, hot-forming SPIF, or DSIF. Toolpath optimization techniques are generally related to improving accuracy and reducing springback and involve finding different algorithms in order to correct the initial toolpath.

Palumbo et al. [111] studied the behavior of the titanium alloy Ti-6Al-4V, an alloy with a strong anisotropy, during single-point incremental forming. The research was based on the numerical simulation of this procedure using the Abaqus/Explicit analysis program. To conduct the numerical research, the mechanical properties and those measured during the uniaxial tensile test were determined for specimens taken at 0°, 45° and 90°, by means of the Aramis 3D optical measurement system. This allowed for the anisotropy of the titanium alloy, as well as its influence on SPIF, to be highlighted. The results were focused on determining the plastic strains and the thinning and shape errors. In the first step, to analyze the influence of the main input parameters on SPIF, a truncated pyramid was chosen as the geometric model for the part manufactured by SPIF. The yield criterion chosen for the titanium alloy was Hill (1948). A face-centered central composite design (CCD) was built in order to perform the numerical simulations. The two input parameters of the SPIF process considered were the ratio between the punch diameter and the vertical step (D/pz) and the wall angle (θ). Regarding the geometric precision, the authors concluded that the parts with the smallest shape errors are obtained when the ratio between the punch diameter and the vertical step (D/pz) has the highest values. Based on the data obtained from the first stage results, the research continued for a car door shell, with two values for the D/pz ratio: 3 and 16. A higher value of D/pz was observed to lead not only to a better geometric accuracy but also to a reduction of thinning on the parts obtained.

Ambrogio et al. [94] analyzed the influence of the punch speed on the behavior of two materials, Grade 2 and Ti-6Al-4V, during SPIF. Speeds between 5 and 500 mm/min were used for Grade 2 and between 6 and 600 mm/min for the titanium alloy. The vertical pitch used varied for both types of materials from 0.1 to 1 mm. The authors showed that it is possible to increase the speed of the punch and implicitly reduce the manufacturing time (the SPIF process being known to be time consuming) without producing a significant degradation of the surface quality. With the increase of punch speed on the Ti-6Al-4 V alloy, a 2% increase in micro-hardness is observed. However, the authors of the study concluded that using a small vertical step leads to long manufacturing times but also to an increase in temperature due to repeated friction between the punch and the blank.

Sbayti et al. [112] propose an algorithm for optimizing the SPIF process factors (punch diameter, vertical step and friction coefficient) to obtain a denture plate with the purpose of increasing its geometric accuracy. Their research is based on simulation using the finite element method and the material used to make the denture plate is Ti Grade 1. The algorithm presented by the study authors starts with the CAD configuration of the geometric model of the denture plate and continues with toolpath generation, numerical simulation using the finite element method, error measurement and the optimization of the process factors, to reduce the geometric errors of the part. The response surface methodology was used as an optimization method, with two optimization techniques having been applied: the single objective optimization technique and the multiple-objective optimization technique. The Box–Behnken strategy was used to develop the design of experiments. In the optimization process for the process factors, all three major groups of geometric errors that occur in parts manufactured by SPIF were taken into account, namely: the springback error, the pillow effect error and the bending error (Figure 16).

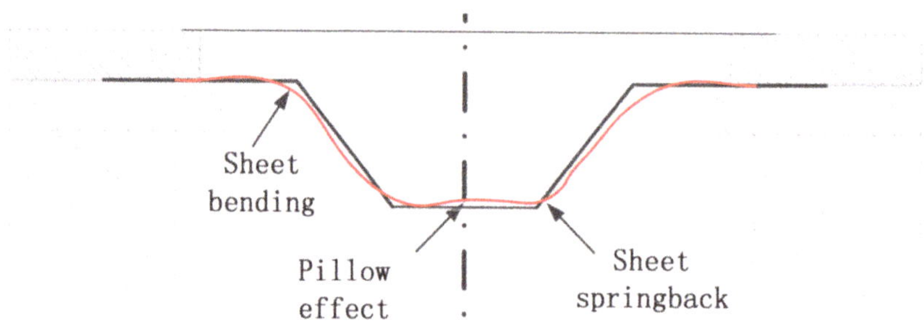

Figure 16. Geometric errors during the SPIF process (reproduced with permission from [113]; copyright © 2007 Elsevier B.V. All rights reserved).

The mathematical expressions for the three types of errors were determined in the form of second-degree polynomials using a variance analysis (ANOVA) [112]. Main effect plots for the punch diameter, vertical step and friction coefficient were also graphically presented. A genetic algorithm (GA) was used for the single objective optimization technique, which allows for error minimization for one of each type of the above-mentioned errors (springback error, pillow effect error and bending error). For the multiple-objective optimization technique, which allows for the simultaneous minimization of the three geometric errors being considered, a combination of GA, global optimum determination by linking and interchanging kindred evaluators algorithm (GODLIKE), and grasshopper optimization algorithm (GOA) was used [113]. The numerical simulations were subsequently validated by experimental tests obtaining the optimal values for the three process factors.

The effect of the tool rotation direction on the formability, surface quality and geometric accuracy of parts made of pure titanium sheets was analyzed by Uheida et al. [96]. The study, based on experimental research, was performed on axisymmetric parts with a variable angle (from 30° to 75°), using a spiral trajectory and a 10 mm diameter punch. Both a "conventional rotation" tool direction, in which the spindle rotation and the feed direction are coincident (both in a clockwise direction), and a "climbing rotation", in which the spindle rotation remains in a clockwise direction while the feed direction is changed to a counter-clockwise direction, were used. To evaluate the accuracy of the parts obtained, 2D contours of the inner surface were measured at different heights with the help of a coordinate measuring machine, which were subsequently used for the 3D reconstruction of the geometric model. The cross-sectional profiles obtained for the two variants are presented in Figure 17. Once the model was reconstructed, it was compared with the CAD model in order to determine the springback errors. It has been found that both types of tool rotation direction lead to parts with a lower height than desired and that a correction of the trajectory in the sense of an "overform" is necessary to reduce this error. However, the shape deviation that appears in proximity to the cone-opening flanges, also called bending error at SPIF, is instead influenced by the tool rotation direction with the smallest errors found to occur in parts manufactured using the "conventional" tool rotation direction.

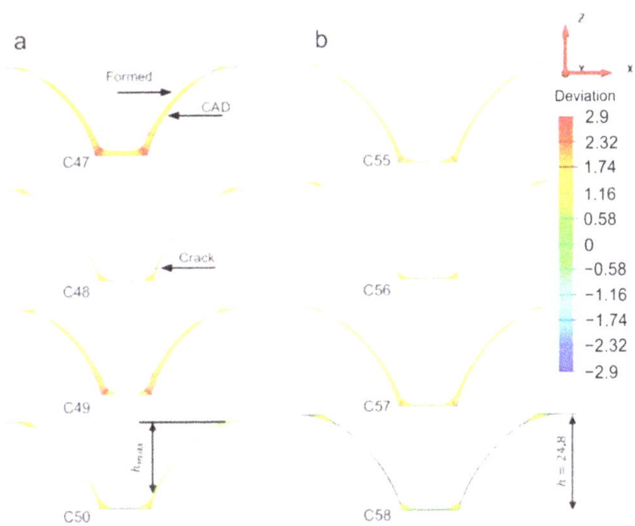

Figure 17. Deviation of the cross-sectional profiles with (**a**) climbing rotation and (**b**) conventional rotation at different heights in the part (reproduced with permission from [96]; copyright © 2018, The Author(s); this is an open-access article distributed under the terms of the Creative Commons CC BY license, which permits unrestricted use, distribution and reproduction in any medium, provided the original work is properly cited).

A combined study—numerical, using the finite element method, and experimental, to improve the accuracy of titanium parts manufactured by electric hot incremental forming—was presented by Fan and Gao [93]. One of the most-used titanium alloys, Ti-6Al-4V, an alloy with very low elongation and high strength that is processed by forming only at high temperatures, was used in this study. Due to the pronounced elasticity of this alloy, the springback errors are higher than with SPIF manufacturing of other types of materials [114]. The first stage consisted of simulation during incremental forming at elevated temperature for a pyramid-type frustum part with a wall angle of 50°, using the MSC-Marc program for the numerical simulation. The electric hot incremental forming process is particular in that the temperature is not uniform on the whole part but produces local heating in the area immediately close to the punch. This feature was introduced in the simulation as the heat transfer rate and was determined following the analysis of the equivalent thermal strain, equivalent plastic strain, and stresses. The authors of the study concluded that, unlike SPIF manufacturing of other metals or alloys in which the predominant deformation was the shear deformation, electric hot incremental forming presents three types of deformation: bending deformation in the first stage, shear deformation in the middle stage and reverse bending in the last stage. Another conclusion drawn by the authors was that the equivalent plastic strain only appears starting with the fourth vertical step due to the low temperature and then increases rapidly. This is the reason the geometric accuracy of the part is influenced by the first trajectories and the reverse bending is influenced by the plastic strain. In order to reduce the reverse bending, the authors propose the use of two point incremental forming together with electric hot incremental forming.

Valoppi et al. [115] conducted a comparative study between electrically-assisted cumulative double-sided incremental forming (E-ADSIF) and electrically-assisted mixed double-sided incremental forming (E-MDSIF), in forming Ti-6Al-4V titanium alloy. The experimental layout used is shown in Figure 18.

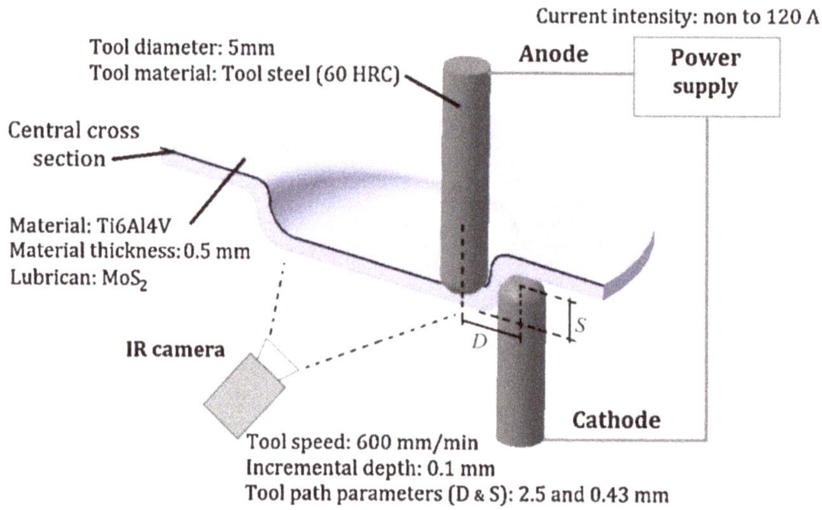

Figure 18. The experimental layout used to study the differences between electrically-assisted cumulative double-sided incremental forming (E-ADSIF) and electrically-assisted mixed double-sided incremental forming (E-MDSIF) (reproduced with permission from [115]; copyright © 2016 College International pour la Recherche en Productique).

A double curvature part was the geometry chosen for the study, and the vertical step used was of 0.1 mm. The main purpose of the paper [115] was to study the influence of electricity on the formability of titanium alloy, but the geometric accuracy and roughness of the parts obtained was also analyzed. For E-ADSIF, the authors used different intensities of electric current, ranging from 40 A to 120 A, and for the E-MDSIF they used the optimal intensity obtained in the case of E-ADSIF. In order to estimate the geometric accuracy, the parts were measured with a non-contact laser instrument and were compared with the ideal geometry. Thus, for E-ADSIF the best results in terms of accuracy were obtained at a current intensity of 50 A. However, at the same current intensity of 50 A, but instead using E-MDSIF, the accuracy is significantly improved.

Another paper that analyzes the behavior of the Ti-6Al-4V alloy during hot single-point incremental forming is presented by Saidi et al. [116]. The novelty factor of this study is that the authors are replacing the expensive laser heating systems that perform local heating on the forming area with a cheaper alternative based on heat cartridges. This variant allows for the heating of the entire surface of the blanksheet, leading to increased formability. In parallel with experimental research, numerical simulations were also performed using the finite element method, with the authors recording small differences (of up to 2%) between the experimental and numerical results. In order to highlight the geometric accuracy of the cone-shaped parts when a frustum wall angle of 55° is obtained, the part profiles were measured and compared with the ideal profiles. The areas with the largest geometric deviations were the bending zone (the connection area between the flange and the wall element); the wall area, where the springback effect occurs; and the bottom of the part, where the pillow effect occurs. The maximum deviation value appeared in the bending zone, and the maximum springback value appeared in the area where the maximum thickness reduction was also recorded.

Behera and Ou [117] analyzed the influence of stress-relieving heat treatment on the geometric accuracy and surface topography of titanium parts manufactured by incremental forming. In the case of pure titanium and titanium alloys, residual stresses appeared and remained during forming operations, unfavorably affecting the subsequent mechanical behavior of the part. Because of this, it is essential to remove these residual stresses by applying the stress-relieving heat treatment. The aforementioned study used titanium Grade 1 with a 0.5 mm thickness. Following SPIF manufacturing and the application of heat treatment (before or after the forming process), the parts obtained were subjected to X-ray diffraction in order to highlight the effect of the stress-relieving heat treatment, as well as being subjected to an analysis using a 3D interferometer in order to study the surface topography. The authors observed the appearance on the surface of the part of an "orange peel effect", which is characteristic of titanium, but also a worsening of the roughness by 0.5 μm on the free surface of the part (the surface that does not come into contact with the punch). The geometric accuracy of the parts was analyzed both before and after the trimming process, which is necessary in order to extract the useful area of the part. The conclusions of the study show that applying the stress-relieving heat treatment before SPIF leads to both an improvement of the surface roughness, as well as to an increase in the geometric accuracy of the parts obtained. Additionally, the value of the residual stresses is lower in the case of the parts that were subjected to stress-relieving heat treatment than in the case of those that have not undergone this treatment at all.

Han et al. [73] developed a digital manufacturing technology for a skull prosthesis based on the fact that creating medical implants and a prosthesis requires an individualized manufacturing model, with each model having its own geometry. Thus, starting with the images from Computed Tomography (CT) and using software for the reconstruction of CAD models based on Non-Uniform Rational B-Splines (NURBS), the authors obtained the geometry of the skull prosthesis. Based on this geometry, the authors went on to generate the tool path using industrial software, followed by the optimization of the punch trajectory in order to reduce errors and springback.

Another paper that studies the technological variants of manufacturing of cranial implants by SPIF on the basis of an analytic hierarchy process (AHP) is that of Racz et al. [118]. For the analytic hierarchy process, they chose four types of manufacturing technologies, including three related to incremental forming, namely, conventional SPIF, hot SPIF and DSIF. Seven criteria were chosen for evaluation using the AHP: formability, microstructure, degree of control, roughness, energy consumption, accuracy and production time. They used three types of tool path for the SPIF manufacturing of implants: two circular, one of which had vertical feed points on the same generatrix, the second of which had other vertical feed points indexed at different angles, and a third spiral trajectory. A profilometer was used to analyze the geometric accuracy of the implants obtained, and the main strains and thickness reduction were determined by means of an optical system. Following this analysis, the authors found the spiral trajectory to be the optimal option for the manufacturing of complex parts such as cranial implants.

The dimensional accuracy of implants made of medical grade titanium and manufactured by SPIF was analyzed by Behera et al. [119]. They also proposed a method to improve the geometric accuracy by compensating the tool path. The authors considered the division of the geometric model into features of great importance, especially in the case of complex shaped parts with double curvature, such as cranial implants. The algorithm they presented contains, in summary, the following steps: the uncompensated toolpath is created starting with the CAD model of the part; the part is manufactured with this tool path and is then measured by means of an optical measurement system; the measurement leads to obtaining a point cloud, which is subsequently transformed into a mesh, and this mesh is compared to the initial CAD model, thus obtaining the deviations on the entire surface of the part. The accuracy file, an STL file, is then obtained. A software product created by the authors of the study identifies the features in this file and then links this file to a set of files called the training set CAD model. The linked data set is then exported to a data file for generating an accuracy response surface using multivariate adaptive regression splines (MARS). The compensated tool path is then obtained. In the study cited, the authors applied the abovementioned algorithm for both an ellipsoid part and a cranial plate. The results obtained for the cranial plate are presented in Figure 19. Using this algorithm led to obtaining a mean deviation of ±0.5 mm, a significant improvement compared to that of the cranial plate manufactured with the uncompensated toolpath.

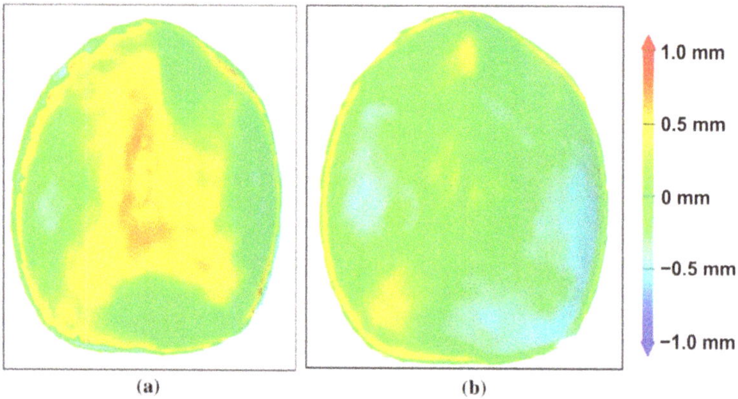

Figure 19. The accuracy plot of a cranial implant manufactured by SPIF with (**a**) an uncompensated toolpath and (**b**) a compensated toolpath (reproduced with permission from [119]; copyright © 2015, Springer-Verlag London; this is an open-access article distributed under the terms of the Creative Commons CC BY license, which permits unrestricted use, distribution and reproduction in any medium, provided the original work is properly cited).

A very interesting study, also based on the finite element method, was developed by Fiorentino et al. [120]. The study uses iterative learning control to increase the accuracy of Ti Grade 2 alloy parts manufactured by SPIF. Iterative learning control uses the following algorithm: the punch trajectory is determined, the part is manufactured, it is measured and the errors are determined. Once these errors are known, the punch trajectory is corrected in order to reduce the geometric error. This algorithm is then cyclically repeated until convergence, which is represented by obtaining a minimum imposed geometric error. The algorithm used by the iterative learning control is presented in Figure 20. The authors focused on parts with planar and curvilinear geometries with wall inclinations of (0°, 29° and 38°) and a depth of 30 mm. They also conducted a comparative study between three types of materials: aluminum Al 1050A, steel DC04 and titanium Grade 2 alloy. The iterative learning control algorithm was applied to the finite element method, and the validation of the results was done by experimental research on a CNC machine. Of all the three types of materials, the largest geometric errors (±1.1 mm) were obtained for titanium Grade 2 due to the springback phenomenon, which is most pronounced in the case of titanium alloys.

Improving the accuracy of industrial parts manufactured by SPIF is desirable, but improving the accuracy of medical implants is an extremely important goal. Thus, Duflou et al. [66] published some of the first concerns related to increasing the geometric accuracy of a frontal orbit cranial implant made of Grade 2. They additionally presented a comparative study between the conventional implant technologies and the single-point incremental forming process, also naming the advantages and disadvantages of the different types of technologies. The steps required to manufacture an implant by SPIF are: carrying out a CT scan of the injured skull, preparing a clay model of the skull, building a CAD file of the implant based on reverse engineering and producing the final 3D CAD model of the implant integrated into the blank sheet. The strategy used to make the implant is a multi-step tool path that allows for the formability of the material to be increased by redistributing the material. To evaluate the geometric accuracy of the orbit cranial implant, they scanned the entire implant following the SPIF processing, noting that there are some overformed areas, such as the backing plate, as well as underformed areas. For this reason, to compensate for the trajectory, an uncompensated trajectory was chosen for a model made of AA 1050 and was translated proportionally with the deviation obtained and multiplied by a scale factor equal to 0.7. By using this strategy, the geometric accuracy increased significantly, with the average deviation decreasing from 1.19 to 0.377 mm.

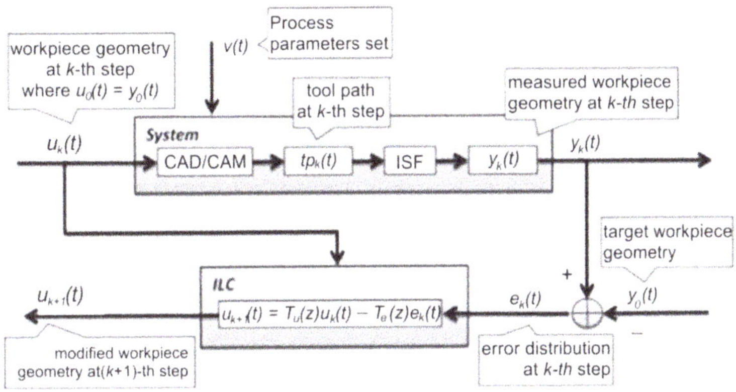

Figure 20. The algorithm used for increasing part accuracy: $u_k(t)$—part geometry, $u_{k+1}(t)$—part geometry (corrected), $v_k(t)$—process parameters set, $tp_k(t)$—toolpath, $y_k(t)$ —obtained geometry, $y_d(t)$—target geometry, $e_k(t)$—error map and $T_e(z)$—correction weight (reprinted with permission from [120]; copyright © 2016, Springer-Verlag France).

Li and Mo [121] studied springback in double curvature parts made of Grade 4 titanium alloy and manufactured by SPIF, based on a numerical simulation using the finite element method. As an example, the authors chose to use SPIF to manufacture a part of an aircraft skin. Using the geometry of the part, the authors inserted an addition that will later be removed by trimming in order to reduce springback. To estimate the springback, the authors presented the nodal displacement in three different sections of the part (two at the ends of the part and one in the middle), following the forming process before and after the appearance of springback. The conclusion of the study is that the plastic deformation value in the case of titanium parts is low, while the springback has particularly high values, especially in the middle area of the part and slightly lower towards the ends of the part. Additionally, the vertical feed values must be as low as possible in order to reduce the springback value and increase the accuracy.

Najafabady and Ghaei [102] conducted a study on the geometric accuracy, surface quality and hardness of titanium alloy parts produced by SPIF. There are other studies related to hot SPIF of titanium alloy, but the novelty brought by the authors is provided by the fact that the local heating due to the electrical contact is replaced by an electric heater. The main advantage is the use of alternating current electricity instead of direct current, with the alternating current allowing the use of cheaper and safer equipment for the operator. The authors studied the main areas where, due to the elasticity of the titanium alloy, the maximum deviations occur. A deviation analysis for a truncated cone geometry is presented in Figure 21. In this type of manufacturing, with the use of an electric heater, the maximum deviation occurs at the connection area between the flanges and the vertical walls, i.e., exactly in the starting area of the trajectory, due to the fact that in this area the blanksheet fails to reach the optimum temperature. After several vertical steps, this inconvenience disappears and the geometric accuracy increases.

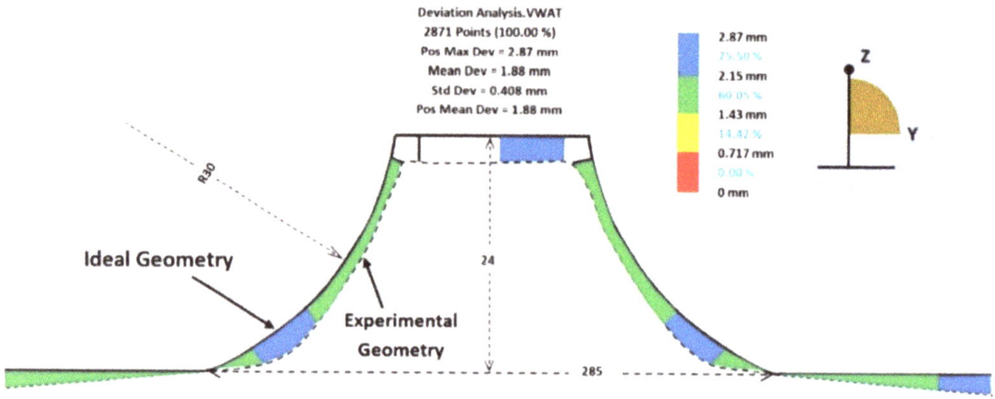

Figure 21. Deviation analysis for a truncated cone geometry (reprinted with permission from [102]; copyright © 2016, Springer-Verlag London).

Another paper refers to SPIF manufacturing accompanied by heating of the blanksheet of the Ti-6Al-4V titanium alloy [105]. Unlike other studies in which heating only occurs locally, near the forming area, the authors of this paper heated the whole blanksheet and kept the temperature constant throughout the entire punch movement. The main objective of the study was to improve the geometric accuracy by means of two approaches: using a trajectory that is subjected to a correction process in order to eliminate springback errors and reducing the deflection of the sheet along the perimeter part by adding an addendum to the shape of the part (in fact a connection between the tilted walls of the part and the undeformed, flat area). The authors studied the influence of part temperature and the trajectory on geometric accuracy. When considering the influence of temperature, it was found that, even if the shape deviation follows the same pattern, its value has a different influence on different areas of the part. Increasing the temperature leads to a large variation in wall angle, while decreasing the temperature leads to the reduction of deviations in the flat/low curvature areas. The trajectory correction was achieved by applying an intelligent process model that has the ability to predict the springback value and apply it to the initial CAD model, which thus modifies the new trajectory being obtained on this corrected geometric model. The use of this algorithm led to significant reduction of springback in the titanium part, as we can observe in Figure 22.

Grimm et al. [122] aimed to reduce springback during incremental forming of titanium alloys by electrically assisting this deformation process. They identified the residual stresses that remain inside the material after the forming process as the main cause of springback in the case of material-forming processes in general and of SPIF in particular. The authors also classified the factors influencing springback, namely, part geometry, part material and part thickness. Having taken all these into consideration, the authors centered their study on truncated pyramid shaped parts made of thick AMS-T-9046 titanium sheets with a thickness of 0.508 mm. Three test parts were initially manufactured, without the use of electricity, and these were then measured and compared with the ideal CAD models for the determination of springback. The same methodology was used for the manufacturing of parts in the presence of electricity, with several paths for applying electricity. The general conclusion was that the use of electrically assisted SPIF led to a significant reduction in springback in the case of AMS-T-9046 titanium sheets but also that the path the electricity takes in each part has an important influence on the reduction of springback.

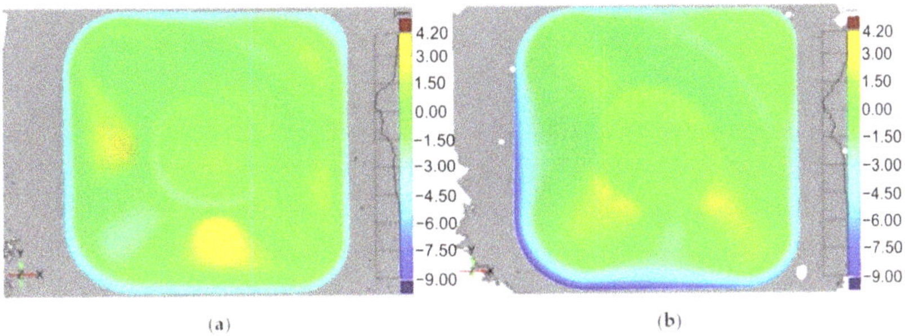

Figure 22. Accuracy analysis for (**a**) an uncompensated trajectory and (**b**) for a trajectory compensated with the intelligent modelling (reproduced with permission from [105]; copyright © 2015, The Authors; this is an open-access article distributed under the terms of the Creative Commons CC BY license, which permits unrestricted use, distribution and reproduction in any medium, provided the original work is properly cited).

7. Surface Quality

7.1. Forming Strategy

Large-scale waviness (Figure 23) created by the tool path is considered a weak point for ISF. The problem of poor surface quality of the SPIFed parts due to large scale waviness created by the tool path can be overcome in SPIF by optimization of the forming strategy. Waved impression roughness is the major surface quality problem of SPIF since the roughness value of the waved impression is larger than with the friction trace [123,124].

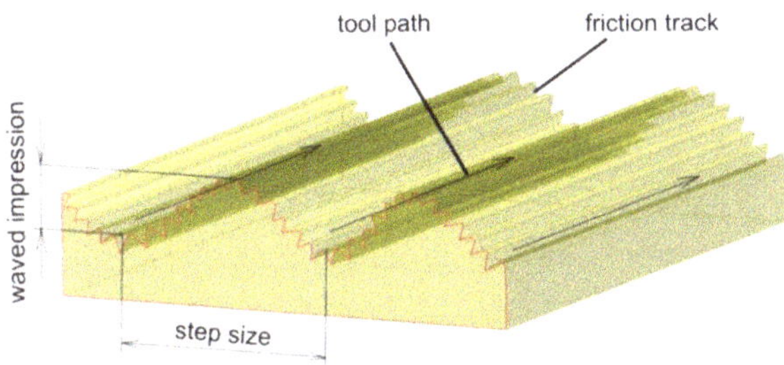

Figure 23. Surface of the drawpiece across the tool-path.

Recently, many studies have focused on applying different strategies to improve the geometric accuracy of finished parts made by SPIF, i.e., neural network strategies [125,126], iterative algorithms methodology [127,128] and the transfer functions approach [129].

Ortiz et al. [105] proposed two approaches to improving the geometric accuracy of typical aerospace Ti-6Al-4V parts produced by hot SPIF. In the first approach, it was proposed to skip the overforming deviations associated with the deflection of the sheet along the perimeter of the part based on a design improvement. In the second approach, an IPM was adopted that counteracts deviations associated with the elastic deformations of the component after springback. The IPM approach leads to an accuracy improvement of the tetragonal pyramid of up to 49%. Veeraajay [130] experimentally studied the effect of the feed rate, incremental step depth, spindle speed and tool feed on the surface roughness of Ti-6Al-4V truncated cones. Response surface methodology (RSM) was used to design the experiments, and ANOVA was performed to find the factor that significantly affects the surface finish of drawpieces. It was found that with an increasing feed rate, the surface roughness decreases from 1.05 mm to 0.9 mm. A better surface finish is achieved by decreasing the step depth.

Szpunar et al. [131] applied a central composite design to determine the mathematical model of the effect of tool feed rate, spindle speed and step size on the surface roughness of truncated cones of Grade 2 commercially pure titanium. It was found that samples formed with high values of spindle speed showed poor surface qualities. They also found that the direction of tool rotation in relation to the feed direction is one of the crucial parameters influencing the risk of cracking (Figure 24) [131]. Step down showed the strongest influence on surface roughness, which is concordant with the results of the Box–Behnken design analyzed by Yao et al. [127].

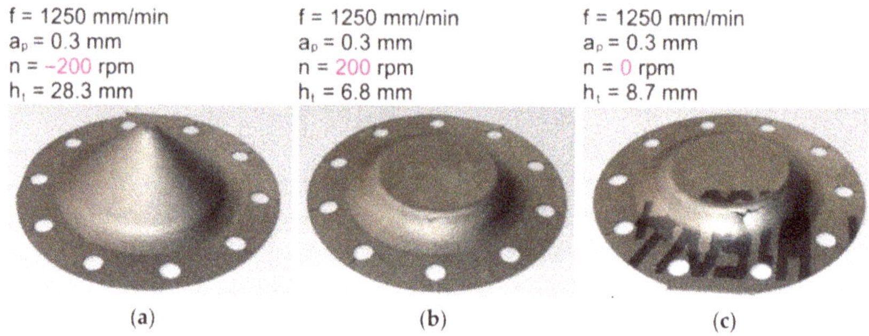

Figure 24. Grade 2 drawpieces formed at a constant step size of 0.3 mm, tool feed rate of 1250 mm/min and rotational speeds: (**a**) −200 rpm (clockwise direction), (**b**) 200 rpm (anticlockwise direction) and (**c**) freely rotatable tool).

A decision-making method based upon an analytic hierarchy process was applied by Racz et al. [118] to optimize the SPIF parameters. The surface roughness of the SPIFed Ti-6Al-4V cranioplasty plates was directly influenced by the vertical step, a decrease in the vertical step leading to a decrease in the roughness value [118].

The radius of the tool tip is one of the most important parameters that influences the surface finish of Grade 1 and Ti-6Al-4V sheets [132]. The higher the feed rate is, the poorer the surface quality of the workpiece is (Figure 25). Mohanraj and Elangovan [133] found that the negative incremental forming method with a sliding motion of the tool provides a better surface finish of Ti-6Al-4V. Ambrogio et al. [94] investigated the effect of feed rate with respect to the surface quality of formed Grade 2 and Ti-6Al-4V sheets in order to develop a high-speed SPIF machine. Preliminary analyses show that the impact of high-speed forming on the forming accuracy and surface roughness was not significant.

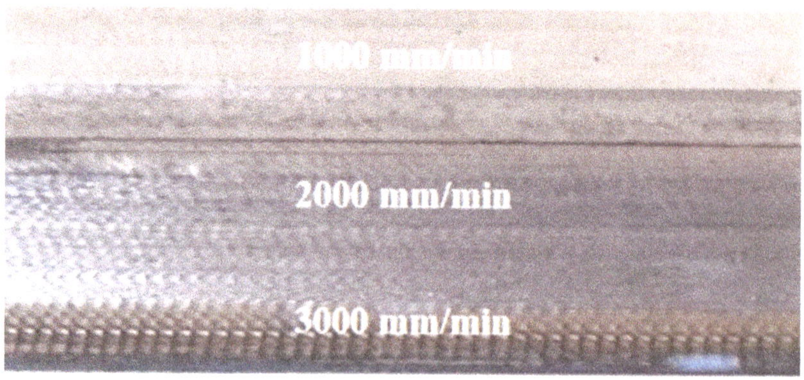

Figure 25. Images showing the effect of the feed rate on the surface quality of the sheet (reproduced with permission from [132]; copyright © 2019 The Author(s). Published by Elsevier B.V.).

7.2. Friction Conditions

Titanium alloys have low wear resistance, high coefficient of friction (COF), a tendency to galling and relatively low hardness [134,135]. Therefore, the improvement of friction conditions is a key factor for surface finish [136,137]. The tribological properties of titanium and its alloys, in particular the tendency to seize during cold and hot forming operations, are intensified by high contact loads. Another significant problem occurring during the forming of titanium is the formation of sticking on the surface of tools (galling), which significantly affects the quality of manufactured products [138–140]. The unfavorable phenomena accompanying the SPIF and TPIF of titanium alloys can be limited by appropriate surface treatment of the tools and the use of appropriately selected technological lubricants [141,142]. Due to the fact that most of the lubricants used in sheet metal forming (SMF) processes belong to the group of synthetic oils, in recent years research has been conducted on replacing mineral oils with biodegradable vegetable oils [143,144], including those with the addition of nanopowders [145–147]. Trzepieciński et al. [144] tested the tribological properties of Ti-6Al-4V titanium alloy sheets in a strip drawing test using five kinds of vegetable oils: palm oil, rapeseed oil, olive oil, sunflower oil and soybean oil. It was found that increasing the load at a constant value of kinematic viscosity of lubricant reduces the value of the coefficient of friction (COF). The lowest value of COF was provided by oil with a low density and at the same time high kinematic viscosity.

Appropriate lubrication is essential to prevent the formation of an oxide layer on the inner surface of the part which promotes sheet fracture and tool wear [148,149]. Due to the high temperature of the sheet in hot SPIF, liquid lubricants cannot be used and therefore solid lubricants should be used instead [92,150].

A poor friction condition could lead to severe scratch, which can damage the implant surface. Moreover, heavy-duty contact conditions can lead to a temperature increase in the contact area. In particular, for titanium and titanium alloy, the adhesion phenomenon is more likely to happen under improper contact conditions [136]. Changing the contact conditions and improving the lubrication are currently the two main approaches adopted: modification of the tool and selection of lubricants with appropriate properties adapted to the processing conditions. Table 1 shows the most important lubricants used during SPIF or TPIF of titanium and Ti-based alloys. It was found that the molybdenum disulphide MoS_2 grease is the most popular lubricant when forming this group of materials.

Table 1. Typical lubricants used in SPIF of selected materials.

Lubricant	Forming Conditions	Material	Shape of Drawpiece	Reference
OKS 280 lubricant	elevated temperature	Ti-6Al-4V	car door shell	[109]
MOS_2	elevated temperature	TiAl2Mn1.5	tetragonal pyramid	[92]
MoS_2	elevated temperature	Ti-6Al-4V	conical frustum/tetragonal pyramid	[83]
graphite	elevated temperature	Ti-6Al-4V	conical frustum/tetragonal pyramid	[83]
boron nitride spray	elevated temperature	Ti-6Al-4V	conical frustum/tetragonal pyramid	[83]
MoS_2	elevated temperature	Ti-6Al-4V	conical frustum	[77]
MoS_2, graphite powder, graphite-based and copper-based anti-seize lubricants	elevated temperature	Ti-6Al-4V	conical frustum	[150]
MoS_2	elevated temperature	Ti-6Al-4V	tetragonal pyramid	[100]
MoS_2	elevated temperature	Ti-6Al-4V	conical frustum	[86,93,102]
ROCOL copper-based anti-seize compound	elevated temperature	Ti-6Al-4V	conical frustum	[101]
boron nitride	elevated temperature	Ti-6Al-4V	tetragonal pyramid	[104,105]
OKS 280 lubricant	elevated temperature	Ti-6Al-4V	flat components with grooves	[106]
OKS 280 lubricant	elevated temperature	Ti-6Al-4V	truncated cone	[107]
MoS_2	elevated temperature	Ti-6Al-4V	truncated cone	[151]
carbon-based, dry-film lubricant (Berulit 935)	elevated temperature	Ti-6Al-4V	asymmetric component	[87]
dry graphite	elevated temperature	Ti-6Al-4V	flat component with longitudinal pockets	[89]

Table 1. *Cont.*

Lubricant	Forming Conditions	Material	Shape of Drawpiece	Reference
MoS_2	elevated temperature	Grade 2 Ti-6Al-4V	conical frustum	[94–96]
Rocol RTD compound	cold-forming conditions	Grade 1	cranial plate	[68]
MoS_2	cold-forming conditions	pure titanium (grade not specified)	conical frustum	[152]
chlorine-containing forming oil	cold-forming conditions	Grade 1	denture plate	[112]
Nuto 46 hydraulic oil	cold-forming conditions	Grade 2	clavicle implant	[74]
ceramic grease WEICON ASW 040P	cold-forming conditions	Ti Grade 2	facial implant	[71]
MoS_2	elevated temperature	Ti Grade 2	conical frustum	[153]
MoS_2	cold-forming conditions	Pure titanium (grade not specified)	conical frustum	[154]

Vahdani et al. [150] examined MoS_2, graphite powder, and graphite-based and copper-based anti-seize lubricants. They experimentally investigated the effects of four different lubricants together with the influence of vertical pitch, feed rate and electric current on the surface finish of Ti-6Al-4V sheet in resistance SPIF using a forming tool with a cooling system that they had developed (Figure 26). It was found that the lubricant type has the most effect on the formability followed by the electric current. The best surface finish is obtained when using a graphite-based compound, owing to the more uniform distribution of the graphite particles within the grease.

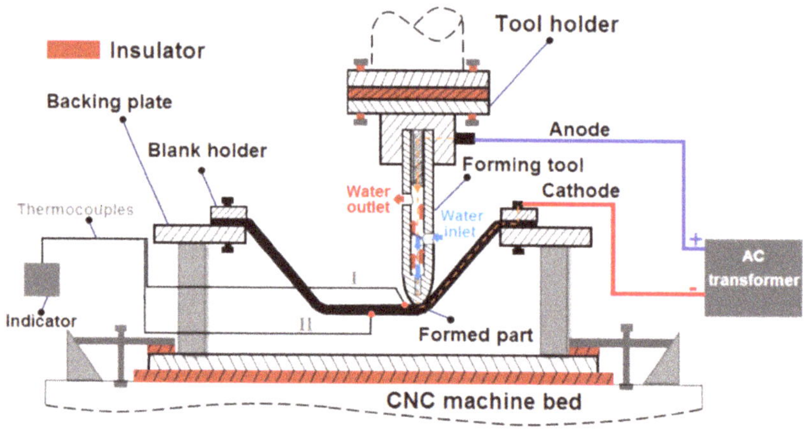

Figure 26. Schematic diagram of resistance SPIF (reprinted with permission from [150]; copyright © 2019, The Indian Institute of Metals—IIM).

Graphite is a good high-temperature lubricant, but when used in EHIF, electric sparks appeared between the tool and sheet metal, which may cause the creation of micro-pits on the surface of the sheet metal and tool [83].

MoS_2 is often dispersed in metal matrices like copper and nickel to achieve enhanced strength of the lubricant for good wear resistance. Fan et al. [83] successfully formed Ti-6Al-4V alloy using EHIF at 500–600 °C with a small amount of oxidation. They confirmed that a self-lubricating Ni disulphide metal matrix composite is suitable for electric hot incremental forming at high temperature.

Hussain et al. [154] investigated the effect of forming parameters on the contact conditions in SPIF of conical drawpieces made from pure titanium. The specimen that was formed by using the MoS_2 powder and grease showed the best surface quality (no striations on the specimen surface or no metal peeling), and no metal sticking to the tool tip was observed. When the blank was formed without using any lubricant, the blank material adhered to the tool tip and the specimen surface peeled off (Figure 27).

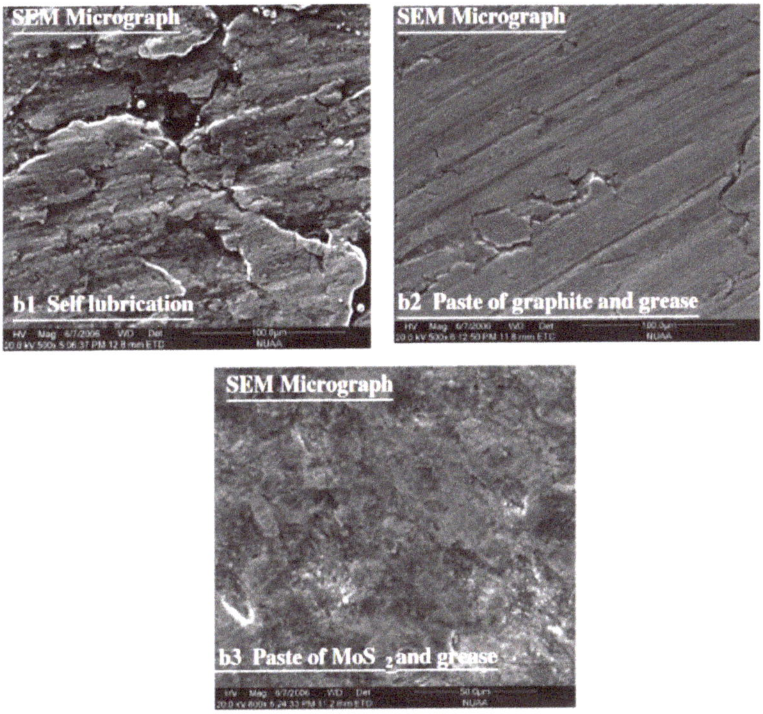

Figure 27. The effect of friction conditions on the surface quality of truncate cone drawpieces formed from pure titanium (reprinted with permission from [154]; copyright © 2007, Springer-Verlag London Limited).

8. Numerical Modeling

Supporting the manufacturing processes with finite element-based programs, although quite an expensive process, may contribute to increasing the efficiency of implementing new products in the market. Both large and small companies are most interested in the possibilities for numerical modeling and simulation of manufacturing processes. The virtual models allow one to modernize existing production systems or to implement the manufacturing process without the need to manufacture real tools. Finite element simulations are most efficient for the prediction of elastic-plastic deformations of metallic sheets. Finite element method can be used to predict forming defects, clarify the forming characteristics and improve the forming process. However, due to complicated contact problems involved in the SPIF, efficient approaches for modeling the ISF are still lacking [155].

Many scholars have conducted significant research into the numerical analysis of incremental sheet forming of titanium and Ti-based alloys. Mohanraj and Elangovan [133] performed experimental work and numerical analysis of Ti-6Al-4V incremental sheet forming, considering the spindle speed, tool diameter, feed rate and step size, to study the geometric accuracy and thinning of an aerospace component with complex shape. Simulation results were found to be in agreement with the experimental results, thus showing the applicability of the process in eco-friendly products in low-volume production.

Palumbo et al. [111] conducted FE-based simulations of the SPIF process of non-axisymmetric truncated pyramids with the aim of investigating the effect of both draw angle and tool/step size ratio, taking into account the anisotropic behavior of sheets of

Ti-6Al-4V titanium alloy. The analysis of thinning maps and shape errors highlighted that the tool/step size parameter plays a key role in SPIF. The numerical analysis conducted in ABAQUS/Explicit program revealed that large values of the draw angle determined dangerous thinning. Sbayti et al. [156] analyzed hot incremental forming of titanium alloy Ti-6Al-4V denture base by finite element-based simulations. The effects of the forming temperature on the material failure and more particularly on the geometric accuracy of the final product were investigated. It was found that numerical modeling of ISF of denture base by using Ti-6Al-4Vsheets is a feasible solution and it clearly shows the potential for real medical application. Naranjo et al. [157] presented results of the SPIF of Ti-6Al-4V conical drawpieces at different temperatures. Ansys Workbench was used as the numerical program. Based on the recent paper [158], two mesh sizes of hexahedral-type elements were considered for FEM modeling in order to optimize the resolution of the procedure and the computing time. Modeling the Ti-6Al-4V material allowed one to reach a high level of agreement among the forces obtained experimentally and by simulation. It was found that simulation results depend highly on the mesh element size, and a low size must be selected. Moreover, mesh in the corners of squared pyramid makes the simulation process more difficult. Abdeljefi et al. [159] proposed a numerical approach of the SPIF process of titanium T40 sheet by using a simple elastic-plastic material model in Abaqus/implicit© software. The effect of the proposed model on the forming forces and the final geometry was investigated. Saidi et al. [116] investigated the warm incremental process of Ti-6Al-4V sheet based on the use of heat cartridges. The objective was to demonstrate that our low-cost heating system can be used in forming a limit angle similar to that obtained with expensive laser heating. The numerical results were in good agreement with the experimental ones. Piccininni et al. [160] conducted numerical simulations with Tsay–Hill formulation and the strain rate dependent isotropic plasticity law in order to design the SPIF processes of Grade 2 titanium sheets. Finite element models created for simulating both forming processes were revealed to be robust and accurate since a good correlation between numerical and experimental thickness distributions could be obtained. Naranjo et al. [161] tested different plastic behavior hypotheses using a Ti6Al4V alloy to compare the results obtained by FEM with the Ansys program to the experimental tests run to measure the forces applied to the tool. For low strains, a high degree of agreement between the experimentally measured forces and those obtained by simulation was achieved. However, for large strains, most models overpredicted forces obtained experimentally due to the cracks that appeared on the bottom of the cones before the failure occurred.

Most of the studies that refer to FEM for SPIF processes have been traditionally carried out using software with explicit-type solvers like Abaqus [71,162] or LS-Dyna [163,164], or even with software developed especially for this purpose, as in Lagamine [165]. Numerical modeling is used to predict temperature distribution, sheet thickness change, springback, material flow, stress and strain distribution and predictions of forming limit diagrams. FE-based numerical simulation has become an indispensable tool for the prediction of wrinkling and crack occurrence during SPIF. The use of simulation programs requires a thorough knowledge of material behavior and FE programming to develop a suitable numerical analysis for a forming process. It is undeniable that the FE simulations are a huge help in reducing the waste of time and material toward a correct and proper process design. In the future, most numerical research work must be focused on the development of a macroscopic constitutive model based on a physical mechanism, grain-twinning interactions and a crystal plasticity model considering the grain–grain interaction [34].

9. Conclusions

Single-point incremental forming plays a key role in transferring state-of-the-art knowledge on manufacturing into innovative solutions of low-volume customized biomedical implants and components in the aerospace industry. This comprehensive review of ISF methods used on titanium and titanium alloys allows the following conclusions to be drawn.

The use of SPIF for biomedical applications is basically limited to the production of implants tailored to individual human needs, mainly dental prosthetics and reconstructive medicine. During the production of custom implants, the SPIF and TPIF methods are limited due to the low dimensional accuracy associated with the residual stresses remaining in the formed material. This research area with regard to titanium and its alloys has not been sufficiently explored.

The geometric accuracy attributed to springback effects and poor surface quality are one of the dominant limits in the implementation of SPIF. Correction of the tool trajectory and heat treatment of components after the forming process allows one to reduce the elastic deformation of the material after cut-out operations and would reduce the pillowing effect.

In SPIF, adhesive wear of the tool predominates. There are insufficient investigations on the synergistic effect of tool rotation direction and rotational speed on the flattening of the surface asperities and quantitative tool wear.

The selection of an appropriate lubricant is a key aspect of the forming process of titanium and its alloys, which exhibit unfavorable tribological properties such as high adhesion and a tendency to adhesive wear that is caused by microscopic transfer of material between the workpiece and tool surfaces. These hot spots are especially important during the deformation of titanium alloys, which are generally only formed under elevated temperature conditions. In such conditions, liquid lubricants cannot be used, and therefore solid lubricants, primarily MoS_2, are required.

The reduction of forming traces on the surface of implants can be effectively improved by optimizing the tool trajectory and forming strategy, as well as improving the friction conditions. A potential method is to use tools with a textured surface that provide improved lubrication conditions, or tools with a movable spherical tip that freely rolls over the surface of the component being formed.

Research on the phenomenon of friction in conventional SMF processes has shown that biodegradable edible oils can be considered as an alternative to synthetic oils. It seems that this may potentially be a field of research on SPIF carried out in cold-forming conditions that is worth developing.

Due to its crystalline structure, the most commonly used titanium alloy, Ti-6Al-4V, is basically only formed at an elevated temperature. A high temperature, on the one hand, helps to increase the formability of the material but at the same time increases the tendency of the titanium alloy to undergo a severe plastic deformation process (galling). There is insufficient research to determine the balance between SPIF temperature and the quality of the surface finish of the components.

One of the ways to increase the formability of titanium alloys is to treat them at elevated temperatures. Many approaches have been developed for introducing temperature into a workpiece such as electrical heating, laser heating, friction heating or combined electrical and friction heating. The advantage of SPIF with laser heating is lower springback of the formed component. It is easier to control temperature with this method than with friction-assisted SPIF.

The influence of SPIF parameters (i.e., step size, radius of tool tip, feed rate and tool rotational speed) on the surface quality and formability of titanium and its alloys is not identical for materials of different thickness. In many studies based on DoE or Taguchi optimization, sheet thickness is a parameter that is omitted. Future work should focus on exploring such exceptional conditions.

Although titanium and titanium alloys have many desirable characteristics, they have problematic areas as well. These include their affinity to oxygen, hydrogen pickup, susceptibility to certain types of chemical attack, and damage generation arising from the elastic and plastic anisotropy of the hexagonal crystal structure [17]. This makes titanium alloys, especially the most popular Ti-6Al-4V alloy, extremely difficult to deform at ambient temperature. Combinations of working and heat treatment can generate highly preferred crystal orientations, so between the coherent phase boundaries with transformation shears and the anisotropy of the thermal expansion coefficient, substantial residual stresses are common even after a stress relief anneal [3,12,17]. To fully optimize the material property design for particular applications, control of texture and microstructure with SPIF strategies requires predictive understanding of the rules for variant selection during the $\alpha \rightarrow \beta \rightarrow \alpha$ transformations, which are not yet well established [17].

Apart from the knowledge of the microstructure of the material and the related formability, attention should be paid to the strong synergistic effect of forming parameters on the surface finish of the drawpiece and the possibility of obtaining a component with a given degree of deformation. The most critical factors of SPIF process affecting the sheet formability are step size, tool rotational speed, feed rate, tool diameter, sheet thickness and wall angle. Understanding how these parameters influence output quantities (i.e., surface finish, material formability and geometric accuracy) still requires clarity and optimization. First of all, attention should be paid to general observations: (i) the direction of tool rotation in relation to the feed direction is one of the key SPIF parameters influencing the possibility of receiving a titanium drawpiece without the risk of cracking, (ii) samples formed with high values of spindle speed showed poor surface qualities and (iii) the rotation speed is dominant process parameter determining the thermal effect and forming force. A major factor affecting forming forces and quality of the surface finish of drawpieces is the step size. By increasing the spindle speed, a reduction in forming forces was observed [131]. In terms of the surface quality of the formed components, scratches caused by the forming tool can be effectively reduced through the optimization of the toolpath strategy and improvement of lubrication conditions. Proper lubrication is a key factor in enhancing the surface quality at elevated temperatures. When increasing the forming temperature, proper lubrication is a key factor affecting the surface roughness obtained, as well as the oxidation process.

Current findings on the effect of process parameters on output quantities in incremental sheet forming of titanium and Ti-based alloys are limited and sometimes contradictory, which may be due to the various settings of the process parameters and the various geometries of components. Only in terms of the lubrication of the sheet surface, it has been found that MoS_2 is the most effective lubricant for cold and elevated temperature conditions.

Author Contributions: Conceptualization, T.T.; methodology, V.O. and T.T.; investigation, V.O., T.T., M.S., Ł.C., D.F. and I.S.; data curation, V.O., T.T., M.S., Ł.C., D.F. and I.S.; writing—original draft preparation, V.O., T.T., M.S., Ł.C., D.F. and I.S.; writing—review and editing, T.T. All authors have read and agreed to the published version of the manuscript.

Funding: This research received no external funding.

Institutional Review Board Statement: Not applicable.

Informed Consent Statement: Not applicable.

Data Availability Statement: The data presented in this study are available on request from the corresponding author.

Acknowledgments: Project financed by Lucian Blaga University of Sibiu and Hasso Plattner Foundation research grants LBUS-IRG-2021-07.

Conflicts of Interest: The authors declare no conflict of interest.

References

1. Bedinger, G.M. *Titanium—Mining Engineering*; USGS Publications Warehouse: Washington, DC, USA, 2013.
2. Veiga, C.; Davim, J.P.; Loureiro, A. Review on machinability of titanium alloys: The process perspective. *Rev. Adv. Mater. Sci.* **2013**, *34*, 148–164.
3. Boyer, R.R.; Briggs, R.D. The use of β titanium alloys in the aerospace industry. *J. Mater. Eng. Perform.* **2005**, *14*, 681–685. [CrossRef]
4. Wang, Z.; Liu, L.; Zhang, L.; Sheng, J.; Wu, D.; Yuan, M. Effect of Heat Treatment on the Microstructure and Mechanical Properties of High-Strength Ti-6Al-4V-5Fe Alloy. *Mater. Trans.* **2019**, *60*, 269–276. [CrossRef]
5. Brice, D.A.; Samimi, P.; Ghamarian, I.; Kiu, Y.; Brice, R.M.; Reidy, R.F.; Cotton, J.D.; Kaufman, M.J.; Collins, P.C. Oxidation behavior and microstructural decomposition of Ti-6Al-4V and Ti-6Al-4V-1B sheet. *Corros. Sci.* **2016**, *112*, 338–346. [CrossRef]

6. Ma, K.; Zhang, R.; Sun, J.; Liu, C. Oxidation Mechanism of Biomedical Titanium Alloy Surface and Experiment. *Int. J. Corros.* **2020**, *2020*, 1678615. [CrossRef]
7. Bocchetta, P.; Chen, L.-Y.; Tardelli, J.D.C.; Reis, A.C.D.; Almeraya-Calderón, F.; Leo, P. Passive Layers and Corrosion Resistance of Biomedical Ti-6Al-4V and β-Ti Alloys. *Coatings* **2021**, *11*, 487. [CrossRef]
8. Aliofkhazrael, M.; Ali, N.; Chipara, M.; Laidani, N.B.; De Hosson, J.T.M. *Handbook of Modern Coating Technologies. Fabrication Methods and Functional Properties*; Elsevier: Cambridge, MA, USA, 2021.
9. Wang, K.; Kopec, M.; Chang, S.; Qu, B.; Liu, J.; Politis, D.J.; Wang, L.; Liu, G. Enhanced formability and forming efficiency for two-phase titanium alloys by Fast light Alloys Stamping Technology (FAST). *Mater. Des.* **2020**, *194*, 108948. [CrossRef]
10. Elias, C.N.; Meyers, M.A.; Valiev, R.Z.; Monteiro, S.N. Ultrafine grained titanium for biomedical applications: An overview of performance. *J. Mater. Res. Technol.* **2013**, *2*, 340–350. [CrossRef]
11. Shahmir, H.; Langdon, T. An evaluation of the hexagonal close-packed to face-centered cubic phase transformation in a Ti-6Al-4V alloy during high-pressure torsion. *Mater. Sci. Eng. A* **2017**, *704*, 212–217. [CrossRef]
12. Kolli, R.P.; Devaraj, A. A Review of Metastable Beta Titanium Alloys. *Metals* **2018**, *8*, 506. [CrossRef]
13. Wood, J.R.; Russo, P.A.; Welter, M.F.; Crist, E.M. Thermomechanical processing and heat treatment of Ti–6Al–2Sn–2Zr–2Cr–2Mo–Si for structural application. *Mater. Sci. Eng. A* **1998**, *243*, 109–118. [CrossRef]
14. Gloria, A.; Montanari, R.; Richetta, M.; Varone, A. Alloys for Aeronautic Applications: State of the Art and Perspectives. *Metals* **2019**, *9*, 662. [CrossRef]
15. Oshida, Y. Materials Classification. In *Bioscience and Bioengineering of Titanium Materials*, 2nd ed.; Yoshida, Y., Ed.; Elsevier: Amsterdam, The Netherlands, 2013; pp. 9–34.
16. Huang, S.; Zhao, Q.; Wu, C.; Lin, C.; Zhao, Y.; Jia, W.; Mao, C. Effects of β-stabilizer elements on microstructure formation and mechanical properties of titanium alloys. *J. Alloys Compd.* **2021**, *867*, 160085. [CrossRef]
17. Bieler, T.R.; Trevino, R.M.; Zeng, L. *Alloys: Titanium. Encyclopedia of Condensed Matter Physics*; Elsevier: Amsterdam, The Netherlands, 2005; pp. 65–76.
18. Sbayti, M.; Ghiotti, A.; Bahloul, R.; Belhadjsalah, H.; Bruschi, S. Finite Element Analysis of hot Single Point Incremental forming of hip prostheses. *Matec. Web Conf.* **2016**, *80*, 14006. [CrossRef]
19. Inagaki, I.; Takechi, T.; Shirai, Y.; Ariyasu, N. Application and Features of Titanium for the Aerospace Industry. *Nippon Steel Sumimoto Met. Tech. Rep.* **2014**, *106*, 22–27.
20. Kim, Y.W. Ordered intermetallic alloys: Part III: Gamma titanium aluminides. *J. Miner. Met. Mater. Soc.* **1994**, *46*, 30–39. [CrossRef]
21. Liu, S.; Shin, C. Additive manufacturing of Ti6Al4V alloy: A review. *Mater. Des.* **2019**, *164*, 107552. [CrossRef]
22. Huang, R.; Riddle, M.; Graziano, D.; Warren, J.; Das, S.; Nimbalkar, S.; Cresko, J.; Masanet, E. Energy and emissions saving potential of additive manufacturing: The case of lightweight aircraft components. *J. Clean. Prod.* **2016**, *135*, 1559–1570. [CrossRef]
23. Sousa, R. Incremental Sheet Forming Technologies. Available online: https://doi.org/10.1016/B978-0-12-803581-8.04055-8 (accessed on 16 October 2021).
24. Advantages & Disadvantages of Casting Process. Available online: https://bansengtrade.com/news/advantages-disadvantages-of-casting-process.html (accessed on 16 October 2021).
25. Kumar, A.R.; Roy, S.; Joshi, K.K.; Sahoo, A.K.; Das, R.K. Machining of Ti-6Al-4V ELI alloy: A brief review. *IOP Conf. Ser. Mater. Sci. Eng.* **2018**, *390*, 012112.
26. Oldenberger, E.L.; Oldenburg, M.; Thilderkvist, P.; Stoehr, T.; Lechler, J.; Merklein, M. Tool development based on modelling and simulation of hot sheet metal forming of Ti-6Al-4V titanium alloy. *J. Mater. Process. Technol.* **2011**, *211*, 1324–1335. [CrossRef]
27. Abdelkader, W.B.; Bahloul, R.; Arfa, H. Numerical Investigation of the Influence of some Parameters in SPIF Process on the Forming Forces and Thickness Distributions of a Bimetallic Sheet CP-Titanium/Low-carbon Steel Compared to an Individual Layer. *Procedia Manuf.* **2020**, *47*, 1319–1327. [CrossRef]
28. Hagan, E.; Jeswiet, J. A review of conventional and modern single-point sheet metal forming methods. *Proc. Inst. Mech. Eng. Part B J. Eng. Manuf.* **2003**, *217*, 213–225. [CrossRef]
29. Milutinović, M.; Lendjel, R.; Baloš, S.; Zlatanović, D.L.; Sevšek, L.; Pepelnjak, T. Characterisation of geometrical and physical properties of a stainless steel denture framework manufactured by single-point incremental forming. *J. Mater. Res. Technol.* **2021**, *10*, 605–623. [CrossRef]
30. Cristino, V.A.; Magrinho, J.P.; Centeno, G.; Silva, M.B.; Martins, P.A.F. Theory of single point incremental forming of tubes. *J. Mater. Process. Technol.* **2021**, *287*, 116659. [CrossRef]
31. Najm, S.M.; Paniti, I. Experimental Investigation on the Single Point Incremental Forming of AlMn1Mg1 Foils using Flat End Tools. *IOP Conf. Ser. Mater. Sci. Eng.* **2018**, *448*, 012032. [CrossRef]
32. Sbayti, M.; Bahloul, R.; Belhadjsalah, H.; Zemzemi, F. Optimization techniques applied to single point incremental forming process for biomedical application. *Int. J. Adv. Manuf. Technol.* **2018**, *95*, 1789–1804. [CrossRef]
33. Trzepieciński, T.; Oleksik, V.; Pepelnjak, T.; Najm, S.M.; Paniti, I.; Maji, K. Emerging Trends in Single Point Incremental Sheet Forming of Lightweight Metals. *Metals* **2021**, *11*, 1188. [CrossRef]
34. Trzepieciński, T. Recent developments and trends in sheet metal forming. *Metals* **2020**, *10*, 779. [CrossRef]
35. Preferred Reporting Items for Systematic Reviews and Meta-Analyses. Available online: http://prisma-statement.org/ (accessed on 16 August 2021).

36. Schreiber, R.G.; Schaeffer, L. Manufacture of absorber fins for solar collector using incremental sheet forming. *J. Mater. Res. Technol.* **2019**, *8*, 1132–1140. [CrossRef]
37. Bastos, R.N.P.; de Sousa, R.J.A.; Ferreira, J.A.F. Enhancing time efficiency on single point incremental forming processes. *Int. J. Mater. Form.* **2016**, *9*, 653–662. [CrossRef]
38. Wei, H.; Hussain, G.; Shi, X.; Isidore, B.B.L.; Alkahtani, M.; Abidi, M.H. Formability of Materials with Small Tools in Incremental Forming. *Chin. J. Mech. Eng.* **2020**, *33*, 55. [CrossRef]
39. Tera, M.; Breaz, R.E.; Racz, S.G.; Girjob, C.E. Processing strategies for single point incremental forming—A CAM approach. *Int. J. Adv. Manuf. Technol.* **2019**, *102*, 1761–1777. [CrossRef]
40. Najm, S.M.; Paniti, I. Artificial neural network for modeling and investigating the effects of forming tool characteristics on the accuracy and formability of thin aluminum alloy blanks when using SPIF. *Int. J. Adv. Manuf. Technol.* **2021**, *114*, 2591–2615. [CrossRef]
41. Saotome, Y.; Okamoto, T. An in-situ incremental microforming system for three-dimensional shell structures of foil materials. *J. Mater. Process. Tech.* **2001**, *113*, 636–640. [CrossRef]
42. Sekine, T.; Obikawa, T. Micro incremental forming characteristics of stainless foil. *Key Eng. Mater.* **2010**, *447–448*, 346–350. [CrossRef]
43. Hapsari, G.; Hmida, R.B.; Richard, F.; Thibaud, S.; Malécot, P. A procedure for ductile damage parameters identification by micro incremental sheet forming. *Procedia Eng.* **2017**, *183*, 125–130. [CrossRef]
44. Cédric, B.; Pierrick, M.; Sébastien, T. Shape Accuracy Improvement Obtained by μ-SPIF by Tool Path Compensation. *Procedia Manuf.* **2020**, *47*, 1399–1402. [CrossRef]
45. Urtekin, L.; Bozkurt, F.; Çanli, M. Analysis of biomedical titanium implant green parts by X-ray tomography. *J. Mater. Res. Technol.* **2021**, *14*, 277–286. [CrossRef]
46. Kurup, A.; Dhatrak, P.; Khasnis, N. Surface modification techniques of titanium and titanium alloys for biomedical dental applications: A review. *Mater. Today Proc.* **2021**, *39*, 84–90. [CrossRef]
47. Jackson, M.J.; Kopac, J.; Balazic, M.; Bombac, D.; Brojan, M.; Kosel, F. Titanium and titanium alloy applications in medicine. In *Surface Engineered Surgical Tools and Medical Devices*; Jackson, M.J., Ahmed, W., Eds.; Springer: Boston, MA, USA, 2007; pp. 533–576.
48. Yuan, Z.; He, Y.; Lin, C.; Liu, P.; Cai, K. Antibacterial surface design of biomedical titanium materials for orthopedic applications. *J. Mater. Sci. Technol.* **2021**, *78*, 51–67. [CrossRef]
49. Balasubramanian, R.; Nagumothu, R.; Parfenov, E.; Valiev, R. Development of nanostructured titanium implants for biomedical implants—A short review. *Mater. Today Proc.* **2021**, *46*, 1195–1200. [CrossRef]
50. Hwang, Y.-J.; Choi, Y.-S.; Hwang, Y.-H.; Cho, H.-W.; Lee, D.-G. Biocompatibility and Biological Corrosion Resistance of Ti–39Nb–6Zr+0.45Al Implant Alloy. *J. Funct. Biomater.* **2021**, *12*, 2. [CrossRef]
51. Chen, L.-Y.; Cui, Y.-W.; Zhang, L.-C. Recent Development in Beta Titanium Alloys for Biomedical Applications. *Metals* **2020**, *10*, 1139. [CrossRef]
52. Wang, G.; Li, J.; Lv, K.; Zhang, W.; Ding, X.; Yang, G.; Liu, X.; Jiang, X. Surface thermal oxidation on titanium implants to enhance osteogenic activity and in vivo osseointegration. *Sci. Rep.* **2016**, *6*, 31769. [CrossRef]
53. Eremin, A.; Panin, S.; Byakov, A.; Sharkeev, Y. Mechanical behavior of course-grain and ultrafine-grain titanium for biomedical application. *Procedia Struct. Integr.* **2019**, *23*, 233–238. [CrossRef]
54. Kim, J.; Lee, H.; Jang, T.-S.; Kim, D.; Yoon, C.-B.; Han, G.; Kim, H.-E.; Jung, H.-D. Characterization of Titanium Surface Modification Strategies for Osseointegration Enhancement. *Metals* **2021**, *11*, 618. [CrossRef]
55. Gonsior, M.; Borowski, T.; Czarnowska, E.; Sanak, M.; Kustosz, R.; Ossowski, M.; Wierzchoń, T. Thrombogenicity of Ti(N,C,O) diffusive coating layers developer on titanium alloy as the blood contact surface. *Eur. Cells Mater.* **2010**, *19*, 12–17.
56. Cai, D.; Zhao, X.; Yang, L.; Wang, R.; Qin, G.; Chen, D.F.; Zhang, E. A novel biomedical titanium alloy with high antibacterial property and low elastic modulus. *J. Mater. Sci. Technol.* **2021**, *81*, 13–25. [CrossRef]
57. Castelan, J.; Schaeffer, L.; Daleffe, A.; Fritzen, D.; Salvaro, V.; da Silva, F.P. Manufacture of custom-made cranial implants from DICOM®using 3D printing, CAD/CAM technology and incremental sheet forming. *Eng. Biomed.* **2014**, *30*, 265–273. [CrossRef]
58. Petrovskaya, T.S.; Toropkov, N.E.; Mironov, E.G.; Azarmi, F. 3D printed biocompatible polylactide-hydroxyapatite based material for bone implants. *Mater. Manuf. Process.* **2018**, *33*, 1899–1904. [CrossRef]
59. Bayirli, A.; Ide, S.; Orujalipour, I.; Demir, O.; Dursun, M.; Ates, H. Nanoscale structural investigation on Ti-6Al-4V implants produced by using selective laser melting technique: The importance of production angle. *J. Mater. Res. Technol.* **2019**, *8*, 2796–2801. [CrossRef]
60. Berretta, S.; Evans, K.; Ghita, O. Additive manufacture of PEEK cranial implants: Manufacturing considerations versus accuracy and mechanical performance. *Mater. Des.* **2018**, *139*, 141–152. [CrossRef]
61. Kim, Y.W. Surface modification of Ti dental implants by grit-blasting and micro-arc oxidation. *Mater. Manuf. Process.* **2010**, *25*, 307–310. [CrossRef]
62. Correa, D.R.N.; Rocha, L.A.; Donato, T.A.G.; Sousa, K.S.J.; Grandini, C.R.; Afonso, C.R.M.; Doi, H.; Tsutsumi, Y.; Hanawa, T. On the mechanical biocompatibility of Ti-15Zr-based alloys for potential use as load-bearing implants. *J. Mater. Res. Technol.* **2020**, *9*, 1241–1250. [CrossRef]

63. Wyte, M.; Nootens, S. Cast/forge a new cost effective approach to the manufacture of medical implants using cobalt–28 chromium–6 molybdenum alloy. In Proceedings of the Materials and Processes for Medical Devices Conference, Anaheim, CA, USA, 8–10 September 2003; ASM International: Novelty, OH, USA; pp. 320–323.
64. Emmens, W.C.; Sebastiani, G.; van den Boogaard, A.H. The technology of incremental sheet forming—A brief review of the history. *J. Mater. Process. Technol.* **2010**, *210*, 981–997. [CrossRef]
65. Jeswiet, J.; Micari, F.; Hirt, G.; Bramley, A.; Duflou, J.; Allwood, J. Asymmetric single point incremental forming of sheet metal. *Ann. CIRP Manuf. Technol.* **2005**, *54*, 623–649. [CrossRef]
66. Duflou, J.R.; Behera, A.K.; Vanhove, H.; Bertol, L.S. Manufacture of accurate titanium cranio-facial implants with high forming angle using single point incremental forming. *Key Eng. Mater.* **2013**, *549*, 223–230. [CrossRef]
67. Göttmann, A.; Korinth, M.; Schäfer, V.; Araghi, B.T.; Bambach, M.; Hirt, G. Manufacturing of individualized cranial implants usingtwo point incremental sheet metal forming. In *Future Trends in Production Engineering*; Schuh, G., Neugebauer, R., Uhlmann, E., Eds.; Springer: Berlin/Heidelberg, Germany, 2013; pp. 287–295.
68. Lu, B.; Ou, H.; Shi, S.Q.; Long, H.; Chen, J. Titanium based cranial reconstruction using incremental sheet forming. *Int. J. Mater. Form.* **2016**, *9*, 361–370. [CrossRef]
69. Lu, B.; Xu, D.K.; Liu, R.Z.; Ou, H.; Long, H.; Chen, J. Cranial reconstruction using double side incremental forming. *Key Eng. Mater.* **2015**, *639*, 535–542. [CrossRef]
70. Eksteen, P.D.; Van der Merwe, A.F. Incremental sheet forming (ISF) in the manufacturing of titanium based plate implants in thebio-medical sector. In Proceedings of the International Conference on Computers & Industrial Engineering (CIE 42), Cape Town, South Africa, 16–18 July 2012; pp. 569–575.
71. Araújo, R.; Teixeira, P.; Montanari, L.; Reis, A.; Silva, M.B.; Martins, P.A. Single point incremental forming of a facial implant. *Prosthet. Orthot. Int.* **2013**, *38*, 369–378. [CrossRef] [PubMed]
72. Ambrogio, G.; De Napoli, L.; Filice, L.; Gagliardi, F.; Muzzupappa, M. Application of incremental forming process for high customized medical product manufacturing. *J. Mater. Process. Technol.* **2005**, *162–163*, 156–162. [CrossRef]
73. Han, F.; Mo, J.H.; Wang, P.; Deng, Y.Z. A digital manufacture technology for skull prosthesis using incremental sheet forming method. *Adv. Mater. Res.* **2010**, *102–104*, 348–352. [CrossRef]
74. Vanhove, H.; Carette, Y.; Vancleef, S.; Duflou, J.R. Production of thin shell clavicle implants through single point incremental forming. *Procedia Eng.* **2017**, *183*, 174–179. [CrossRef]
75. Araújo, R.; Teixeira, P.; Silva, M.B.; Reis, A.; Martins, P.A.F. Single Point Incremental Forming of a Medical Implant. *Key Eng. Mater.* **2013**, *544–557*, 1388–1393. [CrossRef]
76. Ambrogio, G.; Sgambitterra, E.; De Napoli, L.; Gagliardi, F.; Fragomeni, G.; Piccininni, A.; Gugleilmi, P.; Palumbo, G.; Sorgente, D.; La Barbera, L.; et al. Performances analysis of Titanium prostheses manufactured by Superplastic Forming and Incremental Forming. *Procedia Eng.* **2017**, *183*, 168–173. [CrossRef]
77. Honarpisheh, M.; Abdolhoseini, M.J.; Amini, S. Experimental and numerical investigation of the hot incremental forming of Ti-6Al-4V sheet using electrical current. *Int. J. Adv. Manuf. Technol.* **2016**, *83*, 2027–2037. [CrossRef]
78. Mohammadi, A.; Vanhove, H.; Van Bael, A.; Duflou, J.R. Towards accuracy improvement in single point incremental forming of shallow parts formed under laser assisted conditions. *Int. J. Mater. Form.* **2016**, *9*, 339–351. [CrossRef]
79. Saidi, B.; Giraud Moreau, L.; Mhemed, S.; Cheroaut, A.; Adragna, P.A.; Nasri, R. Hot incremental forming of titanium human skull prosthesis by using cartridge heaters: A reverse engineering approach. *Int. J. Adv. Manuf. Technol.* **2019**, *101*, 873–880. [CrossRef]
80. Palumbo, G.; Piccininni, A.; Ambrogio, G.; Sgambitterra, E. Design of custom cranial prostheses combining manufacturing and drop test finite element simulations. *Int. J. Adv. Manuf. Technol.* **2020**, *111*, 1627–1641. [CrossRef]
81. Oleksik, V.; Pascu, A.; Deac, C.; Fleacă, R.; Bologa, O.; Racz, G. Experimental study on the surface quality of the medical implants obtained by single point incremental forming. *Int. J. Mater. Form.* **2010**, *3*, 935–938. [CrossRef]
82. Donachie, M.J. *Titanium: A Technical Guide*; ASM International: Materials Park, OH, USA, 2000.
83. Fan, G.; Sun, F.; Meng, X.; Gao, L.; Tong, G. Electric Hot Incremental Forming of Ti-6Al-4V Titanium Sheet. *Int. J. Adv. Manuf. Technol.* **2010**, *49*, 941–947. [CrossRef]
84. Liu, Z. Heat-Assisted Incremental Sheet Forming: A State-of-the-Art Review. *Int. J. Adv. Manuf. Technol.* **2018**, *98*, 2987–3003. [CrossRef]
85. Sbayti, M.; Bahloul, R.; Belhadjsalah, H. Simulation of the local heating effect on incremental sheet forming process. In Proceedings of the Advances in Mechanical Engineering, Materials and Mechanics, Hammamet, Tunisia, 16–18 December 2019; Kharrat, M., Baccar, M., Dammak, F., Eds.; Springer International Publishing: Cham, Switzerland, 2021; pp. 144–151.
86. Ambrogio, G.; Gagliardi, F.; Chamanfar, A.; Misiolek, W.Z.; Filice, L. Induction Heating and Cryogenic Cooling in Single Point Incremental Forming of Ti-6Al-4V: Process Setup and Evolution of Microstructure and Mechanical Properties. *Int. J. Adv. Manuf. Technol.* **2017**, *91*, 803–812. [CrossRef]
87. Göttmann, A.; Diettrich, J.; Bergweiler, G.; Bambach, M.; Hirt, G.; Loosen, P.; Poprawe, R. Laser-Assisted Asymmetric Incremental Sheet Forming of Titanium Sheet Metal Parts. *Prod. Eng. Res. Dev.* **2011**, *5*, 263–271. [CrossRef]
88. Göttmann, A.; Bailly, D.; Bergweiler, G.; Bambach, M.; Stollenwerk, J.; Hirt, G.; Loosen, P. A Novel Approach for Temperature Control in ISF Supported by Laser and Resistance Heating. *Int. J. Adv. Manuf. Technol.* **2013**, *67*, 2195–2205. [CrossRef]

89. Mosecker, L.; Göttmann, A.; Saeed-Akbari, A.; Bleck, W.; Bambach, M.; Hirt, G. Deformation Mechanisms of Ti6Al4V Sheet Material during the Incremental Sheet Forming with Laser Heating. *Key Eng. Mater.* **2013**, *549*, 372–380. [CrossRef]
90. Ji, Y.H.; Park, J.J. Formability of magnesium AZ31 sheet in the incremental forming at warm temperature. *J. Mater. Proc. Technol.* **2008**, *201*, 354–358. [CrossRef]
91. Gagliardi, F.; Ambrogio, G.; Filice, L. Incremental Forming with Local Induction Heating on Materials with Magnetic and Non-Magnetic Properties. *Procedia Eng.* **2017**, *183*, 143–148.
92. Fan, G.; Gao, L.; Hussain, G.; Wu, Z. Electric hot incremental forming: A novel technique. *Int. J. Mach. Tools Manuf.* **2008**, *48*, 1688–1692. [CrossRef]
93. Fan, G.; Gao, L. Numerical Simulation and Experimental Investigation to Improve the Dimensional Accuracy in Electric Hot Incremental Forming of Ti-6Al-4V Titanium Sheet. *Int. J. Adv. Manuf. Technol.* **2014**, *72*, 1133–1141. [CrossRef]
94. Ambrogio, G.; Gagliardi, F.; Bruschi, S.; Filice, L. On the high-speed Single Point Incremental Forming of titanium alloys. *Cirp. Ann.-Manuf. Technol.* **2013**, *62*, 243–246. [CrossRef]
95. Grün, P.A.; Uheida, E.H.; Lachmann, L.; Dimitrov, D.; Oosthuizen, G.A. Formability of Titanium Alloy Sheets by Friction Stir Incremental Forming. *Int. J. Adv. Manuf. Technol.* **2018**, *99*, 1993–2003. [CrossRef]
96. Uheida, E.H.; Oosthuizen, G.A.; Dimitrov, D.M.; Bezuidenhout, M.B.; Hugo, P.A. Effects of the relative tool rotation direction on formability during the incremental forming of titanium sheets. *Int. J. Adv. Manuf. Technol.* **2018**, *96*, 3311–3319. [CrossRef]
97. Ao, D.; Chu, X.; Gao, J.; Yang, Y.; Lin, S. Experimental Investigation on the Deformation Behaviors of Ti-6Al-4V Sheet in Electropulsing-Assisted Incremental Forming. *Int. J. Adv. Manuf. Technol.* **2019**, *104*, 4243–4254. [CrossRef]
98. Ao, D.; Gao, J.; Chu, X.; Lin, S.; Lin, J. Formability and Deformation Mechanism of Ti-6Al-4V Sheet under Electropulsing Assisted Incremental Forming. *Int. J. Solids Struct.* **2020**, *202*, 357–367. [CrossRef]
99. Asgar, J.; Lingam, R.; Reddy, V.N. Tool Path Influence on Electric Pulse Aided Deformation during Incremental Sheet Metal Forming. *AIP Conf. Proc.* **2013**, *1567*, 840–843.
100. Fan, G.; Gao, L. Mechanical Property of Ti-6Al-4V Sheet in One-Sided Electric Hot Incremental Forming. *Int. J. Adv. Manuf. Technol.* **2014**, *72*, 989–994. [CrossRef]
101. Liu, R.; Lu, B.; Xu, D.; Chen, J.; Chen, F.; Ou, H.; Long, H. Development of Novel Tools for Electricity-Assisted Incremental Sheet Forming of Titanium Alloy. *Int. J. Adv. Manuf. Technol.* **2016**, *85*, 1137–1144. [CrossRef]
102. Najafabady, S.A.; Ghaei, A. An experimental study on dimensional accuracy, surface quality, and hardness of Ti-6Al-4 V titanium alloy sheet in hot incremental forming. *Int. J. Adv. Manuf. Technol.* **2016**, *87*, 3579–3588. [CrossRef]
103. Xu, D.K.; Lu, B.; Cao, T.T.; Zhang, H.; Chen, J.; Long, H.; Cao, J. Enhancement of Process Capabilities in Electrically-Assisted Double Sided Incremental Forming. *Mater. Des.* **2016**, *92*, 268–280. [CrossRef]
104. Ortiz, M.; Penalva, M.; Puerto, M.J.; Homola, P.; Kafka, V. Hot Single Point Incremental Forming of Ti-6Al-4V Alloy. *Key Eng. Mater.* **2014**, *611–612*, 1079–1087. [CrossRef]
105. Ortiz, M.; Penalva, M.; Iriondo, E.; de Lacalle, L.N.L. Accuracy and Surface Quality Improvements in the Manufacturing of Ti-6Al-4V Parts Using Hot Single Point Incremental Forming. *Metals* **2019**, *9*, 697. [CrossRef]
106. Khazaali, H.; Fereshteh-Saniee, F. Application of the Taguchi Method for Efficient Studying of Elevated-Temperature Incremental Forming of a Titanium Alloy. *J. Braz. Soc. Mech. Sci. Eng.* **2018**, *40*, 43. [CrossRef]
107. Khazaali, H.; Fereshteh-Saniee, F. Process Parameter Enhancement for Incremental Forming of Titanium Ti-6Al-4V Truncated Cone with Varying Wall Angle at Elevated Temperatures. *Int. J. Precis. Eng. Manuf.* **2019**, *20*, 769–776. [CrossRef]
108. Mohanraj, R.; Elangovan, S. Thermal Modeling and Experimental Investigation on the Influences of the Process Parameters on Warm Incremental Sheet Metal Forming of Titanium Grade 2 Using Electric Heating Technique. *Int. J. Adv. Manuf. Technol.* **2020**, *110*, 255–274. [CrossRef]
109. Palumbo, G.; Brandizzi, M. Experimental Investigations on the Single Point Incremental Forming of a Titanium Alloy Component Combining Static Heating with High Tool Rotation Speed. *Mater. Des.* **2012**, *40*, 43–51. [CrossRef]
110. Khazaali, H.; Fereshteh-Saniee, F. A Comprehensive Experimental Investigation on the Influences of the Process Variables on Warm Incremental Forming of Ti-6Al-4V Titanium Alloy Using a Simple Technique. *Int. J. Adv. Manuf. Technol.* **2016**, *87*, 2911–2923. [CrossRef]
111. Palumbo, G.; Brandizzi, M.; Cervelli, G.; Fracchiolla, M. Investigations about the Single Point Incremental Forming of anisotropic Titanium alloy sheets. *Adv. Mater. Res.* **2011**, *164–264*, 188–193. [CrossRef]
112. Sbayti, M.; Bahloul, R.; Belhadjsalah, H. Efficiency of optimization algorithms on the adjustment of process parameters for geometric accuracy enhancement of denture plate in single point incremental sheet forming. *Neural. Comput. Appl.* **2020**, *32*, 8829–8846. [CrossRef]
113. Micari, F.; Ambrogio, G.; Filice, L. Shape and dimensional accuracy in single point incremental forming: State of the art and future trends. *J. Mater. Process. Technol.* **2007**, *191*, 390–395. [CrossRef]
114. Hussain, G.; Gao, L.; Hayat, N.; Cui, Z.; Pang, Y.C.; Dar, N.U. Tool and lubrication for negative incremental forming of a commercially pure titanium sheet. *J. Mater. Process. Technol.* **2008**, *203*, 193–201. [CrossRef]
115. Valoppi, B.; Egea, A.J.S.; Zhang, Z.X.; Rojas, H.A.G.; Ghiotti, A.; Bruschi, S.; Cao, J. A hybrid mixed double-sided incremental forming method for forming Ti6Al4V alloy. *CIRP Ann.-Manuf. Technol.* **2016**, *65*, 309–312. [CrossRef]
116. Saidi, B.; Moreau, L.G.; Cherouat, A.; Nasri, R. Experimental and numerical study on warm single-point incremental sheet forming (WSPIF) of titanium alloy Ti-6Al-4V, using cartridge heaters. *J. Braz. Soc. Mech. Sci.* **2020**, *42*, 534. [CrossRef]

117. Behera, A.K.; Ou, H. Effect of stress relieving heat treatment on surface topography and dimensional accuracy of incrementally formed grade 1 titanium sheet parts. *Int. J. Adv. Manuf. Technol.* **2016**, *87*, 3233–3248. [CrossRef]
118. Racz, S.G.; Breaz, R.E.; Tera, M.; Girjob, C.; Biris, C.; Chicea, A.L.; Bologa, O. Incremental Forming of Titanium Ti6Al4V Alloy for Cranioplasty Plates-Decision-Making Process and Technological Approaches. *Metals* **2018**, *8*, 626. [CrossRef]
119. Behera, A.K.; Lu, B.; Ou, H. Characterization of shape and dimensional accuracy of incrementally formed titanium sheet parts with intermediate curvatures between two feature types. *Int. J. Adv. Manuf. Technol.* **2016**, *83*, 1099–1111. [CrossRef]
120. Fiorentino, A.; Feriti, G.C.; Ceretti, E.; Giardini, C. Capability of iterative learning control and influence of the material properties on the improvement of the geometrical accuracy in incremental sheet forming process. *Int. J. Mater. Form.* **2018**, *11*, 125–134. [CrossRef]
121. Li, F.Q.; Mo, J.H. Springback analysis of TC4 titanium alloy complex part with double curvature under single point incremental forming process. *Adv. Mater. Res.* **2012**, *468–471*, 1094–1098. [CrossRef]
122. Grimm, T.J.; Roth, J.T.; Ragai, I. Electrically assisted global springback elimination in AMS-T-9046 titanium after single point incremental forming. In Proceedings of the ASME 2016 11th International Manufacturing Science and Engineering Conference, Backsburg, VA, USA, 27 June–1 July 2016; pp. 1–12.
123. Li, X.; Han, K.; Song, X.; Wang, H.; Li, D.; Li, Y.; Li, Q. Experimental and numerical investigation on surface quality for two-point incremental sheet forming with interpolator. *Chin. J. Aeronaut.* **2020**, *33*, 2794–2806. [CrossRef]
124. Liu, Z.; Liu, S.; Li, Y.; Meehan, P.A. Modeling and optimization of surface roughness in incremental sheet forming using a multi-objective function. *Mater. Manuf. Process.* **2014**, *29*, 808–818. [CrossRef]
125. Radu, C.; Cristea, I.; Herghelegiu, E.; Tabacu, S. Improving the accuracy of parts manufactured by single point incremental forming. *Appl. Mech. Mater.* **2013**, *332*, 443–448. [CrossRef]
126. Bahloul, R.; Arfa, H.; Belhajsalah, H. Application of response surface analysis and genetic algorithm for the optimization of single point incremental forming process. *Key Eng. Mater.* **2013**, *554–557*, 1265–1272. [CrossRef]
127. Yao, Z.; Li, Y.; Yang, M.; Yuan, Q.; Shi, P. Parameter optimization for deformation energy and forming quality in single point incremental forming process using response surface methodology. *Adv. Mech. Eng.* **2017**, *9*, 1–15. [CrossRef]
128. Behera, A.; Lauwers, B.; Duflou, J. Tool path generation for single point incremental forming using intelligent sequencing and multi-step mesh morphing techniques. *Int. J. Mater. Form.* **2015**, *8*, 517–532. [CrossRef]
129. Fu, Z.; Mo, J.; Han, F.; Gong, P. Tool path correction algorithm for single-point incremental forming of sheet metal. *Int. J. Adv. Manuf. Technol.* **2013**, *64*, 1239–1248. [CrossRef]
130. Veeraajay, C. Parameter Optimization in Incremental Forming of Titanium Alloy Material. *Trans. Indian Inst. Met.* **2020**, *73*, 2403–2413.
131. Szpunar, M.; Ostrowski, R.; Trzepieciński, T.; Kaščák, L. Central Composite Design Optimisation in Single Point Incremental Forming of Truncated Cones from Commercially Pure Titanium Grade 2 Sheet Metals. *Metals* **2021**, *14*, 3634.
132. Peter, I.; Fracchia, E.; Canale, I.; Maiorano, R. Incremental sheet forming for prototyping automotive modules. *Procedia Manuf.* **2019**, *32*, 50–58. [CrossRef]
133. Mohanraj, R.; Elangovan, S. Incremental sheet metal forming of Ti-6Al-4V alloy for aerospace application. *Trans. Can. Soc. Mech. Eng.* **2020**, *44*, 56–64. [CrossRef]
134. Kumar, D.; Deepak, K.B.; Muzakkir, S.M.; Wani, M.F.; Lijesh, K.P. Enhancing tribological performance of Ti-6Al-4V by sliding process. *Tribol. Mater. Surf. Interf.* **2018**, *12*, 137–143. [CrossRef]
135. Sreesha, R.B.; Kumar, D.; Chandraker, S.; Agrawal, A. Room temperature sliding wear behavior of Ti6Al4V: A review. *AIP Conf. Proc.* **2021**, *2341*, 040041.
136. Cheng, Z.; Li, Y.; Xu, C.; Liu, Y.; Ghafoor, S.; Li, F. Incremental sheet forming towards biomedical implants: A review. *J. Mater. Res. Technol.* **2020**, *9*, 7225–7251. [CrossRef]
137. Kaur, S.; Ghadirinejad, K.; Oskouei, R.H. An Overview on the Tribological Performance of Titanium Alloys with Surface Modifications for Biomedical Applications. *Lubricants* **2019**, *7*, 65. [CrossRef]
138. Ma, J.; Yang, H.; Li, H.; Wang, D.; Li, G.J. Tribological behaviors between commercial pure titanium sheet and tools in warm forming. *Trans. Nonferrous Met. Soc. China* **2015**, *25*, 2924–2931. [CrossRef]
139. Jozwik, J. Evaluation of Tribological Properties and Condition of Ti6Al4V Titanium Alloy Surface. *Teh. Vjesn. Tech. Gaz.* **2018**, *25*, 170–175.
140. Fridrici, V.; Fouvry, S.; Kapsa, P. *Fretting Wear Behavior of a Titanium Alloy*; Springer: Dordrecht, The Netherlands, 2001; pp. 413–421.
141. Pakhamov, M.A.; Gorlov, D.; Stolyarov, V. Features of wear and friction in titanium. *IOP Conf. Ser. Mater. Sci. Eng.* **2020**, *996*, 012017. [CrossRef]
142. Zhang, H.; Liu, X.; Liu, H.; Wang, Y.; Tang, B. Improvement of the fretting damage resistance of Ti-811 alloy by CuNi multilayer film. *Tribol. Int.* **2011**, *44*, 1488–1494. [CrossRef]
143. Trzepieciński, T.; Lemu, H.G. Recent developments and trends in the friction testing for conventional sheet metal forming and incremental sheet forming. *Metals* **2020**, *10*, 47. [CrossRef]
144. Trzepieciński, T.; Szpunar, M.; Kaščák, L. Modeling of friction phenomena of Ti-6Al-4V sheets based on backward elimination regression and multi-layer artificial neural networks. *Materials* **2021**, *14*, 2570. [CrossRef]

145. Ajmal, M.; Ali, S.A.; Tahir, H.; Shah, M.R.; Saad, M. Synthesis of Emulsion by Using Vegetable Oil Modified by Titanium Dioxide (TiO_2) Nanoparticles: A Peculiar Source for Synthesis of Bio-Based Lubricant and Novel Approach to Enhance the Efficiency of Emulsion. *Eur. J. Lipid Sci. Technol.* **2021**, *123*, 2000280. [CrossRef]
146. Pang, H.; Ngalle, G. Formulation of SiO_2/oil nanolubricant for metal forming using hydrodynamic cavitation. *Proc. Inst. Mech. Eng. Part B J. Eng. Manuf.* **2020**, *234*, 1549–1558. [CrossRef]
147. Chen, Y.; Renner, P.; Liang, H. Dispersion of Nanoparticles in Lubricating Oil: A Critical Review. *Lubricants* **2019**, *7*, 7. [CrossRef]
148. Fellah, M.; Assala, O.; Labaïz, M.; Dekhil, L.; Iost, A. Friction and Wear Behavior of Ti-6Al-7Nb Biomaterial Alloy. *J. Biomater. Nanotech.* **2013**, *4*, 374–384. [CrossRef]
149. Niu, Q.L.; Zheng, X.H.; Ming, W.W.; Chen, M. Friction and Wear Performance of Titanium Alloys against Tungsten Carbide under Dry Sliding and Water Lubrication. *Tribol. Trans.* **2013**, *56*, 101–108. [CrossRef]
150. Vahdani, M.; Mirnia, M.J.; Gorji, H.; Bakhshi-Jooybari, M. Experimental Investigation of Formability and Surface Finish into Resistance Single-Point Incremental Forming of Ti-6Al-4V Titanium Alloy Using Taguchi Design. *Trans. Indian Inst. Met.* **2019**, *72*, 1031–1041. [CrossRef]
151. Gagliardi, F.; Ambrogio, G.; Filice, L. Formability of lightweight alloys by hot incremental sheet forming. *Mater. Des.* **2012**, *34*, 501–508.
152. Hussain, G. Microstructure and mechanical and tribological properties of a lubricant coating for incremental forming of a Ti sheet. *Trans. Mech. Eng.* **2014**, *38*, 423–429.
153. Sornsuwit, N.; Sittisakuljaroen, S.; Sangsai, N.; Suwankan, P. Effect of heat treatment on single point incremental forming for titanium grade 2 sheet. In Proceedings of the 2018 Third International Conference on Engineering Science and Innovative Technology (ESIT), North Bangkok, Thailand, 19–22 April 2018; p. 4.
154. Hussain, G.; Gao, L.; Zhang, Z.Y. Formability evaluation of a pure titanium sheet in the cold incremental forming process. *Int. J. Adv. Manuf. Technol.* **2008**, *37*, 920–926. [CrossRef]
155. Li, Y.; Chen, X.; Liu, Z.; Sun, J.; Li, F.; Li, J.; Zhao, G. A review on the recent development of incremental sheet-forming process. *Int. J. Adv. Manuf. Technol.* **2017**, *92*, 2439–2462. [CrossRef]
156. Sbayti, M.; Bahloul, R.; Belhadjsalah, H. Numerical Modeling of Hot Incremental Forming Process for Biomedical Application. In *Design and Modeling of Mechanical Systems—III. CMSM 2017. Lecture Notes in Mechanical Engineering*; Haddar, M., Chaari, F., Benamara, A., Chouchane, M., Karra, C., Aifaoui, N., Eds.; Springer: Cham, Switzerland, 2018; pp. 881–891.
157. Naranjo, J.; Miguel, V.; Martínez, A.; Coello, J.; Manjabacas, M.C.; Valera, J. Influence of temperature on alloy Ti6Al4V formability during the warm SPIF process. *Procedia Eng.* **2017**, *207*, 866–871. [CrossRef]
158. Naranjoa, J.; Miguela, V.; Martínez-Martínez, A.; Gómez-Lópeza, L.M.; Manjabacas, M.C.; Coello, J. Analysis and simulation of Single Point Incremental Forming by ANSYS®. *Procedia Eng.* **2015**, *132*, 1104–1111. [CrossRef]
159. Abdelkefi, A.; Guinesa, D.; Léotoinga, L.; Thuillier, S. Incremental forming of Titanium T40 sheet: Experimental and numerical investigations. In Proceedings of the 24ème Congrès Français de Mécanique, Brest, France, 26–30 August 2019; pp. 1–7.
160. Piccininni, A.; Gagliardi, F.; Guglielmi, P.; De Napoli, L.; Ambrogio, G.; Sorgente, D.; Palumbo, G. Biomedical Titanium alloy prostheses manufacturing by means of Superplastic and Incremental Forming processes. *Matec. Web Conf.* **2016**, *80*, 15007. [CrossRef]
161. Naranjo, J.; Miguel, V.; Martínez, A.; Coello, J.; Manjabacas, M.C. Analysis of material behavior models for the Ti6Al4V alloy to simulate the Single Point Incremental Forming process. *Procedia Eng.* **2017**, *13*, 307–314. [CrossRef]
162. Nguyen, D.T.; Park, J.G.; Lee, H.J.; Kim, Y.S. Finite element method study of incremental sheet forming for complex shape and its improvement. *Proc. Inst. Mech. Eng. Part B J. Eng. Manuf.* **2010**, *224*, 913–924. [CrossRef]
163. Al-Ghamdi, K.A.; Hussain, G. The pillowing tendency of materials in single-point incremental forming: Experimental and finite element analyses. *Proc. Inst. Mech. Eng. Part B J. Eng. Manuf.* **2015**, *229*, 744–753. [CrossRef]
164. Manco, L.; Filice, L.; Ambrogio, G. Analysis of the thickness distribution varying tool trajectory in single-point incremental forming. *Proc. Inst. Mech. Eng. Part B J. Eng. Manuf.* **2011**, *225*, 348–356. [CrossRef]
165. Henrard, C.; Bouffioux, C.; Eyckens, P.; Sol, H.; Duflou, J.R.; van Houtte, P.; van Bael, A.; Duchêne, L.; Habraken, A.M. Forming forces in single point incremental forming: Prediction by finite element simulations, validation and sensitivity. *Comput. Mech.* **2011**, *47*, 573–590. [CrossRef]

MDPI
St. Alban-Anlage 66
4052 Basel
Switzerland
Tel. +41 61 683 77 34
Fax +41 61 302 89 18
www.mdpi.com

Materials Editorial Office
E-mail: materials@mdpi.com
www.mdpi.com/journal/materials

www.ingramcontent.com/pod-product-compliance
Lightning Source LLC
Chambersburg PA
CBHW040800250926
51850CB00007BA/1040
9783036572826